T0134948

Springer Optimization and Its Applications

Volume 174

Aims and Scope

Optimization has continued to expand in all directions at an astonishing rate. New algorithmic and theoretical techniques are continually developing and the diffusion into other disciplines is proceeding at a rapid pace, with a spot light on machine learning, artificial intelligence, and quantum computing. Our knowledge of all aspects of the field has grown even more profound. At the same time, one of the most striking trends in optimization is the constantly increasing emphasis on the interdisciplinary nature of the field. Optimization has been a basic tool in areas not limited to applied mathematics, engineering, medicine, economics, computer science, operations research, and other sciences.

The series **Springer Optimization and Its Applications (SOIA)** aims to publish state-of-the-art expository works (monographs, contributed volumes, textbooks, handbooks) that focus on theory, methods, and applications of optimization. Topics covered include, but are not limited to, nonlinear optimization, combinatorial optimization, continuous optimization, stochastic optimization, Bayesian optimization, optimal control, discrete optimization, multi-objective optimization, and more. New to the series portfolio include Works at the intersection of optimization and machine learning, artificial intelligence, and quantum computing.

Volumes from this series are indexed by Web of Science, zbMATH, Mathematical Reviews, and SCOPUS.

More information about this series at http://www.springer.com/series/7393

Sergey I. Nikolenko

Synthetic Data for Deep Learning

 Springer

Sergey I. Nikolenko
Synthesis AI
San Francisco, CA, USA

Steklov Institute of Mathematics
St. Petersburg, Russia

ISSN 1931-6828   ISSN 1931-6836   (electronic)
Springer Optimization and Its Applications
ISBN 978-3-030-75180-7   ISBN 978-3-030-75178-4   (eBook)
https://doi.org/10.1007/978-3-030-75178-4

This Springer imprint is published by the registered company Springer Nature Switzerland AG
The registered company address is: Gewerbestrasse 11, 6330 Cham, Switzerland

Preface

Dear reader,

You are holding in your hands... oh, come on, who holds books like this in their hands anymore? Anyway, you are reading this, and it means that I have managed to release one of the first books specifically devoted to the subject of *synthetic data*, that is, data produced artificially with the purpose of training machine learning models. In this preface, let me briefly explain why this subject might deserve your attention.

As we will see in the introductory chapter, machine learning, and especially deep learning, is developing at a breakneck pace these days. Alas, as most exponential growths go, in the real world, there are several constraints that will almost inevitably turn this exponential curve into a sigmoid. One of them is computational: deep learning is currently outpacing Moore's law by far, and this cannot continue indefinitely.

In this book, we primarily deal with another constraint: the amount of available *data*, especially labeled data for supervised learning problems. In many applications, especially computer vision (which is a very big part of this book), manual labeling is excruciatingly hard: imagine labeling for instance segmentation, and don't even try to imagine manual labeling for depth or optical flow estimation.

Synthetic data is a way to prolong the march of progress in these fields: if you have a three-dimensional virtual scene complete with objects of interest, and images for the dataset are produced by rendering, it means that you automatically know which object every pixel belongs to, what are the 3D relations between them, and so on, and so forth. Producing new data becomes very cheap, and labeling becomes free, which is the main attraction of synthetic data. In the book, we will give a broad overview of synthetic data currently used for various machine learning endeavours (mostly, but not exclusively, computer vision problems) and directions in which synthetic data can be further improved in the future.

Naturally, this approach comes with its own problems. Most such problems can be united under the term of *domain transfer*: if we produce synthetic data for machine learning models, we plan to train the models on the synthetic domain, but the final goal is almost always to apply them on real data, i.e., on a completely different domain. A large part of the book will be devoted to ways to cope with the domain transfer problem, including domain randomization, synthetic-to-real refinement, model-based domain adaptation, and others.

We will also discuss another important motivation for synthetic data: *privacy* concerns. In many sensitive applications, datasets theoretically exist but cannot be released to the general public. Synthetic data can help here as well, and we will consider ways to create anonymized datasets with differential privacy guarantees.

But in order to have a meaningful discussion of synthetic data for deep learning and synthetic-to-real domain adaptation, we need a firm grasp of the main concepts of modern machine learning. That includes deep neural networks, especially convolutional networks and their most important architectures for computer vision problems, generative models, especially generative adversarial networks and their loss functions and architectures, and much more. Therefore, the book contains several introductory chapters that present deep neural networks and the corresponding optimization problems and algorithms, neural architectures for computer vision, and deep generative models. I do not claim that this book is a suitable introductory textbook on deep learning in general (this would require a far longer text), but I hope that a somewhat prepared reader will be able to find something of interest in these introductory chapters and use them as reference material for the rest of the book.

Finally, a word of gratitude. This book could not appear without the help of *Synthesis AI*, where I currently serve as Head of AI, and especially the CEO of *Synthesis AI*, Yashar Behzadi. Yashar, many thanks for your support and patience! I also personally thank Alex Davydow and Rauf Kurbanov who read the manuscript and made insightful suggestions for improvement.

St. Petersburg, Russia
February 2021

Sergey I. Nikolenko

Contents

Acronyms

All acronyms are listed below and the corresponding definitions are also introduced and explained in the text of the book; this list is provided purely for reference purposes.

AAE	Adversarial autoencoder
AdaIN	Adaptive instance normalization
AI	Artificial intelligence
ALV	Autonomous land vehicle
ANN	Artificial neural network
BEGAN	Boundary equilibrium generative adversarial network
BERT	Bidirectional encoder representations from Transformers
BN	Batch normalization
CGI	Computer-generated imagery
CNN	Convolutional neural network
COCO	Common objects in context
CPU	Central processing unit
CV	Computer vision
DA	Domain adaptation
DBN	Deep belief network
DCGAN	Deep convolutional generative adversarial network
DL	Deep learning
DP	Differential privacy
DQN	Deep Q-network
EBGAN	Energy-based generative adversarial network
EM	Expectation–maximization
FCN	Fully convolutional network
FPN	Feature pyramid network
FUNIT	Few-shot unsupervised image-to-image translation
FVBN	Fully visible belief network
GAN	Generative adversarial network

GD	Gradient descent
GPT	Generative pretrained transformer
GPU	Graphics processing unit
IAF	Inverse autoregressive flow
ILSVRC	ImageNet large-scale visual recognition challenge
KL	Kullback–Leibler
LSGAN	Least squares generative adversarial network
LSTM	Long short-term memory
MADE	Masked autoencoder for distribution estimation
MAF	Masked autoregressive flow
MUNIT	Multimodal unsupervised image-to-image translation
NAG	Nesterov accelerated gradient
NAS	Neural architecture search
NIPS	Neural information processing systems
NLP	Natural language processing
OCR	Optical character recognition
PATE	Private aggregation of teacher ensembles
PCA	Principal component analysis
QA	Question answering
ReLU	Rectified linear unit
RL	Reinforcement learning
RNN	Recurrent neural network
ROS	Robot operating system
RPN	Region proposal network
SGD	Stochastic gradient descent
SLAM	Simultaneous localization and mapping
SSD	Single-shot detector
SVM	Support vector machine
UAV	Unmanned aerial vehicle
VAE	Variational autoencoder
VGG	Visual geometry group
VQA	Visual question answering
WGAN	Wasserstein generative adversarial network
YOLO	You only look once

Chapter 1
Introduction: The Data Problem

Machine learning has been growing in scale, breadth of applications, and the amounts of required data. This presents an important problem, as the requirements of state-of-the-art machine learning models, especially data-hungry deep neural networks, are pushing the boundaries of what is economically feasible and physically possible. In this introductory chapter, we show and illustrate this phenomenon, discuss several approaches to solving the data problem, introduce the main topic of this book, *synthetic data*, and outline a plan for the rest of the book.

1.1   Are Machine Learning Models Hitting a Wall?

Machine learning is hot, and it has been for quite some time. The field is growing exponentially fast, with new models and new papers appearing every week, if not every day. Since the deep learning revolution, for about a decade, deep learning has been far outpacing other fields of computer science and arguably even science in general. Analysis of the submissions from *arXiv*[1], the most popular preprint repository in mathematics, physics, computer science, and several other fields, shows how deep learning is taking up more and more of the papers. For example, the essay [117] cites statistics that show how:

- the percentage of papers that use deep learning has grown steadily over the last decade within most computer science categories on *arXiv*;
- and moreover, categories that feature high deep learning adoption are growing in popularity compared to other categories.

---

[1] https://arxiv.org/.

© The Author(s), under exclusive license to Springer Nature Switzerland AG 2021
S. I. Nikolenko, *Synthetic Data for Deep Learning*, Springer Optimization
and Its Applications 174, https://doi.org/10.1007/978-3-030-75178-4_1

DATA ⟶ ANNOTATION ⟶ TRAINING ⟶ DEPLOYMENT

Fig. 1.1 The structure of a machine learning project.

The essay [117] was published in 2018, but the trend continues to this day: the number of papers on deep learning is growing exponentially, each individual subfield of deep learning is getting more and more attention, and all of this does not show any signs of stopping. We are living on the third hype wave of artificial intelligence, and nobody knows when and even if it is going to crash like the first two (we will briefly discuss them in Section 2.1).

Still, despite all of these advances, the basic pipeline of using machine learning for a given problem remains mostly the same, as shown in Figure 1.1:

- first, one has to collect raw *data* related to the specific problem and domain at hand;
- second, the data has to be *labeled* according to the problem setting;
- third, machine learning models *train* on the resulting labeled datasets (and often also *validate* their performance on subsets of the datasets set aside for testing);
- fourth, after one has trained the model, it needs to be *deployed* for inference in the real world; this part often involves deploying the models in low-resource environments or trying to minimize latency.

The vast majority of the thousands of papers published in machine learning deal with the "Training" phase: how can we change the network architecture to squeeze out better results on standard problems or solve completely new ones? Some deal with the "Deployment" phase, aiming to fit the model and run inference on smaller edge devices or monitor model performance in the wild.

Still, any machine learning practitioner will tell you that it is exactly the "Data" and (for some problems especially) "Annotation" phases that take upwards of 80% of any real data science project where standard open datasets are not enough. Will these 80% turn into 99% and become a real bottleneck? Or have they already done so? Let us find out.

For computer vision problems, the labeling required is often very labor-intensive. Suppose that you want to teach a model to count the cows grazing in a field, a natural and potentially lucrative idea for applying deep learning in agriculture. The basic computer vision problem here is either *object detection*, i.e., drawing bounding boxes around cows, or *instance segmentation*, i.e., distinguishing the silhouettes of cows. To train the model, you need a lot of photos with labeling like the one shown in Fig. 1.2 (segmentation on Fig. 1.2a; object detection on Fig. 1.2b).

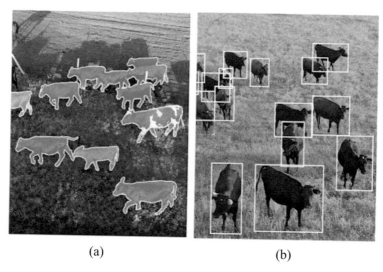

(a) (b)

Fig. 1.2 Sample labeling for standard computer vision problems: (a) instance segmentation; (b) object detection.

Imagine how much work it is to label a photo like this by hand. Naturally, people have developed tools to help partially automate the process. For example, a state-of-the-art labeling tool (see, e.g., [829]) will suggest a segmentation candidate produced by some general-purpose model, and the human annotator is only supposed to fix the mistakes of this model by clicking on individual pixels that are segmented incorrectly. But it might still take minutes per photo, and the training set for a standard segmentation or object detection model should have thousands or even tens of thousands of such photos. This adds up to human-years and hundreds of thousands, if not millions, of dollars spent on labeling only.

There exist large open datasets for many different problems, segmentation and object detection included. But as soon as you need something beyond the classes, settings, and conditions that are already well covered in these datasets, you are out of luck; for instance, *ImageNet* does have cows, but not shot from above with a drone.

And even if the dataset appears to be tailor-made for the problem at hand, it may contain dangerous biases. For example, suppose that the basic problem is *face recognition*, a classic computer vision problem with large-scale datasets freely available [37, 38, 366, 542]; there also exist synthetic datasets of people, and we will discuss them in Section 6.6. But if you want to recognize faces "in the wild", you need a dataset that covers all sorts of rotations for the faces, while standard datasets mostly consist of frontal photos. For example, the work [1017] shows that the distribution of face rotations in IJB-A [460], a standard open large-scale face recognition dataset, is extremely unbalanced; the work discusses how to fill in the gaps in this distribution by producing synthetic images of faces from the IJB-A dataset (see Section 10.3, where we discuss this work in detail).

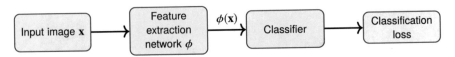

Fig. 1.3   General architecture of a simple classification network.

To sum up: current systems are data-intensive, data is expensive, and we are hitting the ceiling of where we can go with already available or easily collectible datasets, especially with complex labeling. So what's next? How can we solve the data problem? Is machine learning heading towards a brick wall? Hopefully not, but it will definitely take additional effort. Over the next sections, we discuss what can be done to alleviate the data problem. We will give a very high-level overview of several possible approaches currently explored by machine learning researchers and then make an introduction to the main topic of this book: synthetic data.

1.2   One-Shot Learning and Beyond: Less Data for More Classes

We have already mentioned face recognition systems and have just discussed that computer vision systems generally need a lot of labeled training samples to learn how to recognize objects from a new class. But then how are face recognition systems supposed to work at all? The vast majority of face recognition use cases break down if we require hundreds of photos in different contexts taken for every person we need to recognize. How can a face recognition system hope to recognize a new face when it usually has at most a couple of shots for every new person?

The answer is that face recognition systems are a bit different from regular image classifiers. Any machine learning system working with unstructured data (such as photographs) is basically divided into two parts:

- feature extraction, the part that converts an image into a (much smaller) numerical vector, usually called an *embedding* or a *latent code*, and
- a machine learning model (e.g., a classifier) that uses extracted features to actually solve the problem.

So a regular image classifier consists of a feature extraction network followed by a classification model, as shown in Fig. 1.3; this kind of architecture was used by the first neural networks that brought the deep learning revolution to computer vision such as *AlexNet*, *GoogLeNet*, and others (we will discuss them in Section 3.1). In deep learning, the classifier is usually very simple, in most cases basically equivalent to logistic regression, and feature extraction is the interesting part. Actually, most of the advantages of deep learning come from the fact that neural networks can learn to be much better at feature extraction than anything handcrafted that we humans had been able to come up with.

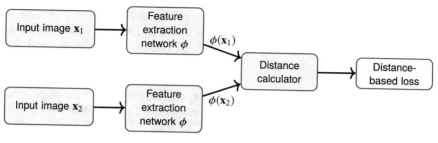

Fig. 1.4 General architecture of a Siamese network.

To train this kind of network, you do indeed need a lot of labeled faces, and to add a new face, you need quite a few examples of the new class. However, this is not the only way. An important direction in the development of modern face recognition systems is related to learning face embeddings (learning feature extraction) in various ways. For example, *FaceNet* [771] learns with a modification of the Siamese network approach, where the target is not a class label but rather the distances or similarities between face embeddings. The goal is to learn embeddings in such a way that embeddings of the same face will be close together while embeddings of different faces will be clearly separated, far from each other.

The general architecture of this approach is shown in Fig. 1.4: feature extractors produce numerical features that are treated as vectors in a Euclidean space, and the loss function is designed to, roughly speaking, bring the vectors corresponding to the same person close to each other and push vectors extracted from images of different people apart.

After the *FaceNet* model shown in Fig. 1.4 has been trained with distance-based metrics in mind, we can use the embeddings to do what is called *one-shot learning*. Assuming that the embedding of a new face will have the same basic properties, we can simply compute the embedding for a new person (with just a single photo as input!) and then do classification by looking for nearest neighbors in the space of embeddings. While this approach has met with some problems specifically for face recognition due to complications in the mining of hard negative examples, and some state-of-the-art face recognition systems are currently trained as classifiers with additional tricks and modified loss functions [188, 915, 966], this approach still remains an important part of the research landscape.

This is a simplified but realistic picture of how one-shot learning systems work. But one can go even further: what if there is no data available at all for a new class? This setting is known as *zero-shot learning*. The problem sounds impossible, and it really is: if all you know are images from "Class 1" and "Class 2", and then you are asked to distinguish between "Class 3" and "Class 4" with no additional information, no amount of machine learning can help you. But in real life, it does not work this way: we usually have some background knowledge about the new classes even if we do not have any images. For example, when we are asked to recognize a

"Yorkshire terrier"[2], we know that it is a kind of dog, and maybe we even have its verbal description that can be lifted, e.g., from *Wikipedia*. With this information, we can try to learn a joint embedding space for both class names and images, and then use the same nearest neighbors approach but look for the nearest label embedding rather than other images (which we do not have for a new class).

This kind of cross-modal zero-shot learning was initiated by Socher et al. [810], who trained a model to recognize objects on images based on knowledge about class labels learned from unsupervised text corpora. Their model learns a joint latent space of word embeddings and image feature vectors. Naturally, this approach will not give the same kind of accuracy as training on a large labeled set of images, but zero-shot learning systems are increasingly successful. In particular, a more recent paper by Zhu et al. [1029] uses a generative adversarial network (GAN) to "hallucinate" images of new classes by their textual descriptions and then extracts features from these hallucinated images; this comes close to the usage of GANs to produce and refine synthetic data that we explore in Chapter 10; see also recent surveys of zero-shot and few-shot learning [911, 916, 917, 949].

Note, however, that one- and zero-shot learning still require large labeled datasets. The difference is that we do not need a lot of images for each new class any more. But the feature extraction network has to be trained on similar labeled classes: a zero-shot approach will not work if you train it on birds and then try to look for a chair based on its textual description. Until we are able to use super-networks trained on every kind of images possible (as we will see shortly, we still have quite a way to go before this becomes possible, if it is even possible at all with our current approaches), this is still a data-intensive approach, although restrictions on what kind of data to use are relaxed.

A middle ground is taken by models that try to generalize from several examples, a field known as *few-shot learning* [916]. Similar to one-shot learning, generalizing from few examples in complex problems is a process that has to be guided by expressive and informative prior distributions—otherwise, the data will simply not be enough. In many ways, this is how human learning works: we usually need a few examples to grasp a novel concept, but never thousands of examples, because the new concept probably fits into the prior framework about the world that we have built over our lifetimes.

To get these prior distributions in a machine learning model, we can use a number of different approaches. We have seen a couple of examples above, and the general classification of one- and few-shot approaches includes at least the following:

- *data augmentation* that extends the small available dataset with transformations that do not change the properties that we need to learn;
- *multitask learning* that trains a model to solve several problems, each of which may have a small dataset, but together these datasets are large enough;
- *embedding learning* that learns latent representations which can be generalized to new examples, as we have discussed above;

---

[2]*ImageNet* [187], the main basic dataset for computer vision models, is very fond of canines: it distinguishes between 120 different dog breeds from around the world.

- *fine-tuning* that updates existing models that have been pretrained on different tasks (possibly unsupervised) with small amounts of new data.

We will encounter all of these techniques in this book.

1.3 Weakly Supervised Training: Trading Labels for Computation

For many problems, obtaining labeled data is expensive but unlabeled data, especially data that is not directly related to the specific problem at hand, is plentiful. Consider again the basic computer vision problems we have talked about: it is very expensive to obtain a large dataset of labeled images of cow herds, but it is much less expensive to get a large dataset of unlabeled such images, and it is almost trivial to simply get a lot of images with and without cows.

But can unlabeled data help? Basic intuition tells that it may not be easy but should be possible. After all, we humans learn from all kinds of random images, and during infancy, we develop an intuition about the world around us that generalizes exceptionally well. Armed with this intuition, we can later in life do one-shot and zero-shot learning with no problem. And the images were never actually labeled, the learning we do can hardly be called supervised. Although it is still a far cry from human abilities (see, e.g., a recent treatise by Francois Chollet [154] for a realistic breakdown of where we stand in terms of artificial general intelligence), there are several approaches being developed in machine learning to make use of all this extra unlabeled data lying around.

First, one can use unlabeled data to produce new labeled data; although the new "pseudolabels" are not guaranteed to be correct, they will still help, and one can revisit and correct them in later iterations. A striking illustration of this approach has appeared in a recent work by Xie et al. [956], where researchers from Google Brain and Carnegie Mellon University applied the following algorithm:

- start from a "teacher" model trained as usual, on a (smaller) labeled dataset;
- use the "teacher" model on the (larger) unlabeled dataset, producing pseudolabels;
- train a "student" model on the resulting large labeled dataset;
- use the trained student model as the teacher for the next iteration, and then repeat the process iteratively.

This is a rather standard approach, used many times before in semi-supervised learning and also known as *knowledge distillation* [129, 293, 345, 541, 603]. But by adding noise to the student model, Xie et al. managed to improve the state-of-the-art results on *ImageNet*, the most standard and beaten-down large-scale image classification dataset. For this, however, they needed a separate dataset with 300 million unlabeled images and a lot of computational power: 3.5 days on a 2048-core Google TPU, on the same scale as *AlphaZero* needed to outperform every other engine in Go and chess [800]; we will shortly see that this kind of computation does not come for free.

Another interesting example of replacing labeled data with (lots of) unlabeled data comes from the problem we already discussed: segmentation. It is indeed very hard to produce labeled data for training segmentation models... but do we have to? Segmentation models from classical computer vision, before the advent of deep learning, do not require any labeled data: they cluster pixels according to their features (color and perhaps features of neighboring pixels) or run an algorithm to cut the graph of pixels into segments with minimal possible cost [657, 839]. Modern deep learning models work better, of course, but it looks like recent advances make it possible to train deep learning models to do segmentation without labeled data as well.

Approaches such as W-Net [948] use unsupervised autoencoder-style training to extract features from pixels and then segment them with a classical algorithm. Approaches such as invariant information clustering [399] develop image clustering approaches and then apply them to each pixel in a convolution-based way, thus transforming image clustering into segmentation. One of the most intriguing lines of work that results in unsupervised clustering uses GANs for image manipulation. The "cut-and-paste" approach [716] works in an adversarial way:

- one network, the mask generator, constructs a segmentation mask from the features of pixels in a detected object;
- then the object is cut out according to this mask and transferred to a different, object-free location on the image;
- another network, the discriminator, now has to distinguish whether an image patch has been produced by this cut-and-paste pipeline or is just a patch of a real image.

The idea is that good segmentation masks will make realistic pasted images, and in order to convince the discriminator, the mask generator will have to learn to produce high-quality segmentation masks. We will discuss this approach and its ramifications for synthetic data in Section 9.3.

Semi-supervised teacher–student training and unsupervised segmentation via cut-and-paste are just two directions out of many that are currently being developed. In these works, researchers are exploring various ways to trade the need for labeled datasets for extra unlabeled data, extra unrelated data, or extra computation, all of which are becoming more and more readily available. Still, this does not completely solve the data problem, and the computational challenges might prove to be insurmountable.

1.4 Machine Learning Without Data: Leaving Moore's Law in the Dust

Interestingly, some kinds of machine learning do not require any external data at all, let alone labeled data. Usually, the idea is that they are able to generate data for themselves, and the catch is that they need *a lot* of generation power.

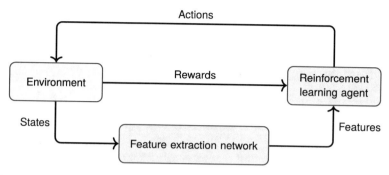

Fig. 1.5 General reinforcement learning flowchart.

The main field where this becomes possible is *reinforcement learning* (RL), where an agent learns to perform well in an interactive environment [788, 831]. An agent can perform actions and receive rewards for these actions from the environment. Usually, modern RL architectures consist of the feature extraction part that processes environment states into features and an RL agent that transforms features into actions and converts rewards from the environment into weight updates; see an illustration in Figure 1.5.

The poster child of these approaches is *AlphaZero* by *DeepMind* [800]. Their original breakthrough was *AlphaGo* [799], a model that beat Lee Sedol, one of the top human Go players, in an official match in March 2016. Long after *DeepBlue* beat Kasparov in chess, professional-level Go was remaining out of reach for computer programs, and *AlphaGo*'s success was unexpected even in 2016. The match between Lee Sedol and *AlphaGo* became one of the most publicized events in AI history and was widely considered as the "Sputnik moment" for Asia in AI, the moment when China, Japan, and South Korea realized that deep learning is to be taken seriously. But *AlphaGo* utilized a lot of labeled data: it had a pretraining step that used a large database of professional games.

AlphaZero takes its name from the fact that it needs zero training data: it begins by knowing only the rules of the game and achieves top results through self-play, actually with a very simple loss function combined with tree search. *AlphaZero* beat *AlphaGo* (and its later version, *AlphaGo Zero*) in Go and one of the top chess engines, *Stockfish*, in chess.

A recent result by the *DeepMind* reinforcement learning team, *MuZero* [770], is even more impressive. *MuZero* is an approach based on *model-based RL*, that is, it builds a model of the environment as it goes and does not know the rules of the game beforehand but has to learn them from scratch; e.g., in chess, it cannot make illegal moves as actions but can consider them in tree search and has to learn that they are illegal by itself. With this additional complication, *MuZero* was able to achieve *AlphaZero*'s skill in chess and shogi and even outperform it in Go. Most importantly, the same model could also be applied to situations with more complex environment

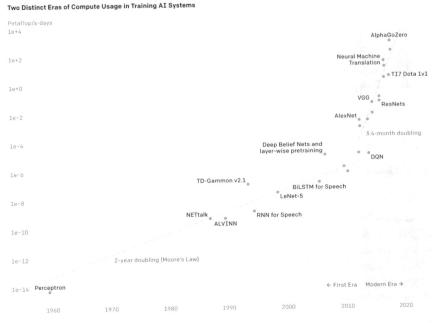

Fig. 1.6 The changing trend in deep learning: a comparison of "Moore's law" in machine learning before and after the rise of deep learning. Chart by *OpenAI* [16].

states, e.g., to computer games in *Atari* environments (a standard benchmark in reinforcement learning).

So what's the catch? Is this the end of the data problem? One drawback is that not every problem can be formulated in terms of RL with no data required. You can learn to play games, i.e., self-contained finite structures where all rules are known in advance. But how do we learn, say, autonomous driving or walking, with a much wider variety of possible situations and individual components of these situations? One possible solution here is to use synthetic virtual environments, and we will discuss them in detail in Chapter 7.

Another, perhaps even more serious, problem is the amount of computation needed for further advances in reinforcement learning. To learn to play chess and Go, *MuZero* used 1000 third-generation Google TPUs to simulate self-play games. This does not tell us much by itself, but here is an interesting observation made by the *OpenAI* team [16], illustrated in Fig. 1.6. They noticed that before 2012, computational resources needed to train state-of-the-art AI models grew basically according to Moore's Law, doubling their computational requirements every two years. But with the advent of deep learning, in 2012–2019, computational resources for top AI model training doubled on average every 3.4 months! This is a huge rate of increase, and, obviously, it cannot continue forever, as the actual hardware computational power growth is only slowing down compared to Moore's Law. Replication of *AlphaZero* experiments has been recently estimated to cost about $35 million at current *Google*

Cloud rates [365]; while the cost of computation is dropping, it does so at a much slower rate than the increase of computation needed for AI.

Thus, one possible scenario for further AI development is that yes, indeed, this "brute force" approach might theoretically take us very far, maybe even to general artificial intelligence, but it would require more computational power than we actually have in our puny Solar System. Note that a similar phenomenon, albeit on a smaller scale, happened with the second wave of hype for artificial neural networks: researchers in the late 1980s had a lot of great ideas about neural architectures (including CNNs, RNNs, RL, and much more), but neither the data nor the computational power was sufficient to make a breakthrough, and neural networks were relegated to "the second best way of doing just about anything"[3].

Still, at present, reinforcement learning represents another feasible way to trade labeled data for computation, as the example of *AlphaGo* blossoming into *AlphaZero* and *MuZero* clearly shows. With this, we finish a brief overview of alternatives and come to the main subject of this book: *synthetic data*.

1.5   Why Synthetic Data?

Let us go back to segmentation, a standard computer vision problem that we already discussed in Section 1.1. How does one produce a labeled dataset for image segmentation? At some point, all images have to be manually processed: humans have to either draw or at least verify and correct segmentation masks. Making the result pixel-perfect is so laborious that it is commonly considered to be not worth the effort. Figure 1.7a–c shows samples from the industry standard *Microsoft Common Objects in Context* (MS COCO) dataset [525]; you can immediately see that the segmentation mask is a rather rough polygon and misses many finer features. It did not take us long to find such rough segmentation maps, by the way; these are some of the first images found by the "dog" and "person" queries.

How can one get a higher quality segmentation dataset? To manually correct all of these masks in the MS COCO dataset would probably cost hundreds of thousand dollars. Fortunately, there is a different solution: *synthetic data*. In the context of segmentation, this means that the dataset developers create a 3D environment with modes of the objects they want to recognize and their surroundings and then render the result. Figure 1.7d–e shows a sample frame from a synthetic dataset called *ProcSy* [449] (we discuss it in more detail in Section 6.5): note how the segmentation map is now perfectly correct. While 3D modeling is still mostly manual labor, this is a one-time investment, and after this investment, one can get a potentially unlimited number of pixel-perfect labeled data: not only RGB images and segmentation maps but also depth images, stereo pairs produced from different viewpoints, point clouds, synthetic video clips, and other modalities.

---

[3]A quote usually attributed to John Denker; see, e.g., [339].

In general, many problems of modern AI come down to insufficient data: either the available datasets are too small or, also very often, even while capturing unlabeled data is relatively easy, the costs of manual labeling are prohibitively high. *Synthetic data* is an important approach to solving the data problem by either producing artificial data from scratch or using advanced data manipulation techniques to produce novel and diverse training examples. The synthetic data approach is most easily exemplified by standard computer vision problems, as we have done above, but it is also relevant in other domains (we will discuss some of them in Chapter 8).

Naturally, other problems arise; the most important of them being the problem of *domain transfer*: synthetic images, as you can see from Figure 1.7, do not look exactly like real images, and one has to make them as photorealistic as possible and/or devise techniques that help models transfer from synthetic training sets to real test sets; thus, domain adaptation becomes a major topic in synthetic data research and in this book as well (see Chapter 10). Note, however, that a common theme in synthetic data research is whether realism is actually necessary; we will encounter this question several times in this book.

We begin with a few general remarks regarding synthetic data. First, note that synthetic data can be produced and supplied to machine learning models on the fly, during training, with software synthetic data generators, thus alleviating the need to ever store huge datasets; see, e.g., Mason et al. [583] who discuss this "on the fly" generation in detail. Second, while synthetic data has been a rising field for some time, I do not know of a satisfactory general overview of synthetic data in machine learning or deep learning, and this was my primary motivation for writing this book. I would like to note surveys that attempt to cover applications of synthetic data [157, 467, 875] and a special issue of the *International Journal of Computer Vision* [253], but I hope that the present work paints a more comprehensive picture.

Third, we distinguish between synthetic data and *data augmentation*; the latter is a set of techniques intended to modify real data rather than create new synthetic data. These days, data augmentation is a crucial part of virtually every computer vision pipeline; we refer to the surveys [792, 914] and especially recommend the *Albumentations* library [104] that has proven invaluable in our practice, but in this survey, we concentrate on synthetic data rather than augmentations (the latter will only be the subject of Section 3.4). Admittedly, the line between them is blurry, and some techniques discussed here could instead be classified as "smart augmentation".

Fourth, we note a natural application of synthetic data in machine learning: testing hypotheses and comparing methods and algorithms in a controlled synthetic setting. Toy examples and illustrative examples are usually synthetic, with a known data distribution so that machine learning models can be evaluated on how well they learn this distribution. This approach is widely used throughout the field, sometimes for entire meta-analyses [80, 491, 797], and we do not dwell on it here; our subject is synthetic data used to transfer to real data rather than direct comparisons between models on synthetic datasets.

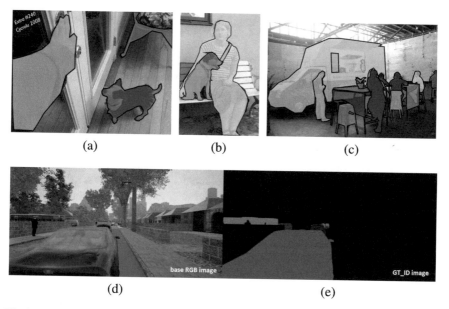

Fig. 1.7 Sample images: (a–c) MS COCO [525] real data samples with ground truth segmentation maps overlaid; (d–e) *ProcSy* [449]: (d) RGB image; (e) ground truth segmentation map.

1.6   The Plan

This book covers three main directions for the use of synthetic data in machine learning; in this section, we introduce all three, giving references to specific parts of the book related to these directions.

1. Using synthetically generated datasets to train machine learning models directly. This is an approach often taken in computer vision, and most of this book is devoted to variations of this approach. In particular, one can:

   - train models on synthetic data with the intention to use them on real data; we discuss this approach through most of Chapters 6, 7, and 8;
   - train (usually generative) models that change (refine) synthetic data in order to make it more suitable for training or adapt the model to allow it to be trained on synthetic data; Chapter 10 is devoted to this kind of models.

2. Using synthetic data to augment existing real datasets so that the resulting hybrid datasets are better suited for training the models. In this case, synthetic data is usually employed to cover parts of the data distribution that are not sufficiently represented in the real dataset, with the main purpose being to alleviate dataset bias. The synthetic data can either:

   - be generated separately with, e.g., CGI-based methods for computer vision (see examples in Chapters 3 and 7)

- or be generated from existing real data with the help of generative models (see Section 10.4).

3. Using synthetic data to resolve privacy or legal issues that make the use of real data impossible or prohibitively hard. This becomes especially relevant for certain specific fields of application, among which we discuss:

   - synthetic data in healthcare, which is not restricted to imaging but also extends to medical records and the like (Section 10.7);
   - synthetic data in finance and social sciences, where direct applications are hard but privacy-related ones do begin to appear (Section 11.5);
   - synthetic data with privacy guarantees: many applications are sensitive enough to require a guarantee of privacy, for example, from the standpoint of the differential privacy framework, and there has been an important line of work that makes synthetic data generation provide such guarantees, which we consider in Chapter 11.

The book is organized as follows. Chapter 2 gives a very brief introduction to deep learning; while one cannot hope to deliver an actual in-depth introductory text in the space of a single chapter, we will nevertheless start with the basics of how the deep learning revolution has transformed machine learning in the late 2000s (Section 2.1), how a neural network is organized (Section 2.3), and how it is trained via various modifications of gradient descent (Sections 2.4 and 2.5).

Next, since computer vision is by far the most common domain for applying synthetic data, we will devote Chapter 3 to deep architectures that are designed for computer vision problems and that will be used throughout other chapters of the book. Section 3.1 introduces the notion of convolutional neural networks, Section 3.2 shows several basic ideas that underlie modern convolutional architectures for image classification, object detection, and segmentation, and Section 3.3 provides a case study of neural architectures for object detection; we cannot hope to cover everything but at least try to take a deep dive into a single topic to illustrate the depth and variability of deep learning approaches in computer vision.

The final, most advanced introductory chapter deals with generative models in deep learning; they are the topic of Chapter 4. We begin by discussing generative models in machine learning in general (Section 4.1), introducing the difference between discriminative and generative models. Next, we discuss Ian Goodfellow's taxonomy of generative models in deep learning and give an overview of tractable density models, including an introduction to normalizing flows (Section 4.2). In Section 4.3, we talk about variational autoencoders as the primary example of approximate explicit density models. Section 4.4 introduces the class of generative models most directly relevant to synthetic data: generative adversarial networks (GAN). Section 4.5 discusses loss functions in modern GANs, Section 4.6 gives a brief overview of some important general GAN-based architectures, and Section 4.7 finishes the chapter with a case study of GAN-based style transfer, a problem which both serves as a good illustration for modern adversarial architectures and is directly relevant to synthetic-to-real domain adaptation.

Chapter 5 is devoted to the early history of synthetic data. It may seem that synthetic data has only been on the rise for the last few years, but we will see that the use of synthetic data dates back to the 1970s, when computer vision took its first steps (Section 5.1), was used as a testbed for experimental comparisons throughout the history of computer vision (Section 5.2), and was used to train one of the first self-driving cars in 1989 (Section 5.3). The final sections of this chapter are devoted to early robotics: we first discuss how synthetic data was used to train robot control systems from their very inception and how it was faced with some reservations and criticism (Section 5.4) and then make a more detailed example of an early robotic navigation system called MOBOT (Section 5.5).

Chapter 6 presents synthetic datasets and results for basic computer vision problems, including low-level problems such as optical flow or stereo disparity estimation (Section 6.2), datasets of basic objects (Section 6.3), basic high-level computer vision problems, including a case study on object detection (Section 6.4) and a discussion of other high-level problems (Section 6.5), human-related synthetic data (Section 6.6), and character and text recognition and visual reasoning problems (Section 6.7). Throughout Chapter 6, we refer to specific architectures whose results have been improved by training on synthetic data and show examples of images from synthetic datasets.

In Chapter 7, we proceed to synthetic datasets that are more akin to full-scale simulated environments, covering outdoor and urban environments (Section 7.2), indoor scenes (Section 7.3), synthetic simulators for robotics (Section 7.4), simulators for autonomous flying vehicles (Section 7.5), and computer games used as simulation environments (Section 7.6). Synthetic simulated environments are an absolute necessity for, e.g., end-to-end reinforcement learning architectures, and we will see examples of situations where they suffice to train a successful RL agent as well as situations where additional work on domain transfer is required.

While synthetic data has been most instrumental in computer vision, there are other domains of application for synthetic data as well, and Chapter 8 is devoted to exactly such domains. In particular, neural programming (Section 8.2) is a field completely reliant on automatically generated training samples: it does not make any sense to force humans to write millions of trivial computer programs. In bioinformatics (Section 8.3), most applications of synthetic data again lie in the domain of medical imaging, but there are fields where one learns to generate other kinds of information, for instance, fingerprints of molecules, and trains subsequent models on this synthetically generated data. Finally, natural language processing (Section 8.4) is not a field where synthetic data has really taken off despite obvious successes in text generation [94, 697, 698]. A computer program that can generate coherent text with predefined properties must be an AI model that captures a lot of useful features about the language, and it is usually more productive to use the model itself than try to learn a different model on its outputs; however, there are some examples of synthetic data in NLP as well.

Chapter 9 discusses research intended to improve synthetic data generation. The notion of *domain randomization* (Section 9.1) means trying to cover as much of the data distribution with synthetic samples as possible, making them maximally different and randomized, with the hope to capture the real data in the support of the synthetic data distribution as well. Section 9.2 discusses current developments in methods for CGI-based generation, including more realistic simulations of real-world sensors (cameras, LiDAR systems, etc.) and more complex ways to define and generate synthetic scenes. Synthetic data produced by "cutting and pasting" parts of real data samples is discussed in Section 9.3, and we end the chapter with a discussion of the direct generation of synthetic data by generative models (Section 9.4). It is rare to see such examples because, similar to natural language processing, a generative model trained to produce high-quality samples from a given domain probably already contains a model that can be used directly or fine-tuned for various applications; still, there are situations where GANs can help directly.

The next part of the book deals with the main research problem of synthetic data, namely *synthetic-to-real domain adaptation*: how can we make a model trained on synthetic data perform well on real samples? After all, the ultimate goal of the entire enterprise is always to apply the model to real data. With this, we get to Chapter 10 that discusses synthetic-to-real domain adaptation itself. There are many approaches to this problem that can be broadly classified into two categories:

- *synthetic-to-real refinement*, where domain adaptation models are used to make synthetic data more realistic (Section 10.1);
- *domain adaptation at the feature/model level*, where the model and/or the training process are adapted rather than the data itself (Section 10.5).

The difference is that with refinement, one usually can extract refined input data: either synthetic samples made "more realistic" or real samples made "more synthetic-like"; with domain adaptation at the model level, the architectures usually just learn to extract common features from both domains. In this chapter, we also discuss case studies of domain adaptation for control and robotics (Section 10.6) and medical imaging (Section 10.7).

Chapter 11 is devoted to the privacy side of synthetic data: can we generate synthetic data which is *guaranteed* not to contain personal information about individual entries from the original dataset? To get such guarantees, we need to venture into *differential privacy*, a field that belongs more to the domain of theoretical cryptography than machine learning. Sections 11.2 and 11.3 introduce differential privacy in general and specifically for deep learning models, Section 11.4 shows how to generate synthetic data with differential privacy guarantees, and Section 11.5 presents a case study about private synthetic data in finance and related fields, in particular, electronic medical records.

In an attempt to look forward, we devote Chapter 12 to directions for further work related to synthetic data that seem most promising. We consider four such directions:

- Section 12.1 considers *procedural generation* of synthetic data, where the data is made more realistic not by low-level refinement but by improving the high-level generation process: for instance, instead of refining the textures of wood and fabric on chairs, we are talking about a more consistent layout of the entire synthetic room's interior;

- in Section 12.2, we introduce the notion of *closing the feedback loop* for synthetic data generation: since the end goal is to improve the performance of models trained on synthetic datasets, maybe we can change the parameters of synthetic data generation in such a way as to directly increase this metric;
- Section 12.3 talks about introducing *domain knowledge* into domain adaptation; specifically, we consider an example where the model contains both a domain-specific generative model designed to produce synthetic images and a bottom-up model that estimates the necessary parameters in an image;
- Section 12.4 shows how domain adaptation models can be improved with additional modalities that are easy to obtain in synthetic datasets; for instance, in computer vision, it is trivial to augment synthetic data with 3D information such as depth maps of surface normals since synthetic data is produced from 3D scenes, so maybe this additional information can help a refiner to make this data even more realistic.

Finally, Section 12.5 concludes the book by drawing some general conclusions about the place of synthetic data in modern AI and possible future work in this direction.

By now, we have seen how the deep learning revolution makes demands on computational power and data that are increasingly hard to satisfy. There are some ways to get around the need for ever growing labeled datasets, but they usually seem to require even more computational resources, which are by now also not so easy to obtain. We have seen that synthetic data is one possible way out of this conundrum. But for the uninitiated, it is still unclear what this "deep learning" is all about, and this is exactly what awaits us in the next chapter.

Chapter 2
Deep Learning and Optimization

Deep learning is currently one of the hottest fields not only in machine learning but in the whole of science. Since the mid-2000s, deep learning models have been revolutionizing artificial intelligence, significantly advancing state of the art across all fields of machine learning: computer vision, natural language processing, speech and sound processing, generative models, and much more. This book concentrates on synthetic data applications; we cannot hope to paint a comprehensive picture of the entire field and refer the reader to other books for a more detailed overview of deep learning [153, 289, 630, 631]. Nevertheless, in this chapter, we begin with an introduction to deep neural networks, describing the main ideas in the field. We especially concentrate on approaches to optimization in deep learning, starting from regular gradient descent and working our way towards adaptive gradient descent variations and state-of-the-art ideas.

2.1   The Deep Learning Revolution

In 2006–2007, machine learning underwent a true revolution that began a new, third "hype wave" for artificial neural networks (ANNs) in particular and artificial intelligence (AI) in general. Interestingly, one can say that artificial neural networks were directly responsible for all three AI "hype waves" in history[1]:

- in the 1950s and early 1960s, Frank Rosenblatt's *Perceptron* [735, 736], which in essence is a very simple ANN, became one of the first machine learning formalisms to be actually implemented in practice and featured in *The New York*

---

[1]For a very comprehensive account of the early history of ANNs and deep learning, I recommend a survey by one of the fathers of deep learning, Prof. Jürgen Schmidhuber [767]; it begins with Newton and Leibniz, whose results, as we will see, are still very relevant for ANN training today.

© The Author(s), under exclusive license to Springer Nature Switzerland AG 2021
S. I. Nikolenko, *Synthetic Data for Deep Learning*, Springer Optimization
and Its Applications 174, https://doi.org/10.1007/978-3-030-75178-4_2

Times, which led to the first big surge in AI research; note that the first mathematical formalizations of neural networks appeared in the 1940s [589], well before "artificial intelligence" became a meaningful field of computer science with the foundational works of Alan Turing [878] and the Dartmouth workshop [587, 611] (see Section 2.3);

- although gradient descent is a very old idea, known and used at least since the early XIX century, only by the 1980s, it became evident that backpropagation, i.e., computing the gradient with respect to trainable weights in the network via its computational graph, can be used to apply gradient descent to deep neural networks with virtually arbitrary acyclic architectures; this idea became common in the research community in the early 1980s [924], and the famous *Nature* paper by Rumelhart et al. [742] marked the beginning of the second coming of neural networks into the popular psyche and business applications.

Both of these hype waves proved to be premature, and neither in the 1960s nor in the 1990s could neural networks live up to the hopes of researchers and investors. Interestingly, by now we understand that this deficiency was as much technological as it was mathematical: neural architectures from the late 1980s or early 1990s could perform very well if they had access to modern computational resources and, even more importantly, modern datasets. But at the moment, the big promises were definitely unfounded; let me tell just one story about it.

One of the main reasons the first hype wave came to a halt was the failure of a large project in no less than machine translation! It was the height of the Cold War, and the US government decided it would be a good idea to develop an automatic translation machine from Russian to English, at least for formal documents. They were excited by the Georgetown–IBM experiment, an early demonstration of a very limited machine translation system in 1954 [381]. The demonstration was a resounding success, and researchers of the day were sure that large-scale machine translation was just around the corner.

Naturally, this vision did not come to reality, and twelve years later, in 1966, the ALPAC (Automatic Language Processing Advisory Committee) published a famous report that had to admit that machine translation was out of reach at the moment and stressed that a lot more research in computational linguistics was needed [620]. This led to a general disillusionment with AI on the side of the funding bodies in the U.S., and when grants stop coming in, researchers usually have to switch to other topics, so the first "AI winter" followed. This is a great illustration of just how optimistic early AI was: naturally, researchers did not expect perfection and would be satisfied with the state of, say, modern Google Translate, but by now we realize how long and hard a road it has been to what Google Translate can do today.

However, in the mid-2000s, deep neural networks started working in earnest. The original approaches to training deep neural networks that proved to work around that time were based on unsupervised pretraining [226]: Prof. Hinton's group achieved the first big successes in deep learning with deep belief networks (DBN), a method where layers of deep architectures are pretrained with the restricted Boltzmann machines, and gradient descent comes only at the very end [344, 347], while in Prof. Bengio's group, similar results on unsupervised pretraining were achieved by stacking autoencoders pretrained on the results of each other [62, 895]. Later, results on activation functions such as ReLU [281], new regularization methods for deep neural networks [387, 816], and better initialization of neural network weights [280] made unsupervised pretraining unnecessary for problems where large labeled datasets are available. These results have changed ANNs from "the second best way" into the method of choice revolutionizing one field of machine learning after another.

The first practical field where deep learning significantly improved state of the art in real-world applications was speech recognition, where breakthrough results were obtained by DBNs used to extract features from raw sound [346]. It was followed closely by computer vision, which we discuss in detail in Chapter 3, and later natural language processing (NLP). In NLP, the key contribution proved to be *word embeddings*, low-dimensional vectors that capture some of the semantic and syntactic properties of words and at the same time make the input dimensions for deep neural networks much more manageable [79, 600, 666]. These word embeddings were processed initially mostly by recurrent networks, but over recent years, the field has been overtaken by architectures based on self-attention: Transformers, BERT-based models, and the GPT family [94, 179, 192, 697, 698, 891]. We will touch upon natural language processing in Section 8.4, although synthetic data is not as relevant for NLP as it is for computer vision.

We have discussed in Section 1.1 that the data problem may become a limiting factor for further progress in some fields of deep learning, and definitely has already become such a factor for some fields of application. However, at present, deep neural networks define state of the art in most fields of machine learning, and progress does not appear to stop. In this chapter, we will discuss some of the basics of deep learning, and the next chapter will put a special emphasis on convolutional neural networks (Section 3.1 and further) because they are the main tools of deep learning in computer vision, and synthetic data is especially important in that field. But let me begin with a more general introduction, explaining how machine learning problems lead to optimization problems, how neural networks represent machine learning models, and how these optimization problems can be solved.

There is one important disclaimer before we proceed. I am writing this in 2020, and deep learning is constantly evolving. While the basic stuff such as the Bayes rule and neural networks as computational graphs will always be with us, it is very hard to say if the current state of the art in almost anything related to machine learning will remain the state of the art for long. Case in point: in the first draft of this book, I wrote that activation functions for individual units are more or less done. ReLU and its leaky variations work well in most problems, you can also try *Swish* found by automated search (pretty exhaustive, actually), and that's it, the future most probably shouldn't surprise us here. After all, these are just unary functions, and the *Swish* paper explicitly says that simpler activation functions outperform more complicated ones [702]. But in September 2020... well, let's not spoil it, see the end of Section 2.3.

That is why throughout this chapter and the next one, I am trying to mostly concentrate on the ideas and motivations behind neural architectures. I am definitely *not* trying to recommend any given architecture because most probably, when you are reading this, the recommendations have already changed. When I say "current state of the art", it's just that the snapshot of ideas that I have attempted to make as up to date as I could, and some of which may have become obsolete by the time you are reading this. The time for comprehensive surveys of deep learning has not yet come. So I am glad that this book is about synthetic data, and all I need from these two chapters is a brief introduction.

2.2 A (Very) Brief Introduction to Machine Learning

Before proceeding to neural networks, let me briefly put them into a more general context of machine learning problems. I usually begin my courses in machine learning by telling students that machine learning is a field of applied probability theory. Indeed, most of machine learning problems can be mathematically formulated as an application of the *Bayes rule*:

$$p\left(\theta \mid D\right) = \frac{p(\theta)p\left(D \mid \theta\right)}{p(D)},$$

where D denotes the data and θ denotes the model parameters. The distributions in the Bayes rule have the following meaning and intuition behind them in machine learning:

- $p\left(D \mid \theta\right)$ is the *likelihood*, i.e., the model itself; the likelihood captures our assumptions about how data is generated in a probability distribution;
- $p(\theta)$ is the *prior* probability, i.e., the distribution of our beliefs about the model parameters *a priori*, before we get any data;
- $p\left(\theta \mid D\right)$ is the *posterior* probability, i.e., the distribution of our beliefs about the model parameters *a posteriori*, after we take available data into account;
- $p(D) = \int p\left(D \mid \theta\right) p(\theta)d\theta$ is the *evidence* or *marginal probability* of the data averaged over all possible values of θ according to the likelihood.

This simple formula gives rise to the mathematical formulations of most machine learning problems. The first problem, common in classical statistics as well, is to find the *maximum likelihood hypothesis*

$$\theta_{ML} = \arg \max_{\theta} p\left(D \mid \theta\right).$$

The second problem is to multiply the likelihood by the prior, getting the posterior

$$p\left(\theta \mid D\right) \propto p\left(D \mid \theta\right) p(\theta),$$

and then find the *maximum a posteriori* hypothesis:

$$\theta_{MAP} = \arg \max_{\theta} p(\theta \mid D) = \arg \max_{\theta} p\left(D \mid \theta\right) p(\theta).$$

These two problems usually have similar structure when considered as optimization problems (we will see that shortly), and most practical machine learning is being done by maximizing either the likelihood or the posterior.

The final and usually the hardest problem is to find the *predictive distribution* for the next data point:

$$p(x \mid D) = \int p(x, \theta \mid D) \, d\theta = \int p(x \mid \theta) \, p(\theta \mid D) \, d\theta.$$

For at least moderately complex model likelihoods, this usually leads to intractable integrals and the need to develop approximate methods. Sometimes, it is this third problem which is called *Bayesian inference*, although the term is applicable as soon as a prior appears.

This mathematical essence can be applied to a wide variety of problems of different nature. With respect to their setting, machine learning problems are usually roughly classified into (Figure 2.1 provides an illustration):

- *supervised learning* problems, where data is given in the form of pairs $D = \{(\mathbf{x}_n, \mathbf{y}_n)\}_{n=1}^{N}$, with $\mathbf{x}_n$ being the nth data point (input of the model) and $\mathbf{y}_n$ being the target variable:

  - in *classification* problems, the target variable $\mathbf{y}$ is categorical, discrete, that is, we need to place $\mathbf{x}$ into one of a discrete set of classes;
  - in *regression* problems, the target variable $\mathbf{y}$ is continuous, that is, we need to predict values of $\mathbf{y}$ given values of $\mathbf{x}$ with as low error as possible;

- *unsupervised learning* problems that are all about learning a distribution of data points; in particular,

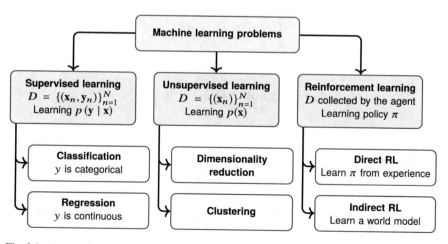

Fig. 2.1   A general taxonomy of machine learning problems.

- *dimensionality reduction* techniques aim to learn a low-dimensional representation that still captures important information about a high-dimensional dataset;
- *clustering* does basically the same but reduces not to a continuous space but to a discrete set of clusters, assigning each **x** from the input dataset with a cluster label; there is a parallel here with the classification/regression distinction in supervised learning;

- *reinforcement learning* problems where the data usually does not exist before learning begins, and an agent is supposed to collect its own dataset by interacting with the environment;

  - agents in *direct reinforcement learning* learn their behaviour policy π directly, either by learning a value function for various states and actions or by parameterizing π and learning it directly via policy gradient;
  - agents in *indirect reinforcement learning* use their experience to build a model of the environment, thus allowing for planning.

There are, of course, intermediate cases and fusions of these problems, the most important being probably *semi-supervised learning*, where a (usually small) part of the dataset is labeled and the other (usually far larger) part is not.

In this book, we will mostly consider supervised learning problems. For example, in computer vision, an image classification problem might be formalized with $\mathbf{x}_n$ being the image (a three-dimensional array of pixels, where the third dimension is the color) and $\mathbf{y}_n$ being a one-hot representation of target classes, i.e., $\mathbf{y}_n = (\,0 \ldots 0\,1\,0 \ldots 0\,)$, where 1 marks the position of the correct answer.

For a simple but already nontrivial example, consider the Bernoulli trials, the distribution of tossing a (not necessarily fair) coin. There is only one parameter here, let's say θ is the probability of the coin landing heads up. The data D is in this case a sequence of outcomes, heads or tails, and if D contains n heads and m tails, the likelihood is

$$p(D \mid \theta) = \theta^n \,(1 - \theta)^m \,.$$

The maximum likelihood hypothesis is, obviously,

$$\theta_{ML} = \arg \max_{\theta} \theta^n \,(1 - \theta)^m = \frac{n}{n + m},$$

but in real life, this single number is clearly insufficient. If you take a random coin from your purse, toss it once, and observe heads, your dataset will have $n = 1$ and $m = 0$, and the maximum likelihood hypothesis will be $\theta_{ML} = 1$, but you will hardly expect that this coin will now *always* land heads up. The problem is that you already have a prior distribution for the coin, and while the maximum likelihood hypothesis is perfectly fine in the limit, as the number of experiments approaches infinity, for smaller samples, the prior will definitely play an important role.

Suppose that the prior is uniform, $p(\theta) = 1$ for $\theta \in [0, 1]$ and 0 otherwise. Note that this is not quite what you think about a coin taken from your purse, you would rather expect a bell-shaped distribution centered at $\frac{1}{2}$. This prior is more suitable for a new phenomenon about which nothing is known *a priori* except that it has two outcomes. But even for that prior, the conclusion will change. First, the posterior distribution is now

$$p(\theta \mid D) = \frac{p(\theta)p(D \mid \theta)}{p(D)} = \begin{cases} \frac{1}{p(D)}\theta^n (1 - \theta)^m, & \text{for } \theta \in [0, 1], \\ 0 & \text{otherwise,} \end{cases}$$

where the normalizing constant can be computed as

$$p(D) = \int p(\theta)p(D \mid \theta)d\theta = \int_0^1 \theta^n (1 - \theta)^m d\theta =$$

$$= \frac{\Gamma(n + 1)\Gamma(m + 1)}{\Gamma(n + m + 2)} = \frac{n!m!}{(n + m + 1)!}.$$

Since the prior is uniform, the posterior distribution is still maximized at the exact same point:

$$\theta_{MAP} = \theta_{ML} = \frac{n}{n + m}.$$

This situation is illustrated in Figure 2.2a that shows the prior distribution, likelihood, and posterior distribution for the parameter θ with uniform prior and the dataset consisting of two heads. The posterior distribution, of course, has the same maximum as the likelihood, at $\theta_{MAP} = \theta_{ML} = 1$.

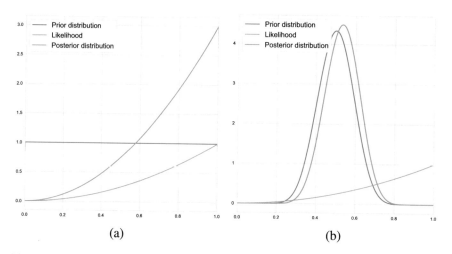

(a)   (b)

Fig. 2.2 Distributions related to the Bernoulli trials: (a) uniform prior, two heads in the dataset; (b) prior Beta(15, 15), two heads in the dataset.

But the predictive distribution will tell a different story because the posterior is maximized at its right border, and the predictions should integrate over the entire posterior. Let us find the probability of this coin landing heads on the next toss:

$$p(\text{heads}|D) = \int_0^1 p(\text{heads}|\theta)p(\theta|D)d\theta = \int_0^1 \frac{\theta^{n+1}(1-\theta)^m}{p(D)}d\theta =$$

$$= \frac{(n+1)!m!}{(n+m+2)!} \cdot \frac{(n+m+1)!}{n!m!} = \frac{n+1}{n+m+2}.$$

In this formula, we have derived what is known as *Laplace's rule of succession*, which shows how to apply Bayesian smoothing to the Bernoulli trials.

Note that in reality, if you take a random coin out of your pocket, the uniform prior would be a poor model for your beliefs about this coin. It would probably be more like the prior shown in Fig. 2.2b, where we show the exact same dataset processed with a non-uniform, informative prior $p(\theta) = \text{Beta}(\theta; 15, 15)$. The beta distribution

$$\text{Beta}(\theta; \alpha, \beta) \propto \theta^{\alpha-1}(1-\theta)^{\beta-1}$$

is the *conjugate prior* distribution for the Bernoulli trials, which means that after multiplying a beta distribution by the likelihood of the Bernoulli trials, we again get a beta distribution in the posterior:

$$\text{Beta}(\theta; \alpha, \beta) \times \theta^n(1-\theta)^m \propto \text{Beta}(\theta; \alpha+n, \beta+m).$$

For instance, in Fig. 2.2b, the prior is $\text{Beta}(\theta; 15, 15)$, and the posterior, after multiplying by θ^2 and renormalizing, becomes $\text{Beta}(\theta; 17, 15)$.

In machine learning, one assumption that is virtually always made is that different data points are produced independently given the model parameters, that is,

$$p(D \mid \theta) = \prod_{n=1}^N p(d_n \mid \theta)$$

for a dataset $D = \{d_n\}_{n=1}^N$. Therefore, it is virtually always a good idea to take logarithms before optimizing, getting the log-likelihood

$$\log p(D \mid \theta) = \sum_{n=1}^N \log p(d_n \mid \theta)$$

and the log-posterior (note that proportionality becomes an additive constant after taking the logarithm)

$$\log p(\theta \mid D) = \log p(\theta) + \sum_{n=1}^N \log p(d_n \mid \theta) + \text{Const},$$

which are usually the actual functions being optimized. Therefore, in complex machine learning models priors usually come in the form of *regularizers*, additive terms that impose penalties on unlikely values of θ.

For another relatively simple example, let us consider *linear regression*, a supervised learning problem of fitting a linear model to data, that is, finding a vector of weights $\mathbf{w}$ such that $y \sim \mathbf{w}^\top \mathbf{x}$ for a dataset of pairs $D = (X, \mathbf{y}) = \{(\mathbf{x}_n, y_n)\}_{n=1}^N$. The first step here is to define the likelihood, that is, represent

$$y = \mathbf{w}^\top \mathbf{x} + \epsilon$$

for some random variable (noise) ϵ and define the distribution for ϵ. The natural choice is to take the normal distribution centered at zero, $\epsilon \sim \mathcal{N}(0, \sigma^2)$, getting the likelihood as

$$p(\mathbf{y} \mid \mathbf{w}, X) = \prod_{n=1}^N p(y_n \mid \mathbf{w}, \mathbf{x}_n) = \prod_{n=1}^N \mathcal{N}(y_n \mid \mathbf{w}^\top \mathbf{x}_n, \sigma^2).$$

Taking the logarithm of this expression, we arrive at the least squares optimization problem:

$$\log p(\mathbf{y} \mid \mathbf{w}, X) = \sum_{n=1}^N \log \mathcal{N}(y_n \mid \mathbf{w}^\top \mathbf{x}_n, \sigma^2) =$$

$$= -\frac{N}{2} \log\left(2\pi\sigma^2\right) - \frac{1}{2\sigma^2} \sum_{n=1}^N \left(y_n - \mathbf{w}^\top \mathbf{x}_n\right),$$

so maximizing $\log p(\mathbf{y} \mid \mathbf{w}, X)$ is the same as minimizing $\sum_{n=1}^N \left(y_n - \mathbf{w}^\top \mathbf{x}_n\right)$, and the exact value of σ^2 turns out not to be needed for finding the maximum likelihood hypothesis.

Linear regression is illustrated in Figure 2.3, with the simplest one-dimensional linear regression shown in Fig. 2.3a. However, even if the data is one-dimensional, the regression does not have to be: if we suspect a more complex dependency than linear, we can express it by extracting *features* from the input before running linear regression.

In this example, the data is generated from a single period of a sinusoid function, so it stands to reason that it should be interpolated well by a cubic polynomial. Figure 2.3b shows the resulting approximation, obtained by training the model

$$y = w_0 + w_1 x + w_2 x^2 + w_3 x^3 + \epsilon,$$

which is equivalent to $y = \mathbf{w}^\top \mathbf{x} + \epsilon$ for $\mathbf{x} = \left(1 \ x \ x^2 \ x^3\right)^\top$, i.e., equivalent to manually extracting polynomial features from x before feeding it to linear regression. In this way, linear regression can be used to approximate much more complex depen-

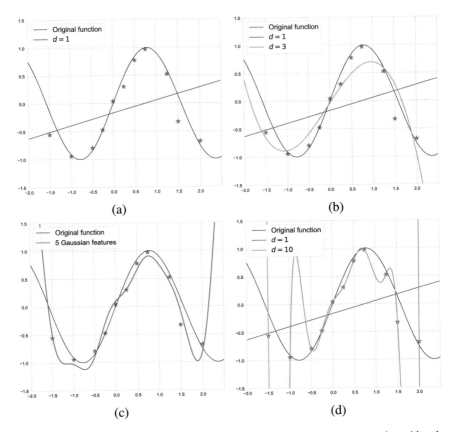

Fig. 2.3 Linear regression: (a) one-dimensional linear regression; (b) linear regression with polynomial features; (c) linear regression with Gaussian features; (d) overfitting in linear regression.

dencies. For example, Figure 2.3c shows the same dataset approximated with five Gaussian features, i.e., features of the form

$$\phi(x; \mu, s) = e^{-\frac{1}{2s}(x-\mu)^2}.$$

In fact, most neural networks that solve regression problems have a linear regression as their final layer, while neural networks for classification problems use a softmax layer, i.e., the logistic regression model. The difference and the main benefit that neural networks are providing is that the features for these simple models implemented at final layers are also learned automatically from data.

With this additional feature extraction, even linear regression can show signs of overfitting, for instance, if the features (components of the vector **x**) are too heavily correlated with each other. The ultimate case of overfitting in linear regression is shown in Fig. 2.3d: if we fit a polynomial of degree $N - 1$ to N points, it will obviously be simply interpolating all of these points, getting a perfect zero error on the training set but providing quite useless predictions, as Fig. 2.3d clearly illustrates.

In this case, we might want to restrict the desirable values of $\mathbf{w}$, for instance say that the values of $\mathbf{w}$ should be "small". This statement, which I have quite intentionally made very vague, can be formalized via choosing a suitable prior distribution. For instance, we could set a normal distribution centered at zero as prior. This time, it's a multi-dimensional normal distribution, and let's say that we do not have preferences with respect to individual features so we assume the prior is round:

$$p(\mathbf{w}) = \mathcal{N}(\mathbf{w} \mid \mathbf{0}, \sigma_0^2 \mathbf{I}).$$

Then we get the following posterior distribution:

$$\log p(\mathbf{w} \mid \mathbf{y}, X) = \sum_{n=1}^{N} \log \mathcal{N}(y_n \mid \mathbf{w}^\top \mathbf{x}_n, \sigma^2) + \log \mathcal{N}(\mathbf{w} \mid \mathbf{0}, \sigma_0^2 \mathbf{I}) =$$

$$= -\frac{N}{2} \log\left(2\pi\sigma^2\right) - \frac{1}{2\sigma^2} \sum_{n=1}^{N} \left(y_n - \mathbf{w}^\top \mathbf{x}_n\right) - \frac{d}{2} \log\left(2\pi\sigma^2\right) - \frac{1}{2\sigma_0^2} \mathbf{w}^\top \mathbf{w}.$$

The maximization problem for this posterior distribution is now equivalent to minimizing

$$\sum_{n=1}^{N} \left(y_n - \mathbf{w}^\top \mathbf{x}_n\right) + \frac{\lambda}{2} \mathbf{w}^\top \mathbf{w}, \quad \text{where} \quad \lambda = \frac{\sigma^2}{\sigma_0^2}.$$

This is known as *ridge regression*. More generally speaking, regularization with a Gaussian prior centered around zero is known as L_2-regularization because as we have just seen, it amounts to adding the L_2-norm of the vector of weights to the objective function.

We will not spend much more time on Bayesian analysis in this book, but note one thing: machine learning problems are motivated by probabilistic assumptions and the Bayes rule, but from the algorithmic and practical standpoint, they are usually *optimization problems*. Finding the maximum likelihood hypothesis amounts to maximizing $p(D \mid \theta)$, and finding the maximum a posteriori hypothesis means to maximize $p(\theta \mid D)$; usually, the main computational challenge in machine learning lies either in these maximization procedures or in finding suitable methods to approximate the integral in the predictive distribution.

Therefore, once probabilistic assumptions are made and formulas such as the ones shown above are worked out, algorithmically machine learning problems usually look like an objective function depending on the data points and model parameters that need to be optimized with respect to model parameters. In simple cases, such as the Bernoulli trials or linear regression, these optimization problems can be worked out exactly. However, as soon as the models become more complicated, optimization problems become much harder and almost always nonconvex.

This means that for complex optimization problems, such as the ones represented by neural networks, virtually the only available way to solve them is to use first-order optimization methods based on gradient descent. Over the next sections, we will consider how neural networks define such optimization problems and what methods are currently available to solve them.

2.3   Introduction to Deep Learning

Before delving into state-of-the-art first-order optimization methods, let us begin with a brief introduction to neural networks in general. As a mathematical model, neural networks actually predate the advent of artificial intelligence in general: the famous paper by Warren McCulloch and Walter Pitts was written in 1943 [589], and AI as a field of science is generally assumed to be born in the works of Alan Turing, especially his 1950 essay *Computing Machinery and Intelligence* where he introduced the Turing test [877, 878]. What is even more interesting, the original work by McCulloch and Pitts already contained a very modern model of a single artificial neuron (perceptron), namely the linear threshold unit, which for inputs $\mathbf{x}$, weights $\mathbf{w}$, and threshold a outputs

$$y = \begin{cases} 1, & \text{if } \mathbf{w}^\top \mathbf{x} \geq a, \\ 0, & \text{if } \mathbf{w}^\top \mathbf{x} < a. \end{cases}$$

This is exactly how units in today's neural networks are structured, a linear combination of inputs followed by a nonlinearity:

$$y = h(\mathbf{w}^\top \mathbf{x}).$$

The *activation function* h is usually different today, and we will survey modern activation functions in a page or two.

The linear threshold unit was one of the first machine learning models actually implemented in software (more like hardware in those times): in 1958, the *Perceptron* device developed by Frank Roseblatt [735] was able to learn the weights from a dataset of examples and actually could receive a 20×20 image as input. The Perceptron was also an important factor in the first hype wave of artificial intelligence. For instance, a *New York Times* article (hardly an unreliable tabloid) devoted to the machine said the following: "The embryo of an electronic computer... learned to differentiate between right and left after fifty attempts in the Navy's demonstration... The service said that it would use this principle to build the first of its Perceptron thinking machines that will be able to read and write. It is expected to be finished in about a year at a cost of $100,000" [858]. Naturally, nothing like that happened, but artificial neural networks were born.

The main underlying idea of the deep neural network is *connectionism*, an approach in cognitive science and neurobiology that posits the emergence of complex behaviour and intelligence in very large networks of simple computational elements [51, 52]. As a movement in both philosophy and computer science, connectionism rose to prominence in the 1980s, together with the second AI "hype wave" caused by deep neural networks. Today, deep learning provides plenty of evidence that complex networks of simple units can perform well in the most complex tasks of artificial intelligence, even if we still do not understand the human brain fully and perhaps strong human-level AI cannot be achieved by simple stacking of layers (to be honest, we don't really know).

An *artificial neural network* is defined by its computational graph. The computational graph is a directed acyclic graph $G = (V, E)$ whose nodes correspond to elementary functions and edges incoming into vertices correspond to their arguments. The source vertices (vertices of indegree zero) represent input variables, and all other vertices represent functions of these variables obtained as compositions of the functions shown in the nodes (for brevity and clarity, I will not give the obvious formal recursive definitions). In the case of neural networks for machine learning, a computational graph usually contains a single sink vertex (vertex of outdegree zero) and is said to compute the function that corresponds to this sink vertex.

Figure 2.4 shows a sample computational graph composed of simple arithmetic functions. The graph shows variables and elementary functions inside the corresponding nodes and shows the results of a node as functions of input variables along its outgoing edge; the variables are artificially divided into "inputs" x and "weights"

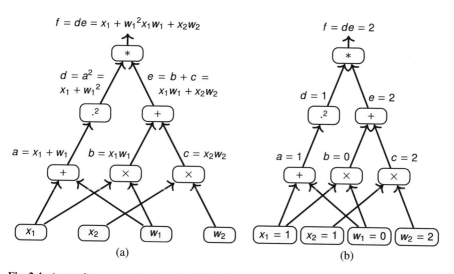

Fig. 2.4 A sample computational graph: (a) function definitions; (b) sample computation for $x_1 = x_2 = 1$, $w_1 = 0$, $w_2 = 2$.

w for illustrative purposes. In this example, the top vertex of the graph computes the function

$$f = (x_1 + w_1)^2 (x_1 w_1 + x_2 w_2).$$

The main idea of using computational graphs is to be able to solve optimization problems with the functions computed by these graphs as objectives. To apply a first-order optimization method such as gradient descent to a function $f(\mathbf{w})$ with respect to its inputs $\mathbf{w}$, we need to be able to do two things:

(1) compute the function $f(\mathbf{w})$ at every point $\mathbf{w}$;
(2) take the gradient $\nabla_{\mathbf{w}} f$ of the objective function with respect to the optimization variables.

The computational graph provides an obvious algorithm for the first task: if we know how to compute each elementary function, we simply traverse the graph from sources (variables) to the sink, computing intermediate results and finally getting the value of f. For example, let us set $x_1 = x_2 = 1$, $w_1 = 0$, $w_2 = 2$; traversing the graph in Fig. 2.4 yields the values shown in Fig. 2.4b:

$$a = x_1 + w_1 = 1, \quad b = x_1 w_1 = 0, \quad c = x_2 w_2 = 2,$$
$$d = a^2 = 1, \quad e = b + c = 2, \quad f = de = 2.$$

As for the second task, there are two possible ways to take the gradients given a computational graph. Suppose that in Fig. 2.4, we want to compute $\nabla_{\mathbf{w}} f$ for $x_1 = x_2 = 1$, $w_1 = 0$, $w_2 = 2$. The first approach, *forward propagation*, is to compute the partial derivatives along with the function values. In this way, we can compute the partial derivatives of each node in the graph with respect to the same variable; Fig. 2.5 shows the results for derivatives with respect to w_1:

$$\frac{\partial a}{\partial w_1} = \frac{\partial w_1}{\partial w_1} = 1, \quad \frac{\partial b}{\partial w_1} = x_1 \frac{\partial w_1}{\partial w_1} = 0, \quad \frac{\partial c}{\partial w_1} = 0,$$
$$\frac{\partial d}{\partial w_1} = 2a \frac{\partial a}{\partial w_1} = 2, \quad \frac{\partial e}{\partial w_1} = \frac{\partial b}{\partial w_1} + \frac{\partial c}{\partial w_1} = 1, \quad \frac{\partial f}{\partial w_1} = d \frac{\partial e}{\partial w_1} + e \frac{\partial d}{\partial w_1} = 1 + 4 = 5.$$

This approach, however, does not scale; it only yields the derivative $\frac{\partial f}{\partial w_1}$, and in order to compute $\frac{\partial f}{\partial w_2}$, we would have to go through the whole graph again! Since in deep learning, the problem is usually to compute the gradient $\nabla_{\mathbf{w}} f$ with respect to a vector of weights $\mathbf{w}$ that could have thousands or even millions of components, either running the algorithm $|\mathbf{w}|$ times or spending the memory equal to $|\mathbf{w}|$ on every computational node is entirely impractical.

That is why in deep learning, the main tool for taking the gradients is the reverse procedure, *backpropagation*. The main advantage is that this time we obtain both derivatives, $\frac{\partial f}{\partial w_1}$ and $\frac{\partial f}{\partial w_2}$, after only a single backwards pass through the graph. Again, the main tool in this computation is simply the chain rule. Given a graph node $v = h(x_1, \ldots, x_k)$ that has children $g_1, \ldots, g_l$ in the computational graph, the backpropagation algorithm computes

$$\frac{\partial f}{\partial w_1} = d\frac{\partial e}{\partial w_1} + e\frac{\partial d}{\partial w_1} = 1 + 4 = 5$$

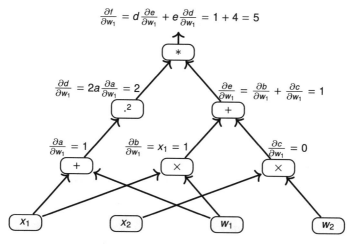

Fig. 2.5 Gradient computation on the graph from Fig. 2.4 for $x_1 = x_2 = 1$, $w_1 = 0$, $w_2 = 2$: forward propagation.

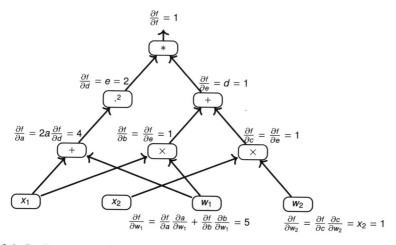

Fig. 2.6 Gradient computation on the graph from Fig. 2.4 for $x_1 = x_2 = 1$, $w_1 = 0$, $w_2 = 2$: backpropagation.

$$\frac{\partial f}{\partial v} = \frac{\partial f}{\partial g_1}\frac{\partial g_1}{\partial v} + \ldots + \frac{\partial f}{\partial g_l}\frac{\partial g_l}{\partial v},$$

where the values $\frac{\partial f}{\partial g_j}\frac{\partial g_j}{\partial v}$ have been obtained in the previous steps of the algorithm and received by the node v from its children, and sends to each of the parents x_i of the node v the value $\frac{\partial f}{\partial v}\frac{\partial v}{\partial x_i}$. The base of the induction here is the sink node, $\frac{\partial f}{\partial f} = 1$, rather than source nodes as before. In the example shown in Figure 2.4, we get the derivatives shown in Figure 2.6:

$$\frac{\partial f}{\partial f} = 1, \qquad\qquad\qquad \frac{\partial f}{\partial d} = e = 2, \qquad\qquad \frac{\partial f}{\partial e} = d = 1,$$

$$\frac{\partial f}{\partial a} = 2a\frac{\partial f}{\partial d} = 4, \qquad\qquad \frac{\partial f}{\partial b} = \frac{\partial f}{\partial e} = 1, \qquad\qquad \frac{\partial f}{\partial c} = \frac{\partial f}{\partial e} = 1,$$

$$\frac{\partial f}{\partial w_1} = \frac{\partial f}{\partial a}\frac{\partial a}{\partial w_1} + \frac{\partial f}{\partial b}\frac{\partial b}{\partial w_1} = 5, \qquad \frac{\partial f}{\partial w_2} = \frac{\partial f}{\partial c}\frac{\partial c}{\partial w_2} = x_2 = 1.$$

A real-life neural network is almost always organized into *layers*. This means that the computational graph of a neural network has subsets of nodes that are incomparable in topological order and hence can be computed in parallel. These nodes usually also have the same inputs and activation functions (or at least the same structure of inputs, like convolutional neural networks that we will consider in Section 3.1), which means that operations on entire layers can be represented as matrix multiplications and componentwise applications of the same functions to vectors.

This structure enables the use of graphics processing units (GPUs) that are specifically designed as highly parallel architectures to handle matrix operations and componentwise operations on vectors, giving speedups of up to 10-50x for training compared to CPU-based implementations. The idea of using GPUs for training neural networks dates back at least to 2004 [639], and convolutional networks were put on GPUs already in 2006 [127]. This idea was quickly accepted across the board and became a major factor in the deep learning revolution: for many applications, this 10-50x speedup was exactly what was needed to bring neural networks into the realm of realistic solutions.

Therefore, one of the first and most natural neural network architectures is the *fully connected* (or *densely connected*) network: a sequence of layers such that a neuron at layer l receives as input activations from *all* neurons at layer $l - 1$ and sends its output to *all* neurons at layer $l + 1$, i.e., each two neighboring layers form a complete bipartite graph of connections.

Fully connected networks are still relevant in some applications, and many architectures include fully connected layers. However, they are usually ill-suited for unstructured data such as images or sound because they scale badly with the number of inputs, leading to a huge number of weights that will almost inevitably overfit. For instance, the first layer of a fully connected network that has 200 neurons and receives a 1024×1024 image as input will have more than 200 million weights! No amount of L_2 regularization is going to fix that, and we will see how to avoid such overkill in Section 3.1. On the other hand, once a network of a different structure has already extracted a few hundred or a couple thousand features from this image, it does make sense to have a fully connected layer or two at the end to allow the features to interact with each other freely, so dense connections definitely still have a place in modern architectures.

Figure 2.7 presents a specific example of a three-layered network together with the backpropagation algorithm worked out for this specific case. On the left, it shows the architecture: input $\mathbf{x}$ goes through two hidden layers with weight matrices $W^{(1)}$ and $W^{(2)}$ (let us skip the bias vectors in this example in order not to clutter notation even further). Each layer also has a nonlinear activation function h, so its outputs are $\mathbf{z}^{(1)} = h\left(W^{(1)}\mathbf{x}\right)$ and $\mathbf{z}^{(2)} = h\left(W^{(2)}\mathbf{z}^{(1)}\right)$. After that, the network has a scalar output $y = h\left(\mathbf{w}^{(3)\top}\mathbf{z}^{(2)}\right)$, again computed with activation function h and weight vector

$\mathbf{w}^{(3)}$ from $\mathbf{z}^{(2)}$, and then the objective function f is a function of the scalar y. The formulas in the middle column show the forward propagation part, i.e., computation along the graph, and formulas on the right show the backpropagation algorithm that begins with $\frac{\partial f}{\partial y}$ and progresses in the opposite direction. Dashed lines on the figure divide the architecture into computational layers, and the computations are grouped inside the dashed lines.

For example, on the second hidden layer, we have the weight matrix $W^{(2)}$, input $\mathbf{z}^{(1)}$ from the layer below, output $\mathbf{z}^{(2)} = h\left(W^{(2)}\mathbf{z}^{(2)}\right)$, and during backpropagation, we also have the gradient $\nabla_{\mathbf{z}^{(2)}} f$ coming from the layer above as part of the induction hypothesis. In backpropagation, this dense layer needs to do two things:

- compute the gradient with respect to its own matrix of weights $W^{(2)}$ so that they can be updated; this is achieved as

$$\nabla_{W^{(2)}} f = h'\left(W^{(2)}\mathbf{z}^{(2)}\right)\left(\nabla_{\mathbf{z}^{(2)}} f\right)\mathbf{z}^{(1)}{}^{\top};$$

note how the dimensions match: $W^{(2)}$ is a 3×4 matrix in this example, $\nabla_{\mathbf{z}^{(2)}} f$ is a 3×1 vector, and $\mathbf{z}^{(1)}$ is a 4×1 vector;
- compute the gradient with respect to its input $\mathbf{z}^{(1)}$ and send it down to the layer below:

$$\nabla_{\mathbf{z}^{(1)}} f = h'\left(W^{(2)}\mathbf{z}^{(2)}\right) W^{(2)}{}^{\top}\nabla_{\mathbf{z}^{(2)}} f.$$

Figure 2.7 shows the rest of the computations, introducing intermediate vectors $\mathbf{o}$ for brevity. As we can see, this algorithm is entirely expressed in the form of matrix operations and componentwise applications of functions, and thus it lends itself easily to GPU-based parallelization.

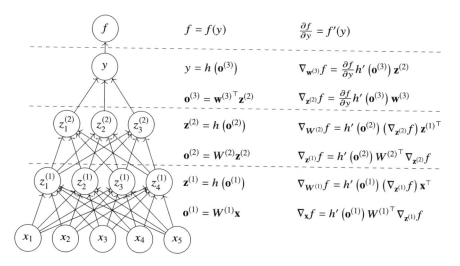

Fig. 2.7 A three-layered fully connected architecture with computations for backpropagation.

The only thing left for this section is to talk a bit more about activation functions. The original threshold activation, suggested by McCulloch and Pitts and implemented by Rosenblatt, is almost never used now: if nothing else, thresholds are hard to optimize by gradient descent because the derivative of a threshold function is everywhere zero or nonexistent. Throughout neural network history, the most popular activation functions had been sigmoids, usually either the logistic sigmoid

$$\sigma(a) = \frac{1}{1 + e^{-a}}$$

or the hyperbolic tangent

$$\tanh(a) = \frac{e^a - e^{-a}}{e^a + e^{-a}}.$$

Several classical activation functions, from threshold to ReLU, are shown in Fig. 2.8.

Research of the last decade, however, shows that one can get a much better family of activation functions (for internal layers of deep networks—you still can't get around a softmax at the end of a classification problem, of course) if one does not restrict it by a horizontal asymptote at least on one side. The most popular activation function in modern artificial networks is the *rectified linear unit* (ReLU)

$$\mathrm{ReLU}(x) = \begin{cases} 0, & \text{if } x < 0, \\ x, & \text{if } x \geq 0 \end{cases}$$

and its variations that do not have a hard zero for negative inputs but rather a slower growing function, for example, the *leaky ReLU* [570]

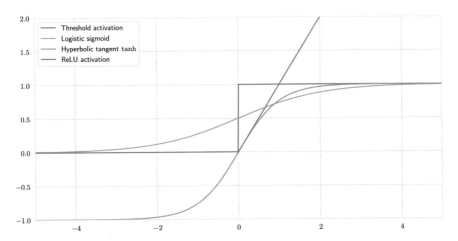

Fig. 2.8 A comparison of activation functions: classical activation functions.

$$\text{LReLU}(x) = \begin{cases} ax, & \text{if } x < 0, \\ x, & \text{if } x > 0 \end{cases}$$

or the *exponential linear unit* [161]

$$\text{ELU}(x) = \begin{cases} \alpha \left(e^x - 1\right), & x < 0, \\ x, & x \geq 0. \end{cases}$$

There is also a smooth variant of ReLU, known as the *softplus* function or [282]:

$$\text{softplus}(x) = \ln \left(1 + e^x\right).$$

You can see a comparison of ReLU variations in Figure 2.9. In any case, all activation functions used in modern neural networks must be differentiable so that gradient descent can happen; it's okay to have kinks on a subset of measure zero, like ReLU does, since one can always just set the derivative to zero at that point.

It is always tempting to try and get a better result just by switching the activation function, but most often it fails: it is usually the last optimization I would advise to actively try. However, if you try many different activation functions systematically and with a wide range of models, it might be possible to improve upon standard approaches.

In 2017, *Google Brain* researchers Ramachandran et al. [702] did exactly this: they constructed a search space of possible activation functions (basically a recursive computational graph with a list of possible unary and binary functions to insert there), used the ideas of *neural architecture search* [1031] to formulate the search for activation functions as a reinforcement learning problem, and made good use of *Google*'s huge computational power to search this space as exhaustively as they

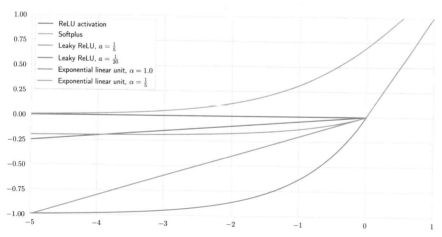

Fig. 2.9 A comparison of activation functions: ReLU variations.

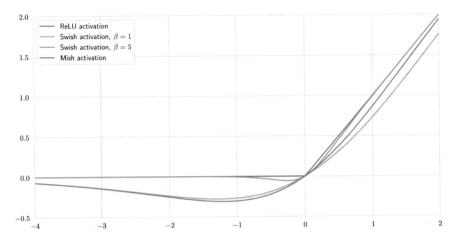

Fig. 2.10 A comparison of activation functions: *Swish* and *Mish*.

could. The results were interesting: in particular, Ramachandran et al. found that complicated activation functions consistently underperform simpler ones. As the most promising resulting function, they highlighted the so-called *Swish* activation:

$$\text{Swish}(x) = x\sigma(\beta x) = \frac{x}{1 + e^{-\beta x}}.$$

Depending on the parameter β, *Swish* scales the range from perfectly linear (when $\beta = 0$) to ReLU (when $\beta \to \infty$). Figure 2.10 shows *Swish* activation and other variations; the most interesting feature of *Swish* is probably the fact that it is not monotone and has a minimum in the negative part of the spectrum.

In 2019, Misra [604] suggested the *Mish* activation, a variation of *Swish*:

$$\text{Mish}(x) = x\tanh(\text{softplus}(x)) = x\tanh\left(\ln\left(1 + e^x\right)\right).$$

Both *Swish* and *Mish* activations have been tested in many applications, including convolutional architectures for computer vision, and they now define state of the art, although good old ReLUs are far from completely replaced.

The text above was written in the summer of 2020. But, of course, this was not the end of the story. In September 2020, Ma et al. [567] presented a new look on *Swish* and ReLUs. They generalized them both by using a smooth approximation of the maximum function:

$$S_\beta(x_1, \ldots, x_n) = \frac{\sum_{i=1}^n x_i e^{\beta x_i}}{\sum_{i=1}^n e^{\beta x_i}}.$$

Let us substitute two functions in place of the arguments: for the hard maximum $\max(g(x), h(x))$, we get its smooth counterpart

$$S_\beta\left(g(x), h(x)\right) = g(x)\frac{e^{\beta g(x)}}{e^{\beta g(x)} + e^{\beta h(x)}} + h(x)\frac{e^{\beta h(x)}}{e^{\beta g(x)} + e^{\beta h(x)}} =$$
$$= g(x)\sigma\left(\beta\left(g(x) - h(x)\right)\right) + h(x)\sigma\left(\beta\left(h(x) - g(x)\right)\right) =$$
$$= \left(g(x) - h(x)\right)\sigma\left(\beta\left(g(x) - h(x)\right)\right) + h(x).$$

Ma et al. call this the *ActivateOrNot* (ACON) activation function. They note that

- for $g(x) = x$ and $h(x) = 0$, the hard maximum is $\mathrm{ReLU}(x) = \max(x, 0)$, and the smooth counterpart is

$$f_{ACON-A}(x, 0) = S_\beta(x, 0) = x\sigma(\beta x),$$

that is, precisely the *Swish* activation function;
- for $g(x) = x$ and $h(x) = ax$ with some $a < 1$, the hard maximum is $\mathrm{LReLU}(x) = \max(x, px)$, and the smooth counterpart is

$$f_{ACON-B} = S_\beta(x, ax) = (1 - a)x\sigma(\beta(1 - a)x) + ax;$$

- both of these functions can be straightforwardly generalized to

$$f_{ACON-C} = S_\beta(a_1 x, a_2 x) = (a_1 - a_2)x\sigma(\beta(a_1 - a_2)x) + a_2 x;$$

in this case, a_1 and a_2 can become learnable parameters, and their intuitive meaning is that they serve as the limits of ACON-C's derivative:

$$\lim_{x\to\infty}\frac{\mathrm{d}f_{ACON-C}}{\mathrm{d}x} = a_1, \qquad \lim_{x\to-\infty}\frac{\mathrm{d}f_{ACON-C}}{\mathrm{d}x} = a_2.$$

Figure 2.11 shows the ACON-C function for different values of a_1, a_2, and β, starting from exactly the *Swish* function and showing the possible variety.

All this has only just happened, and so far it is hard to say whether this idea is going to catch on across many neural architectures or die down quietly. But this is a great example of how hard it is to write about deep learning; we will see such examples in later sections as well.

Let us summarize. We have seen how neural networks are structured as computational graphs composed of simple functions, and how this structure allows us to develop efficient algorithms for computing the gradient of an objective function represented as such a graph with respect to any subset of its variables, in case of neural networks usually with respect to the weights. However, being able to take the gradient is only the first step towards an efficient optimization algorithm. In the next section, we briefly discuss the main first-order optimization algorithms currently used in deep learning.

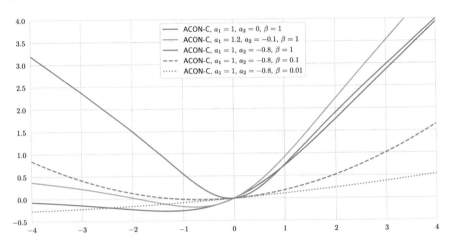

Fig. 2.11 A comparison of activation functions: the ACON-C function with different parameters.

2.4   First-Order Optimization in Deep Learning

In this section, we continue our introduction to deep learning, considering it from the optimization point of view. We have already seen how to compute the gradients, and here we will discuss how to use these gradients to find the local minima of given functions. Throughout this section, we assume that we are given a loss function $f(\mathbf{w}, d)$, where d is a data point and $\mathbf{w}$ is the vector of weights, and the optimization problem in question is to minimize the total loss function over a given dataset D with respect to $\mathbf{w}$:

$$F(\mathbf{w}) = \sum_{d \in D} f(\mathbf{w}, d) \rightarrow_{\mathbf{w}} \min.$$

Algorithm 1 shows the regular "vanilla" gradient descent (GD): at every step, compute the gradient at the current point and move in the opposite direction. In regular GD, a lot depends on the learning rate α. It makes sense that the learning rate should decrease with time, and the first idea would be to choose a fixed schedule for varying α:

- either with linear decay:

$$\alpha = \alpha_0 \left(1 - \frac{t}{T} \right)$$

- or with exponential decay:

$$\alpha = \alpha_0 e^{-\frac{t}{T}}.$$

In both cases, T is called the *temperature* (it does play a role similar to the temperature in statistical mechanics), and the larger it is, the slower the learning rate decays with time.

Algorithm 1: Gradient descent

Initialize $\mathbf{w}_0, k := 1$;
repeat
    $\mathbf{w}_{k+1} := \mathbf{w}_k - \alpha \sum_{d \in D} \nabla_{\mathbf{w}} F(\mathbf{w}_k, d)$;
    $k := k + 1$;
until *a stopping condition is met*;

But these are, of course, just the very first ideas that can be much improved. Optimization theory has a whole field of research devoted to gradient descent and how to find the optimal value of α on any given step. We refer to, e.g., books and surveys [84, 96, 625, 633] for a detailed treatment of this and give only a brief overview of the main ideas.

In particular, basic optimization theory known since the 1960s leads to the so-called *Wolfe conditions* and *Armijo rule*. If we are minimizing $f(\mathbf{w})$, and on step k we have already found the direction $\mathbf{p}_k$ to which we need to move—for instance, in gradient descent, we have $\mathbf{p}_k = \nabla_{\mathbf{w}} f(\mathbf{w}_k)$—the problem becomes

$$\min_\alpha f(\mathbf{w}_k + \alpha \mathbf{p}_k),$$

a one-dimensional optimization problem.

Studying this problem, researchers have found that

- for $\phi_k(\alpha) = f(\mathbf{w}_k + \alpha \mathbf{p}_k)$, we have $\phi_k'(\alpha) = \nabla f(\mathbf{w}_k + \alpha \mathbf{p}_k)^\top \mathbf{p}_k$, and if $\mathbf{p}_k$ is the direction of descent, then $\phi_k'(0) < 0$;
- the step size α must satisfy the Armijo rule:

$$\phi_k(\alpha) \le \phi_k(0) + c_1 \alpha \phi_k'(0) \text{ for some } c_1 \in (0, \tfrac{1}{2});$$

- or even stronger Wolfe conditions, which mean the Armijo rule and, in addition,

$$|\phi_k'(\alpha)| \le c_2 |\phi_k'(0)|,$$

i.e., we aim to reduce the projection of the gradient.

The optimization process now should stop according to a stopping condition with respect to the L_2-norm of the gradient, i.e., when $\|\nabla_{\mathbf{w}} f(\mathbf{w}_k)\|^2 \le \epsilon$ or $\|\nabla_{\mathbf{w}} f(\mathbf{w}_k)\|^2 \le \epsilon \|\nabla_{\mathbf{w}} f(\mathbf{w}_0)\|^2$.

However, first-order methods such as gradient descent begin to suffer if the scale of different variables is different. A classical example of such behaviour is shown in Figure 2.12, where the three plots show gradient descent optimization for three very simple quadratic functions, all in the form of $x^2 + \rho y^2$ for different values of ρ.

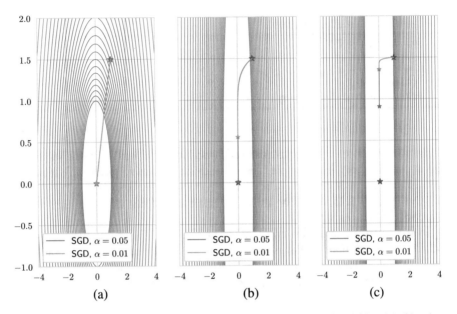

Fig. 2.12 Sample gradient descent optimization with different scale of variables: (a) $f(x, y) = x^2 + y^2$; (b) $f(x, y) = x^2 + \frac{1}{10}y^2$; (c) $f(x, y) = x^2 + \frac{1}{100}y^2$.

The functions are perfectly convex, and SGD (in fact, full GD in this case) should have no trouble at all in finding the optimum. And indeed it doesn't, but as the scale of variables becomes too different, gradient descent slows down to a crawl when it comes to the vertical axis; the plots show how the same learning rates work great for comparable x and y but slow down significantly as they become too different.

Algorithm 2: Stochastic gradient descent with mini-batches

Initialize $\mathbf{w}_0, k := 0$;
repeat
$\quad$ $D_k := \text{Sample}(D)$;
$\quad$ $\mathbf{w}_{k+1} := \mathbf{w}_k - \alpha \sum_{d \in D_k} \nabla_{\mathbf{w}} f(\mathbf{w}_k, d)$;
$\quad$ $k := k + 1$;
until *a stopping condition is met*;

Cases like this are very common in deep learning: for instance, weights from different layers of a deep network certainly might have different scales. Therefore, for machine learning problems, it is much better to use *adaptive* gradient descent algorithms that set the scale for different variables adaptively, depending on the optimization landscape. Naturally, the best way to do that would be to pass to *second-order methods*. Theoretically, we could apply Newton's method here:

$$\mathbf{g}_k = \nabla_{\mathbf{w}} f(\mathbf{w}_k), \quad H_k = \nabla_{\mathbf{w}}^2 f(\mathbf{w}_k),$$

and we get that

$$\mathbf{w}_{k+1} = \mathbf{w}_k - \alpha_k H_k^{-1} \mathbf{g}_k.$$

The Armijo rule is applicable here as well: we should choose α_k such that

$$f(\mathbf{w}_{k+1}) \leq f(\mathbf{w}_k) - c_1 \alpha_k \mathbf{g}_k^\top H_k^{-1} \mathbf{g}_k, \quad \text{where } c_1 \approx 10^{-4}.$$

Using Newton's method to train deep neural networks would be great! Unfortunately, real-life neural networks have a lot of parameters, on the order of thousands or even millions. It is completely impractical to compute and support a Hessian matrix in this case, and even less practical to invert it—note that second-order methods make use of H_k^{-1}.

There exist a wide variety of *quasi-Newton methods* that do not compute the Hessian explicitly but rather approximate it via the values of the gradient. The most famous of them is probably the BFGS algorithm, named after Charles George Broyden, Roger Fletcher, Donald Goldfarb, and David Shanno [237]. The idea is not to compute the Hessian but keep a low-rank approximation and update it via the current value of the gradient. It's a great algorithm, it has versions with bounded memory, and it would also work great to rescale the gradients...

...But a huge Hessian is just the beginning of our troubles. What's more, in the case of machine learning, we cannot really afford gradient descent either! The problem is that the loss function is defined as a sum over the input dataset $\sum_{d \in D} f(\mathbf{w}, d)$, and in reality, it is usually infeasible to go over the entire dataset to make only a single update to the neural network weights. This means that we cannot use the BFGS algorithm and other quasi-Newton methods because we don't have the value of the gradient either.

Therefore, in deep learning, one usually implements *stochastic gradient descent* (SGD), shown in its general form in Algorithm 2: the difference is that on every step, the gradient is computed not over the entire dataset D but over a subsample of the data D_k.

How do we understand stochastic gradient descent formally and how does it fit into optimization theory? Usually, the problem we are trying to solve can be framed as a *stochastic optimization* problem:

$$F(\mathbf{w}) = \mathbb{E}_{q(\mathbf{y})} \left[f(\mathbf{w}, \mathbf{y}) \right] \rightarrow \min_{\mathbf{w}},$$

where $q(\mathbf{y})$ is some known distribution. The basic example here is the minimization of empirical risk:

$$F(\mathbf{w}) = \frac{1}{N} \sum_{i=1}^{N} f_i(\mathbf{w}) = \mathbb{E}_{i \sim U(1,...,N)} \left[f_i(\mathbf{w}) \right] \rightarrow \min_{\mathbf{w}}.$$

Another important example is provided by minimizing the variational lower bound, but this goes beyond the scope of this section.

This formalization makes it clear what mini-batches are from the formal point of view. Averaging over a mini-batch can be thought of simply as an empirical estimate of the stochastic optimization objective function, computed on a subsample:

$$\hat{F}(\mathbf{w}) = \frac{1}{m} \sum_{i=1}^{m} f(\mathbf{w}, \mathbf{y}_i), \quad \hat{\mathbf{g}}(\mathbf{w}) = \frac{1}{m} \sum_{i=1}^{m} \nabla_{\mathbf{w}} f(\mathbf{w}, \mathbf{y}_i),$$

where m denotes the mini-batch size. Basic mathematical statistics tells us that these estimates have a lot of desirable properties: they are unbiased, they always converge to the true value of the expectation (albeit convergence might be slow), and they are easy to compute.

In general, stochastic gradient descent is motivated by these ideas and can be thought of as basically a Monte Carlo variation of gradient descent. Unfortunately, it still does not mean that we can plug these Monte Carlo estimates into a quasi-Newton method such as BFGS: the variance is huge, the gradient on a single mini-batch usually has little in common with the true gradient, and BFGS would not work with these estimates. It is a very interesting open problem to devise stochastic versions of quasi-Newton methods, but it appears to be a very hard problem.

But SGD has several obvious problems even in the first-order case:

- it never goes in the exactly correct direction;
- moreover, SGD does not even have zero updates at the exact point where $F(\mathbf{w})$ is minimized, i.e., even if we get lucky and reach the minimum, we won't recognize it, and SGD with constant step size will never converge;
- since we know neither $F(\mathbf{w})$ nor $\nabla F(\mathbf{w})$ (only their Monte Carlo estimates with huge variances), we cannot use the Armijo rule and Wolfe conditions to find the optimal step size.

There is a standard analysis that can be applied to SGD; let us look at a single iteration of SGD for some objective function

$$F(\mathbf{w}) = \mathbb{E}_{q(\mathbf{y})} \left[f(\mathbf{w}, \mathbf{y}) \right] \to_{\mathbf{w}} \min.$$

In what follows, we denote by $\mathbf{g}_k$ the gradient of F at point $\mathbf{w}_k$, so that

$$\mathbf{w}_{k+1} = \mathbf{w}_k - \alpha_k \hat{\mathbf{g}}_k, \quad \mathbb{E}\left[\hat{\mathbf{g}}_k\right] = \mathbf{g}_k = \nabla F(\mathbf{w}_k).$$

Let us try to estimate the residue of the point on iteration k; denoting by $\mathbf{w}_{\text{opt}}$ the true optimum, we get

$$\|\mathbf{w}_{k+1} - \mathbf{w}_{\text{opt}}\|^2 = \|\mathbf{w}_k - \alpha_k \hat{\mathbf{g}}_k - \mathbf{w}_{\text{opt}}\|^2 =$$
$$= \|\mathbf{w}_k - \mathbf{w}_{\text{opt}}\|^2 - 2\alpha_k \hat{\mathbf{g}}_k^\top (\mathbf{w}_k - \mathbf{w}_{\text{opt}}) + \alpha_k^2 \|\hat{\mathbf{g}}_k\|^2.$$

Taking the expectation with respect to $q(\mathbf{y})$ on iteration k, we get

$$\mathbb{E}\left[\|\mathbf{w}_{k+1} - \mathbf{w}_{\text{opt}}\|^2\right] = \|\mathbf{w}_k - \mathbf{w}_{\text{opt}}\|^2 - 2\alpha_k \mathbf{g}_k^\top (\mathbf{w}_k - \mathbf{w}_{\text{opt}}) + \alpha_k^2 \mathbb{E}\left[\|\hat{\mathbf{g}}_k\|^2\right].$$

And now comes the common step in optimization theory where we make assumptions that are far too strong. In further analysis, let us assume that F is convex; this is, of course, not true in real deep neural networks, but it turns out that the resulting analysis is indeed relevant to what happens in practice, so let's run with it for now. In particular, we can often assume that even in nonconvex optimization, once we get into a neighborhood of a local optimum, the function can be considered to be convex. Specifically, we will use the fact that

$$F(\mathbf{w}_{\text{opt}}) \geq F(\mathbf{w}_k) + \mathbf{g}_k^\top (\mathbf{w}_k - \mathbf{w}_{\text{opt}}).$$

Now let us combine this with the above formula for $\mathbb{E}\left[\|\mathbf{w}_{k+1} - \mathbf{w}_{\text{opt}}\|^2\right]$:

$$\alpha_k (F(\mathbf{w}_k) - F(\mathbf{w}_{\text{opt}})) \leq \alpha_k \mathbf{g}_k^\top (\mathbf{w}_k - \mathbf{w}_{\text{opt}}) =$$

$$= \frac{1}{2}\|\mathbf{w}_k - \mathbf{w}_{\text{opt}}\|^2 + \frac{1}{2}\alpha_k^2 \mathbb{E}\left[\|\hat{\mathbf{g}}_k\|^2\right] - \frac{1}{2}\mathbb{E}\left[\|\mathbf{w}_{k+1} - \mathbf{w}_{\text{opt}}\|^2\right].$$

Next, we take the expectation of the left-hand side and sum it up:

$$\sum_{i=0}^{k} \alpha_i (\mathbb{E}\left[F(\mathbf{w}_i)\right] - F(\mathbf{w}_{\text{opt}})) \leq$$

$$\leq \frac{1}{2}\|\mathbf{w}_0 - \mathbf{w}_{\text{opt}}\|^2 + \frac{1}{2}\sum_{i=0}^{k} \alpha_i^2 \mathbb{E}\left[\|\hat{\mathbf{g}}_i\|^2\right] - \frac{1}{2}\mathbb{E}\left[\|\mathbf{w}_{k+1} - \mathbf{w}_{\text{opt}}\|^2\right] \leq$$

$$\leq \frac{1}{2}\|\mathbf{w}_0 - \mathbf{w}_{\text{opt}}\|^2 + \frac{1}{2}\sum_{i=0}^{k} \alpha_i^2 \mathbb{E}\left[\|\hat{\mathbf{g}}_i\|^2\right].$$

We have obtained a sum of values of the function in different points with weights α_i. Let us now use the convexity assumption:

$$\mathbb{E}\left[F\left(\frac{\sum_i \alpha_i \mathbf{w}_i}{\sum_i \alpha_i}\right) - F(\mathbf{w}_{\text{opt}})\right] \leq$$

$$\leq \frac{\sum_i \alpha_i (\mathbb{E}\left[F(\mathbf{w}_i)\right] - F(\mathbf{w}_{\text{opt}}))}{\sum_i \alpha_i} \leq \frac{\frac{1}{2}\|\mathbf{w}_0 - \mathbf{w}_{\text{opt}}\|^2 + \frac{1}{2}\sum_{i=0}^{k} \alpha_i^2 \mathbb{E}\left[\|\hat{\mathbf{g}}_i\|^2\right]}{\sum_i \alpha_i}.$$

Thus, we have obtained a bound on the residue for some intermediate value in a linear combination of $\mathbf{w}_i$; this is also a common situation in optimization theory, and again, in practice, it usually turns out that there is no difference between the mean and the last point, or the last point $\mathbf{w}_K$ is even better.

In other words, we have found that if the initial residue is bounded by R, i.e., $\|\mathbf{w}_0 - \mathbf{w}_{\text{opt}}\| \leq R$, and if the variance of the stochastic gradient is bounded by G, i.e., $\mathbb{E}\left[\|\hat{\mathbf{g}}_k\|^2\right] \leq G^2$, then

$$\mathbb{E}\left[F(\hat{\mathbf{w}}_k) - F(\mathbf{w}_{\text{opt}})\right] \leq \frac{R^2 + G^2 \sum_{i=0}^{k} \alpha_i^2}{2 \sum_{i=0}^{k} \alpha_i}.$$

This is the main formula in the theoretical analysis of stochastic gradient descent. In particular, for a constant step size $\alpha_i = h$, we get that

$$\mathbb{E}\left[F(\hat{\mathbf{w}}_k) - F(\mathbf{w}_{\text{opt}})\right] \leq \frac{R^2}{2h(k+1)} + \frac{G^2 h}{2} \rightarrow_{k \to \infty} \frac{G^2 h}{2}.$$

Let us summarize the behaviour of SGD that follows from the above analysis:

- SGD comes to an "uncertainty region" of radius $\frac{1}{2}G^2 h$, and this radius is proportional to the step size;
- this means that the faster we walk, the faster we reach the uncertainty region, but the larger this uncertainty region will be; in other words, it makes sense to reduce the step size as optimization progresses;
- SGD converges quite slowly: it is known that the full gradient for convex functions converges at rate $O(1/k)$, and SGD has a convergence rate of only $O(1/\sqrt{k})$;
- on the other hand, the rate of convergence for SGD is also $O(1/k)$ when it is far from the uncertainty region, it slows down only when we have reached it;
- but all of this still depends on G, which in practice we cannot really estimate reliably, and this is also an important point for applications of Bayesian analysis in deep learning.

Algorithm 3: Stochastic gradient descent with momentum

Initialize $\mathbf{w}_0, \mathbf{u}_0 := 0, k := 0$;
repeat
    $D_k := \text{Sample}(D)$;
    $\mathbf{u}_{k+1} := \gamma \mathbf{u}_k + \alpha \sum_{d \in D_k} \nabla_{\mathbf{w}} f(\mathbf{w}_k, d)$;
    $\mathbf{w}_{k+1} := \mathbf{w}_k - \mathbf{u}_{k+1}$;
    $k := k + 1$;
until *a stopping condition is met*;

All of the above means that we need some further improvements: plain vanilla SGD may be not the best way, there is no clear answer as to how to change the learning rate with time, and the problem of rescaling the gradients in an adaptive way still remains open and important.

Fortunately, there are plenty of improvements that do exactly that. We again refer to [84, 96, 625, 633] for classical optimization techniques and proceed to the approaches that have proven particularly fruitful for optimization in deep learning.

Algorithm 4: Nesterov accelerated gradient

Initialize $\mathbf{w}_0, \mathbf{u}_0 := 0, k := 0$;
repeat
　　$D_k := \text{Sample}(D)$;
　　$\mathbf{u}_{k+1} := \gamma \mathbf{u}_k + \alpha \sum_{d \in D_k} \nabla_{\mathbf{w}} f(\mathbf{w}_k - \gamma \mathbf{u}_k, d)$;
　　$\mathbf{w}_{k+1} := \mathbf{w}_k - \mathbf{u}_{k+1}$;
　　$k := k + 1$;
until *a stopping condition is met*;

2.5 Adaptive Gradient Descent Algorithms

As we have seen in the previous section, basic gradient descent is infeasible in the case of deep neural networks, and stochastic gradient descent needs one to be careful about the choice of the learning rate and, most probably, requires some rescaling along different directions as well. Here, we discuss various ideas in first-order optimization, some classical and some very recent, that have proven to work well in deep learning.

The first idea is the *momentum method*: let us think of the current value of $\mathbf{w}$ as a material point going down the landscape of the function F that we are minimizing, and let us say that, as in real Newtonian physics, this material point carries a part of its momentum from one time moment to the next. In discrete time, it means that in step k we are preserving a part of the previous update $\mathbf{u}_{k-1}$, as shown in Algorithm 3. In real situations, the momentum decay parameter γ is usually close to 1, e.g., $\gamma = 0.9$ or even $\gamma = 0.999$. The momentum method has been a staple of deep learning since at least the mid-1980s; it was proposed for neural networks in the famous *Nature* paper by Rumelhart, Hinton, and Williams [742].

The momentum method is often combined with another important heuristic, *implicit updates*. In regular SGD, using implicit updates means that the gradient is computed not at the point $\mathbf{w}_k$ but at the next point $\mathbf{w}_{k+1}$:

$$\mathbf{w}_{k+1} := \mathbf{w}_k - \alpha \sum_{d \in D_k} \nabla_{\mathbf{w}} f(\mathbf{w}_{k+1}, d).$$

This makes it an implicit equation rather than explicit and can be thought of as the stochastic form of the proximal gradient method. In classical optimization theory, using implicit updates often helps with numerical stability.

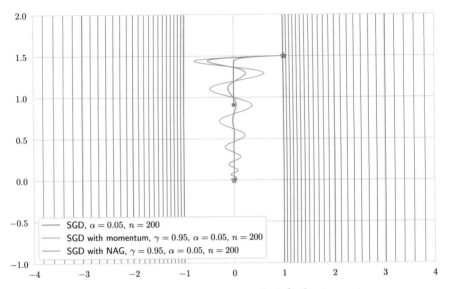

Fig. 2.13 Gradient descent optimization for $F(x, y) = x^2 + \frac{1}{100}y^2$: the effect of momentum.

Applied to the computation of momentum, implicit updates yield the *Nesterov accelerated gradient* (NAG) method, named after its inventor Yurii Nesterov [624]. In this approach, instead of computing the gradient at the point $\mathbf{w}_k$, we first apply the momentum update (after all, we already know that we will need to move in that direction) and then compute the gradient at the point $\mathbf{w}_k + \gamma \mathbf{u}_k$. In this way, the updates are still explicit but numerical stability is much improved, and Nesterov's result was that this version of gradient descent converges faster than the usual version. We show the Nesterov accelerated gradient in Algorithm 4.

Figure 2.13 shows how momentum-based methods solve the problem that we saw in Fig. 2.12. In Fig. 2.13, we consider the same problematic function $F(x, y) = x^2 + \frac{1}{100}y^2$ and show the same number of iterations for every method. Now both regular momentum and Nesterov accelerated gradient converge much faster and in fact have enough time to converge while regular SGD is still crawling towards the optimum. Note how the Nesterov accelerated gradient is more stable and does not oscillate as much as basic momentum-based SGD: this is exactly the stabilization effect of the "lookahead" in computing the gradients.

To move further, note that so far, the learning rate was the same along all directions in the vector $\mathbf{w}$, and we either set a global learning rate α with some schedule of decreasing or chose it with the Armijo rule along the exact chosen direction. Modern adaptive variations of SGD use the following heuristic: let us move faster along the components of $\mathbf{w}$ that change F slowly and move slower when we get to a region of rapid changes in F (which usually means that we are in the vicinity of a local extremum).

The first approach along these lines was *Adagrad* proposed in 2011 [209]. The idea was to keep track of the total accumulated gradient values in the form of their sum of squares; this is vectorized in the form of a diagonal matrix G_k whose diagonal elements contain sums of partial derivatives accumulated up to this point:

Algorithm 5: Adagrad

Initialize $\mathbf{w}_0$, $G_0 := 0$, $k := 0$;
repeat
   | $D_k :=$ Sample(D);
   | $\mathbf{g}_k := \sum_{d \in D_k} \nabla_{\mathbf{w}} f(\mathbf{w}_k, d)$;
   | $G_{k+1} := G_k + \text{diag}(\mathbf{g}_k)$;
   | $\mathbf{w}_{k+1} := \mathbf{w}_k - \frac{\alpha}{\sqrt{G_{k+1}} + \epsilon} \mathbf{g}_{k+1}$;
   | $k := k + 1$;
until *a stopping condition is met*;

$$G_{k,ii} = \sum_{l=1}^{k} \frac{\partial F_l}{\partial \mathbf{w}_i}, \quad \text{where } F_l(\mathbf{w}) = \sum_{d \in D_l} f(\mathbf{w}, d).$$

Adagrad is summarized in Algorithm 5. The learning rate now becomes adaptive: when the gradients along some direction i become large, the sum of their squares $G_{k,ii}$ also becomes large, and gradient descent slows down along this direction. Thus, one does not have to manually tune the learning rate anymore; in most implementations, the initial learning rate is set to $\alpha = 0.01$ or some other similar constant and left with no change.

However, the main problem of *Adagrad* is obvious as well: while it can slow descent down, it can never let it pick the pace back up. Thus, if the slope of the function F becomes steep in a given direction but then flattens out again, *Adagrad* will keep going very slowly along this direction. The fix for this problem is quite straightforward: instead of a sum of squares of the gradients, let's use an exponential moving average.

Algorithm 6: RMSprop

Initialize $\mathbf{w}_0$, $G_0 := 0$, $k := 0$;
repeat
   | $D_k :=$ Sample(D);
   | $\mathbf{g}_k := \sum_{d \in D_k} \nabla_{\mathbf{w}} f(\mathbf{w}_k, d)$;
   | $G_{k+1} := \gamma G_k + (1 - \gamma) \text{diag}(\mathbf{g}_k)$;
   | $\mathbf{w}_{k+1} := \mathbf{w}_k - \frac{\alpha}{\sqrt{G_{k+1}} + \epsilon} \mathbf{g}_{k+1}$;
   | $k := k + 1$;
until *a stopping condition is met*;

The first attempt at this is the *RMSprop* algorithm, proposed by Geoffrey Hinton in his *Coursera* class but, as far as I know, never officially published. It replaces the sum of squares of gradients $G_{k+1} := G_k + \text{diag}(\mathbf{g}_k)$ with a formula that computes the exponential moving average:

$$G_{k+1} := \gamma G_k + (1 - \gamma) \text{diag}(\mathbf{g}_k);$$

Hinton suggested to use $\gamma = 0.9$. We show *RMSprop* in Algorithm 6.

But there is one more, slightly less obvious problem. If you look at the final update rule in *RMSprop*, or actually at the update rule in any of the stochastic gradient descent variations we have considered so far, you can notice that the measurement units in the updates don't match! For instance, in the vanilla SGD, we update

$$\mathbf{w}_{k+1} := \mathbf{w}_k - \alpha \nabla_{\mathbf{w}} F_k(\mathbf{w}_k, d),$$

which means that we are subtracting from $\mathbf{w}$ the partial derivatives of F_k with respect to $\mathbf{w}$. In other words, if $\mathbf{w}$ is measured in, say, seconds and $f(\mathbf{w}, d)$ is measured in meters, we are subtracting meters per second from seconds, hardly a justified operation from the physical point of view! In mathematical terms, this means that the scale of these vectors may differ drastically, leading to mismatches and poor convergence.

Adagrad and *RMSprop* change the units but the problem remains: we are now dividing the gradient update by a square root of the sum of squared gradients, so instead of meters per second we are now subtracting a dimensionless value—hardly a big improvement. Note that in second-order methods, this problem does not arise; in Newton's method, the update rule is $\mathbf{w}_{k+1} = \mathbf{w}_k - \alpha_k H_k^{-1} \mathbf{g}_k$; in the example above, we would get

$$\text{seconds} := \text{seconds} - \alpha \left(\frac{\text{meters}}{\text{second}^2} \right)^{-1} \frac{\text{meters}}{\text{second}},$$

and now the measurement units match nicely.

Algorithm 7: Adadelta

Initialize $\mathbf{w}_0$, $G_0 := 0$, $H_0 := 0$, $\mathbf{u}_0 := 0$, $k := 0$;
repeat
    $D_k := \text{Sample}(D)$;
    $\mathbf{g}_k := \sum_{d \in D_k} \nabla_{\mathbf{w}} f(\mathbf{w}_k, d)$;
    $H_{k+1} := \rho H_k + (1 - \rho) \text{diag}(\mathbf{u}_k)$;
    $G_{k+1} := \gamma G_k + (1 - \gamma) \text{diag}(\mathbf{g}_k)$;
    $\mathbf{u}_{k+1} := \frac{\sqrt{H_{k+1} + \epsilon}}{\sqrt{G_{k+1} + \epsilon}} \mathbf{g}_k$;
    $\mathbf{w}_{k+1} := \mathbf{w}_k - \mathbf{u}_{k+1}$;
    $k := k + 1$;
until *a stopping condition is met*;

Algorithm 8: Adam

Initialize $\mathbf{w}_0$, $G_0 := 0$, $\mathbf{m}_0 := 0$, $\mathbf{v}_0 := 0$, $\mathbf{u}_0 := 0$, $k := 1$;
repeat
$\quad D_k := \text{Sample}(D)$;
$\quad \mathbf{g}_k := \sum_{d \in D_k} \nabla_{\mathbf{w}} f(\mathbf{w}_k, d)$;
$\quad \mathbf{m}_k := \beta_1 \mathbf{m}_{k-1} + (1 - \beta_1) \mathbf{g}_k$;
$\quad \mathbf{v}_k := \beta_2 \mathbf{v}_{k-1} + (1 - \beta_2) \mathbf{g}_k^2$;
$\quad \hat{\mathbf{m}}_k := \frac{\mathbf{m}_k}{1 - \beta_1^k}$, $\hat{\mathbf{v}}_k := \frac{\mathbf{v}_k}{1 - \beta_2^k}$;
$\quad \mathbf{u}_k := \frac{\alpha}{\sqrt{\hat{\mathbf{v}}_k} + \epsilon} \hat{\mathbf{m}}_k$;
$\quad \mathbf{w}_{k+1} := \mathbf{w}_k - \mathbf{u}_k$;
$\quad k := k + 1$;
until *a stopping condition is met*;

To fix this problem without resorting to second-order methods, the authors of *Adadelta* [989] propose to add another exponential moving average, this time of the weight updates themselves, adding it to the numerator and thus arriving at the correct measurement units for the update. In other words, in *Adadelta*, we are rescaling the update with respect to the values of the weights, keeping track of the average weights:

$$\mathbf{w}_{k+1} := \mathbf{w}_k - \mathbf{u}_{k+1} = \mathbf{w}_k - \frac{\sqrt{H_{k+1} + \epsilon}}{\sqrt{G_{k+1} + \epsilon}} \mathbf{g}_k,$$

where

$$H_{k+1} = \rho H_k + (1 - \rho) \operatorname{diag}(\mathbf{u}_k),$$

that is, H_k accumulates the weight updates from previous steps. In this way, the updates are properly rescaled, and the measurement units are restored to their proper values.

But that's not the end of the line. The next algorithm, *Adam* (Adaptive Moment Estimation) [454], in many applications remains the algorithm of choice in deep learning up to this day. It is very similar to *Adadelta* and *RMSprop*, but *Adam* also stores an average of the past gradients (that is, an exponential moving average as usual), which acts as a kind of momentum for its updates.

Formally, this means that *Adam* has two parameters for the decay of updates, β_1 and β_2, keeps two exponential moving averages, $\mathbf{m}_k$ for gradients and $\mathbf{v}_k$ for their squares, and computes the update similar to *RMSProp* but with momentum-based $\mathbf{m}_k$ instead of just $\mathbf{g}_k$. Another feature is that since $\mathbf{m}_k$ and $\mathbf{v}_k$ are initialized by zeros, they are biased towards zero, and the authors correct for this bias by dividing over $(1 - \beta_i^k)$:

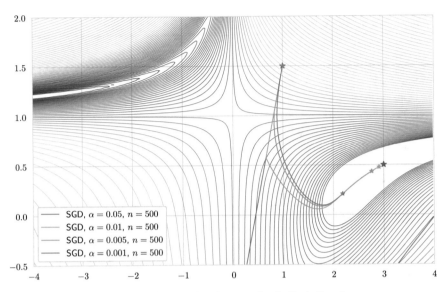

Fig. 2.14 Gradient descent with different learning rates for the Beale function.

$$\mathbf{m}_k = \beta_1 \mathbf{m}_{k-1} + (1 - \beta_1)\mathbf{g}_k,$$
$$\mathbf{v}_k = \beta_2 \mathbf{v}_{k-1} + (1 - \beta_2)\mathbf{g}_k^2.$$
$$\mathbf{u}_k = \frac{\alpha}{\sqrt{\frac{\mathbf{v}_k}{1-\beta_2^k}} + \epsilon} \frac{\mathbf{m}_k}{1 - \beta_1^k}.$$

We give a full description in Algorithm 8.

When *Adam* appeared, it quickly took the field of deep learning by storm. One of its best selling features was that it needed basically no tuning of the hyperparameters: the authors, Diederik Kingma and Jimmy Ba, recommended $\beta_1 = 0.9$, $\beta_2 = 0.999$, $\epsilon = 10^{-8}$, and these values work just fine for the vast majority of practical cases. Basically, by now *Adam* is the default method of training deep neural networks, and practitioners turn to something else only if *Adam* fails for some reason.

Before proceeding to a brief overview of other approaches and recent news, let me give an example of all these algorithms in action. For this example, I have chosen a standard function that is very common in examples like this; this is the Beale function

$$F(x, y) = (1.5 - x + xy)^2 + \left(2.25 - x + xy^2\right)^2 + \left(2.625 - x + xy^3\right)^2.$$

It is a simple and continuous but nonconvex function that has an interesting optimization landscape. All optimization algorithms were run starting from the same point $(1, \frac{3}{2})$, and the Beale function has a single global minimum at $(3, \frac{1}{2})$, which is our main goal.

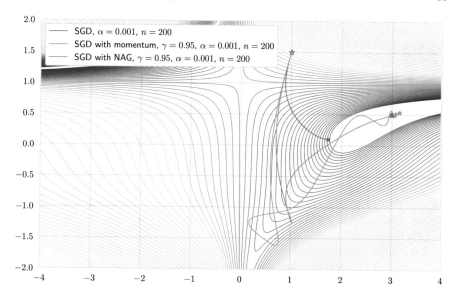

Fig. 2.15 Momentum-based methods for the Beale function.

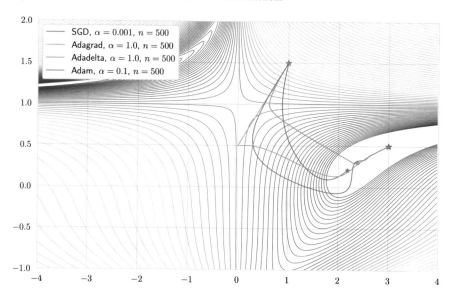

Fig. 2.16 Adaptive gradient descent methods for the Beale function.

Experimental results are shown in Figures 2.14, 2.15, and 2.16. Figure 2.14 shows that even in complex optimization landscapes, one usually can find a learning rate for the basic SGD that would work well. The problem is that this value is far from obvious: for instance, in this case, we see that overshooting the best learning rate (which appears to be around 0.01) even a little can lead to divergence or other undesirable behaviour: note how the plot with $\alpha = 0.05$ quickly gets out of hand. Figure 2.15 shows how momentum-based methods work: we see that for the same initial learning rate and the same number of iterations, SGD with momentum and SGD with Nesterov accelerated gradients find the optimum much faster. We also see the stabilization effect of NAG again: SGD with Nesterov momentum overshoots and oscillates much less than SGD with regular momentum. Finally, Fig. 2.16 shows the behaviour of adaptive gradient descent algorithms; in this specific example, *Adagrad* and *Adam* appear to work best although this two-dimensional example does not really let the adaptive approaches shine.

There have been attempts to explain what is going on with *Adam* and why it is so good. In particular, Heusel et al. in an influential paper [340] showed that stochastic optimization with *Adam* can be described as the dynamics of a heavy ball with friction (HBF), that is, *Adam* follows the differential equation for an HBF in Newtonian mechanics. Averaging over past gradients helps the "ball" (current value of **w**) get out of small regions with local minima and prefer large "valleys" in the objective function landscape. Heusel et al. use this property to help a GAN generator avoid mode collapse (we will talk much more about GANs in Chapter 4), but the remark is fully general and applies to *Adam* optimization in any context.

There have also been critiques of *Adam* and similar approaches. In another influential paper, Wilson et al. [927] demonstrate that

- when the optimization problem has a lot of local minima, different adaptive algorithms can converge to different minima even from the same starting point;
- in particular, adaptive methods can overfit, i.e., find non-generalizing local solutions;
- and all of this happens not only in the theoretical worst case, which would be natural and fine with us, but in practical examples.

Wilson et al. conclude that adaptive gradient descent methods are not really advantageous over standard SGD and advise to use SGD with proper step size tuning over *Adam* and other algorithms.

Therefore, researchers have continued to search for the holy grail of a fast and universal adaptive gradient descent algorithm. Since *Adam* was presented in 2014, there have been a lot of attempts to improve adaptive gradient descent algorithms further. Since this is not the main subject of the book and since *Adam* still remains mostly the default, I will not give a comprehensive survey of these attempts but will only mention in passing a few of the most interesting ideas.

First of all, *Adam* itself has received several modifications:

- the original *Adam* paper [454] proposed *Adamax*, a modification based on the L_∞-norm instead of L_2-norm for scaling the gradients; in this variation, $\mathbf{m}_k$ is computed as above, and instead of $\mathbf{v}_k$ the scaling is done with

$$\mathbf{v}_k^\infty = \max\left(\beta_a \mathbf{v}_{k-1}, |\mathbf{g}_k|\right),$$

and $\mathbf{v}_k^\infty$ is used instead of $\hat{\mathbf{v}}_k$ in Algorithm 8 (initialization bias does not occur in this case);

- *AMSGrad* [708] is a very similar idea: the authors present an example of a simple problem where the original *Adam* does not converge and fix this by normalizing the running average of the gradient with a maximum of all $\mathbf{v}_t$ up to this point instead of the exponential moving average $\mathbf{v}_t$; in Algorithm 8, it means that we let

$$\mathbf{u}_k := \frac{\alpha}{\sqrt{\mathbf{v}_k'} + \epsilon} \hat{\mathbf{m}}_k$$

for $\mathbf{v}_k' = \max\left(\mathbf{v}_{k-1}', \mathbf{v}_k\right)$, where max is understood componentwise and $\mathbf{v}_k$ is defined exactly as in Algorithm 8;

- *Nadam* [204] is the modification of *Adam* that uses the Nesterov momentum instead of regular momentum for $\mathbf{m}_k$; expanding one step back, the *Adam* update rule can be written as

$$\mathbf{w}_{k+1} = \mathbf{w}_k - \frac{\alpha}{\sqrt{\hat{\mathbf{v}}_k} + \epsilon} \left(\frac{\beta_1 \mathbf{m}_{k-1}}{1 - \beta_1^k} + \frac{(1 - \beta_1)\mathbf{g}_k}{1 - \beta_1^k}\right) \approx$$

$$\approx \mathbf{w}_k - \frac{\alpha}{\sqrt{\hat{\mathbf{v}}_k} + \epsilon} \left(\beta_1 \hat{\mathbf{m}}_{k-1} + \frac{(1 - \beta_1)\mathbf{g}_k}{1 - \beta_1^k}\right)$$

(approximate becase we do not distinguish between $1 - \beta_1^k$ and $1 - \beta_1^{k-1}$ in the denominator), and now we can replace the bias-corrected estimate $\hat{\mathbf{m}}_{k-1}$ with the current estimate $\hat{\mathbf{m}}_k$, thus changing regular momentum into Nesterov's version:

$$\mathbf{w}_{k+1} = \mathbf{w}_k - \frac{\alpha}{\sqrt{\hat{\mathbf{v}}_k} + \epsilon} \left(\beta_1 \hat{\mathbf{m}}_k + \frac{(1 - \beta_1)\mathbf{g}_k}{1 - \beta_1^k}\right);$$

- *QHAdam* (quasi-hyperbolic Adam) [565] replaces both momentum estimators in *Adam*, $\mathbf{v}_k$ and $\mathbf{m}_k$, with their quasi-hyperbolic versions, i.e., with weighted averages between plain SGD and momentum-based *Adam* updates; the update rule in *QHAdam* looks like

$$\mathbf{w}_{k+1} = \mathbf{w}_k - \alpha \frac{(1 - \nu_1)\,\mathbf{g}_k + \nu_1 \hat{\mathbf{m}}_k}{(1 - \nu_2)\,\mathbf{g}_k^2 + \nu_2 \hat{\mathbf{v}}_k},$$

where $\hat{\mathbf{m}}_k$ and $\hat{\mathbf{v}}_k$ are defined as in Algorithm 8 and ν_1, ν_2 are new constants;

- the critique of Wilson et al., combined with much faster convergence of *Adam* during the initial stages of the optimization process, led to the idea of switching from *Adam* to SGD at some strategic point during the training process [447].

AdamW [556, 557] is probably one of the most interesting *Adam* variations. It goes back to the 1980s, to the original L_2 regularization method for neural networks which was *weight decay* [322]:

$$\mathbf{w}_{k+1} = (1 - \beta)\mathbf{w}_k - \alpha\nabla_{\mathbf{x}_k} F_k,$$

where β is the weight decay rate and α is the learning rate. In this formulation, the weights are brought closer to zero. Naturally, it was immediately noted (right in the original paper [322]) that this approach is completely equivalent to changing the objective function F_k:

$$F_k^{\text{reg}}(\mathbf{w}_k) = F_k(\mathbf{w}_k) + \frac{\beta}{2}\|\mathbf{w}_k\|_2^2$$

or even directly changing the gradient:

$$\nabla_{\mathbf{w}_k} F_k^{\text{reg}} = \nabla_{\mathbf{w}_k} F_k + \beta\mathbf{w}_k.$$

But while this equivalence holds for plain vanilla SGD, it does not hold for adaptive variations of gradient descent! The idea of *AdamW* is to go back to the original weight decay and "fix" *Adam* so that the equivalence is restored. The authors show how to do that without losing efficiency, by changing *Adam* updates only very slightly. The only change compared to Algorithm 8 is that now the update $\mathbf{u}_k$ is given by

$$\mathbf{u}_k := \frac{\alpha\hat{\mathbf{m}}_k}{\sqrt{\hat{\mathbf{v}}_k} + \epsilon} + \lambda\mathbf{w}_k$$

instead of adding $\lambda\mathbf{w}_k$ to $\mathbf{g}_k$ as it would happen if *Adam* was straightforwardly applied to a regularized objective function.

It has been shown in [556, 557] that *AdamW* has several important beneficial properties. Apart from improved generalization (in some experiments), it is better than the original *Adam* in decoupling the hyperparameters. This means that the best values of hyperparameters such as initial learning rate α and regularization parameter λ do not depend on each other and thus can be found with independent trials, which makes hyperparameter tuning much easier.

Despite this wide variety of first-order optimization algorithms and their variations, the last word has not yet been said in optimization for deep learning. As often as now, new ideas that at first glance might revolutionize the field fade into obscurity after the original experiments are not confirmed in wider research and engineering practice. One good example of such an idea is *super-convergence* [806], the idea that it is beneficial to change the learning rate with a cyclic schedule, increasing it back to large values from time to time in order to provide additional regularization and improve generalization power. The original experiments were extremely promising, and the idea of curriculum learning has a long and successful history in deep learning [63] (we will actually return to this idea in a different context, in particular, in Section 6.4). But the "super" in "super-convergence" has not really proven to be true across a wide variety of situations. The idea of increasing the learning rate back has been added to the toolbox of deep learning practicioners, but cyclic learning rates have not become the staple of deep learning.

2.6 Conclusion

To sum up, in this section, we have seen the main ideas that have driven the first-order optimization as applied to deep neural networks over recent years. There has been a lot of progress in adaptive gradient methods: apart from classical momentum-based approaches, we have discussed the recently developed optimization methods that adapt their learning rates differently to different weights. By now, researchers working in applied deep learning mostly treat the optimization question as tentatively solved, using *Adam* or some later variation of it such as *AdamW* by default and falling back to SGD if *Adam* proves to get stuck in local minima too much. However, new variations of first-order adaptive optimization methods continue to appear, and related research keeps going strong.

Second-order methods or their approximations such as quasi-Newton optimization methods remain out of reach. It appears that it would be very hard indeed to develop their variations suitable for stochastic gradient descent with the huge variances inherent in optimizing large datasets by mini-batches. But there are no negative results that I know in this direction either, so who knows, maybe the next big thing in deep learning will be a breakthrough in applying second-order optimization methods or their approximations to neural networks.

I would like to conclude by noting some other interesting directions of study that so far have not quite led to new optimization algorithms but may well do so. In the latest years, researchers have begun to look at deep neural networks and functions expressed by them as objects of research rather than just tools for approximation or optimization. In particular, there have been interesting and enlightening recent studies of the optimization landscape in deep neural networks. I'd like to highlight a few works:

- Li et al. [514] study how the learning rate influences generalization and establishes a connection between the learning rates used in training and the curriculum of which patterns the model "learns" first;
- Huang et al. [370] show that a real-life neural network's optimization landscape has plenty of bad minima that have near-perfect training set accuracy but very bad generalization (test set accuracy); the authors describe this landscape as a "minefield" but find that SGD somehow "miraculously" avoids the bad minima and finds a local minimum with good generalization properties;

- Keskar et al. [446] and He et al. [326] study the landscape of the loss functions commonly used in deep learning, find that "flat" local minima have better generalization properties than "sharp" ones, and discuss which training modes are more likely to fall into flat or sharp local minima;
- Chen et al. [132] put some of this theory into practice by showing how to *deform* the optimization landscape in order to help the optimizer fall into flat minima;
- Izmailov et al. [391] note that better (wider) local optima can be achieved by a very simple trick of *stochastic weight averaging*, where the weights at several points along the SGD trajectory are combined together; this is a simplification of the previously developed *fast geometric ensembling* trick [263];
- Nakkiran et al. [618] discuss the *double descent* phenomenon, where performance gets worse before getting better as the model size increases; double descent had been known for some time, but Nakkiran et al. introduce *effective model complexity*, a numerical measure of a training procedure that might explain double descent;
- Wilson and Izmailov [928, 929] considers the same effects from the Bayesian standpoint, explaining some mysteries of deep learning from the point of view of Bayesian inference and studying the properties of the prior over functions that are implied by regularization used in deep learning.

These are just a few examples, there are many more. This line of research appears to be very promising and actually looks like it is still early in development. Therefore, I expect exciting new developments in the nearest future in this direction.

In the next chapter, we proceed from general remarks about deep learning and optimization to specific neural architectures. We will consider in more detail the field of machine learning where synthetic data is used most widely and with the best results: computer vision.

Chapter 3
Deep Neural Networks for Computer Vision

Computer vision problems are related to the understanding of digital images, video, or similar inputs such as 3D point clouds, solving problems such as image classification, object detection, segmentation, 3D scene understanding, object tracking in videos, and many more. Neural approaches to computer vision were originally modeled after the visual cortex of mammals, but soon became a science of their own, with many architectures already developed and new ones appearing up to this day. In this chapter, we discuss the most popular architectures for computer vision, concentrating mainly on ideas rather than specific models. We also discuss the first step towards synthetic data for computer vision: data augmentation.

3.1 Computer Vision and Convolutional Neural Networks

Computer vision is one of the oldest and most important subfields of artificial intelligence. In the early days of AI, even leading researchers believed that computer vision might prove to be easy enough: Seymour Papert, one of the fathers of AI, initially formulated several basic computer vision problems as a student project in "an attempt to use our summer workers effectively" [654] (see also Section 5.1). But it soon became apparent that computer vision is actually a much more ambitious endeavour, and despite decades of effort and progress the corresponding problems are still not entirely solved.

One of the most important advances in the study of the visual cortex was made by David H. Hubel and Torsten N. Wiesel who, in their Nobel Prize-winning collaboration, were the first to analyze the activations of individual neurons in the visual cortex of mammals, most famously cats [376, 377, 925]. They studied the early layers of the visual cortex and realized that individual neurons on the first layer of processing react to simple shapes while neurons of the second layer react to certain combinations of first layer neurons. For example, one first layer neuron might

© The Author(s), under exclusive license to Springer Nature Switzerland AG 2021
S. I. Nikolenko, *Synthetic Data for Deep Learning*, Springer Optimization
and Its Applications 174, https://doi.org/10.1007/978-3-030-75178-4_3

react to a horizontal line in its field of view (called a *receptive field*, a term that also carried over to artificial intelligence), and another first layer neuron might be activated by a vertical line. And if these two neurons are activated at the same time, a second layer neuron might react to a cross-like shape appearing in its receptive field by implementing something close to a logical AND (naturally, I'm simplifying immensely but that's the basic idea). In other words, first layer neurons pick up on very simple features of the input, and second layer neurons pick up on combinations of first layer neurons. Hubel and Wiesel were wise enough not to go much farther than the first two layers because signal processing in the brain becomes much more complicated afterwards. But even these initial insights were enough to significantly advance artificial intelligence...

The basic idea of *convolutional neural networks* (CNN) is quite simple. We know, e.g., from the works of Hubel and Wiesel that a reasonable way to process visual information is to extract simple features and then produce more complicated features as combinations of simple ones. Simple features often correspond to small receptive fields: for instance, we might want a first layer neuron to pick up a vertical gradient in a window of size 5×5 pixels. But then this feature extraction should be applied equally to *every* 5×5 window, that is, instead of training a neural network to perform the same operation for each window across, say, a 1024×1024 image we could apply *the same learnable transformation* to every window, with shared weights. This idea works as a structural regularizer, saving an immense number of weights in the CNN compared to an equivalent fully connected network. Mathematically, this idea can be expressed as a convolution between the input and the small learnable transformation, hence the name.

Figure 3.1 illustrates the basic idea of a convolutional network: a 5×5 input image is broken down into 3×3 windows, and each window is passed through the same small neural network, getting a vector of features as a result. After this transformation, the 5×5 image becomes a 3×3 output in terms of width and height.

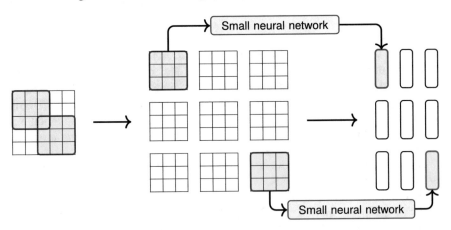

Fig. 3.1 The basic idea of a convolutional neural network; blue and red colors follow the transformations of two 3×3 windows, and the same small neural network is applied to the other windows as well.

A *convolutional layer* is actually just one way to implement this idea, albeit the most popular one by far. It is a layer defined by a set of learnable *filters* (or *kernels*) that are convolved across the entire input tensor. Convolutions can come in any dimension, but the most popular and intuitive ones for computer vision are two-dimensional convolutions, so we will use them in our examples. In this case, the input is a three-dimensional tensor width × height × channels (a grayscale image has one channel, a color image three, and intermediate representations inside a neural network can have arbitrarily many), and the convolution is best thought of as a four-dimensional tensor of dimension

$$\text{input channels} \times \text{width} \times \text{height} \times \text{output channels}.$$

Figure 3.2 shows a toy numerical example of a convolutional layer consisting of a linear convolution with a $3 \times 3 \times 2$ tensor of weights and a ReLU activation, applied to a $5 \times 5 \times 1$ input image.

For the first "real" example, let us consider the Sobel operator, a classical computer vision tool dating back to 1968 [809]. It is a discrete computation of the image gradient, used for edge detection in classical computer vision. For our example, we note that the main components of the Sobel operator are two 3×3 convolutions with matrices

$$S_x = \begin{pmatrix} 1 & 0 & -1 \\ 2 & 0 & -2 \\ 1 & 0 & -1 \end{pmatrix}, \qquad S_y = \begin{pmatrix} 1 & 2 & 1 \\ 0 & 0 & 0 \\ -1 & -2 & -1 \end{pmatrix}.$$

Basically, S_x shows the horizontal component of the image gradient and S_y shows the vertical component.

If we take an image such as a handwritten digit from the MNIST dataset as shown in Fig. 3.3a, and apply a convolution with matrix S_x, we get the result shown in Fig. 3.3b. The result of convolving with matrix S_y is shown in Fig. 3.3c. In this example, we see how convolutions with small matrices give rise to meaningful feature

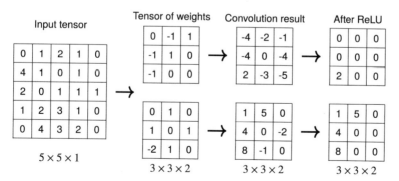

Fig. 3.2 Sample application of a convolutional layer consisting of a linear convolution with a $3 \times 3 \times 2$ tensor of weights and a ReLU activation.

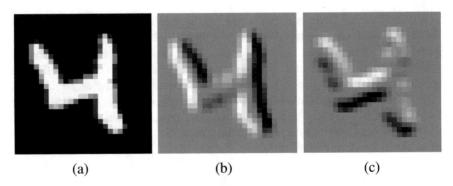

(a)   (b)   (c)

Fig. 3.3 Sample application of convolutions: (a) original handwritten digit; (b) convolution with matrix S_x, the horizontal component of the image gradient; (c) convolution with matrix S_y, the vertical component of the image gradient.

extraction: the meaning of the features shown in Fig. 3.3b, c is quite clear, and they indeed reflect the variation of the input image's pixel intensities in the corresponding direction. Note that in a neural network, both convolutions would be treated as a single tensor of dimension $3 \times 3 \times 2$, and the result of applying it to an image of size 28×28 would be a tensor with two channels (feature maps) shown in Fig. 3.3b, c. Note that in this example, the width and height of the output stay at 28 instead of being reduced by 1 on every side because the results are computed with *padding*, i.e., an extra row and column of zeroes is added on every side of the input; this is a common trick in convolutional networks used when it is desirable to leave the input size unchanged.

In a trainable neural network, weights in the matrices S_x and S_y would not be set in advance but would represent weights that need to be learned with gradient descent, as we discussed in the previous chapter. Actually, the first introduction of convolutions to artificial neural networks happened a long time ago, when even backpropagation had not been universally accepted. This was the *Neocognitron* developed by Kuni-hiko Fukushima in the late 1970s [248–250]. The *Neocognitron* was a pioneering architecture in many respects: it was a deep neural network in times when deep networks were almost nonexistent, it had feature extraction from small windows of the input—precisely the idea of convolutional networks—it was training in an unsu-pervised fashion, learning to recognize different kinds of patterns presented, and it actually already had ReLU activations—all this in the 1970s! The *Neocognitron* is widely regarded as a predecessor to CNNs, and although it did take a few years to adapt all these ideas into modern CNN architectures, they actually appeared in the 1980s pretty much in their modern form.

In a classical convolutional network such as *LeNet* [501], convolutional layers are usually interspersed with nonlinear activation functions (applied componentwise) and *pooling* layers that reduce the dimension. The most popular is the *max-pooling* layer that does not have any trainable parameters and simply covers the input tensor with windows (often 2×2) and chooses the largest value of every feature in each

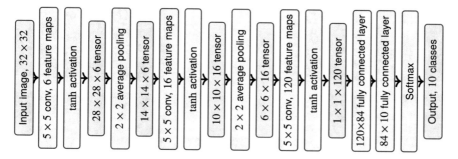

Fig. 3.4 The *LeNet*-5 architecture [501].

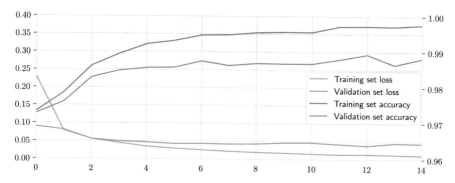

Fig. 3.5 The *LeNet*-5 training process on MNIST dataset of handwritten digits.

window. The basic intuition is that we want the features to correspond to certain properties that extend from smaller to larger windows; for example, if there is a cat present in a 128×128 window in the image, there is also a cat in every 256×256 window containing it. Max-pooling also induces a lot of sparsity that helps keep the computations more efficient.

For an extended example, let us implement and study the (slightly modified) *LeNet*-5 network operating on 32×32 grayscale images (we will be using MNIST handwritten digits), a simple convolutional architecture shown in Figure 3.4. In the figure, layers that perform transformations are shown as rectangles filled in green, and dimensions of current feature maps are shown as rectangles filled in blue. As you can see, each 5×5 convolution reduces the width and height of the input tensor by 4 because there is no padding here, and each 2×2 pooling layer (average pooling in the case of *LeNet*) halves the input dimensions.

Figure 3.5 shows the learning process of this network, trained on the MNIST dataset with *Adam* optimizer and batch size 32. As you can see, the loss function (average cross-entropy between the correct answers and network predictions) on the training set decreases steadily, but the loss function on the held-out validation set is far from monotone. The best result on the validation set is achieved in this experiment

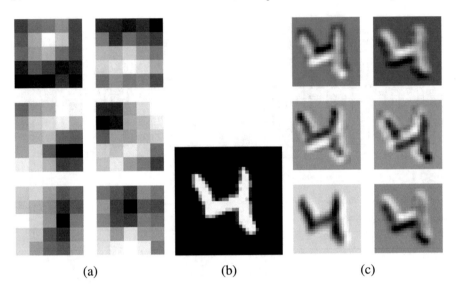

(a)   (b)   (c)

Fig. 3.6 A view into the first layer of *LeNet*-5: (a) weights of the six convolutions; (b) sample input image; (c) activations produced by convolutions from (a) on the image from (b).

(specific results might change after a restart with random re-initialization of the weights) after 12 epochs of training.

Figure 3.6 shows the first layer of the resulting trained network. It shows the weights of the six 5×5 convolutions trained on the first layer in Fig. 3.6a and the results of applying them to a sample digit shown in Fig. 3.6b (the same as in Fig. 3.3, only padded with zeroes to size 32×32) are shown in Fig. 3.6c. You can see that the first layer has learned to extract simple geometric features, in many ways similar to the image gradients shown in Fig. 3.3.

Modern networks used in computer vision employ very similar convolutional layers. They usually differ from *LeNet* in that they use ReLU activation functions or its variations rather than sigmoidal activations (recall the discussion in Section 2.1). The ReLU nonlinearity was re-introduced into deep learning first for Boltzmann machines in [616] and widely popularized in deep convolutional networks by *AlexNet* (see Section 3.2 below).

Over the last decade, CNNs have been dominating computer vision and have been rising in popularity in many other fields of machine learning. For example, in 2014–2016 one-dimensional CNNs were becoming increasingly crucial for natural language processing (NLP), supplementing and even replacing recurrent networks [423, 1010]; after 2017, the best results in NLP were produced by architectures based on self-attention such as Transformer, especially BERT and GPT families [192, 697, 891]. But BERT-like models are still often used to pretrain word embeddings, and embeddings are then processed by CNNs and/or RNNs to solve downstream tasks.

However, there still remain valid criticisms even for the basic underlying idea of convolutional architectures. Two of the most important criticisms deal with the *lack of translational invariance* and *loss of geometry* along a deep CNN. Lack of translational invariance means that units in a CNN that are supposed to produce a given feature (say, recognize a cat's head on a photo) might not activate if the head is slightly moved or rotated. In machine learning practice, translational invariance in CNNs is usually achieved by extensive data augmentation that always includes simple geometric transformations such as re-cropping the image, rescaling by a factor close to 1, and so on (we will discuss augmentations in detail in Section 3.4). However, achieving full translational invariance by simply extending the dataset is far from guaranteed and appears extremely wasteful: if we are dealing with images we already know that translational invariance should be in place, why should we learn it from scratch for every single problem in such a roundabout way?

The "loss of geometry" problem stems from the fact that standard convolutional architectures employ pooling layers that propagate information about low-level features to high-level feature maps with smaller resolutions. Therefore, as the signal travels from bottom to top layers, networks progressively lose sight of the exact locations where features have originated. As a result, it is impossible for a high-level feature to activate on specific geometric interrelations between low-level features, a phenomenon sometimes called the "Picasso problem": a convolutional feature can look for two eyes, nose, and mouth on a face but cannot ensure that these features are indeed properly positioned with respect to each other. This is because pooling layers, while they are helpful to reduce the redundancy of feature representation in neural networks, prevent overfitting, and improve the training process, and at the same time represent a fixed and very crude way of "routing" low-level information to high-level features.

These two problems have been pointed out already in 2014 by Geoffrey Hinton [343]. An attempt to alleviate these problems has led Hinton to develop a new approach to architectures that perform feature composition: *capsule networks*. In a capsule network, special (trainable) routing coefficients are used to indicate which low-level features are more important for a given high-level feature, and the features (capsules) themselves explicitly include the orientations and mutual positions of features and explicitly estimate the likelihood of the resulting composite feature [247, 349, 452, 748]. As a result, translational invariance is built in, routing is dynamic, capsule networks have much fewer parameters than CNNs, and the entire process is much more similar to human vision than a CNN: capsules were designed with cortical columns in mind.

However, at present it still appears too hard to scale capsule networks up to real-world problems: computational tricks developed for large-scale CNNs do not help with capsule networks, and so far they struggle to scale far beyond MNIST and similar-sized datasets. Therefore, all modern real-life computer vision architectures are based on CNNs rather than capsules or other similar ideas, e.g., other equivariant extensions such as spherical CNNs [163] or steerable CNNs [164], and applications of capsule networks are only beginning to appear [227]. Thus, we will not be discussing these alternatives in detail, but I did want to mention that the future of computer vision may hold something very different from today's CNNs.

3.2 Modern Convolutional Architectures

In this section, we give an overview of the main ideas that have brought computer vision to its current state of the art. We will go over a brief history of the development of convolutional architectures during the deep learning revolution, but will only touch upon the main points, concentrating on specific important ideas that each of these architectures has brought to the smörgåsbord of CNNs that we have today. For a more detailed introduction, we refer to [153, 225, 631] and other sources (Fig. 3.7).

MCDNN. The deep learning revolution in computer vision started during 2010–2011, when recent advances in deep learning theory and the technology of training and using neural networks on highly parallel graphical processors (GPUs) allowed training much deeper networks with much more units than before. The first basic problem that was convincingly solved by deep learning was image classification. In 2011, a network by Dan Cireşan from Jürgen Schmudhuber's group won a number of computer vision competitions [159]. In particular, this network was the first to achieve superhuman performance in a practically relevant computer vision problem, achieving a mere 0.56% error in the IJCNN Traffic Sign Recognition Competition, while the average human error on this dataset was 1.16% [160].

Architecturally, Ciresan's network, called *Multi-Column Deep Neural Network* (MCDNN), is a committee of deep convolutional networks with max-pooling. It showcases several important ideas:

- MCDNN uses a basic convolutional architecture very similar to the *LeNet* family of networks (so we do not show a separate figure for it), but it was one of the first to consistently use max-pooling instead of average-pooling or other variations;
- the architecture contains several identical networks trained on differently preprocessed inputs, where preprocessing variations include different combinations of color normalization and contrast adjustment; thus, MCDNN was already showing

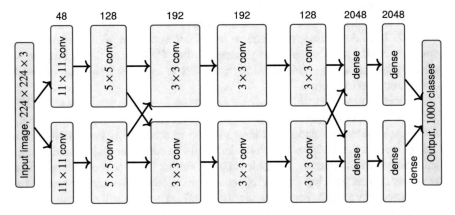

Fig. 3.7 The *AlexNet* architecture [477].

the power of *data augmentation* for computer vision, a theme that remains crucial to this day and that represents one of the motivations for synthetic data.

AlexNet. However, MCDNN operated on very small images, cutting out traffic sign bounding boxes of size 48 × 48 pixels. The development of large-scale modern architectures that could deal with higher resolution images started with *AlexNet* [477], a network developed by Alex Krizhevsky in Prof. Hinton's group (see Fig. 3.7 for an illustration). With 8 trainable layers, *AlexNet* became one of the first successful truly deep convolutional networks. It was introduced at the ImageNet Large Scale Visual Recognition Challenge (ILSVRC) in 2012, where *AlexNet* beat all competitors with an unprecedented margin: two submitted versions of *AlexNet* had test set errors (measured as classification accuracy for top-5 guesses) about 15–16%, while the nearest competitor could only achieve an error of 26%[1]! Architecturally, *AlexNet* again introduced several new ideas:

- it introduced and immediately popularized ReLU activations as nonlinearities used in convolutional layers; previously, tanh activations had been most often used in convolutional networks;
- it emphasized the crucial role of data augmentation in training neural networks for computer vision problems; we will discuss the case of *AlexNet* in detail in Section 3.4;
- it was one of the first large-scale networks to consistently use dropout for additional regularization;
- finally, it was one of the first neural networks to feature model parallelization: the model was distributed between two GPUs; back in 2012, it was a real engineering feat, but since then it has become a standard feature of deep learning frameworks such as *PyTorch* or *Tensorflow*.

AlexNet's resounding success marked the start of a new era in computer vision: since 2012, it has been dominated by convolutional neural architectures. CNNs have improved and defined state of the art in almost all computer vision problems: image classification, object detection, segmentation, pose estimation, depth estimation, object tracking, video processing, and many more. We will talk in more detail about object detection and segmentation architectures in Section 3.3. For now, the important part is that they all feature a convolutional backbone network that performs feature extraction, often on several layers simultaneously: bottom layers (nearest to input) of a CNN extract local features and can produce high-resolution feature maps, while features extracted on top layers (nearest to output) have large receptive fields, generalize more information, and thus can learn to have deeper semantics, but lose some of the geometry along the way (we have discussed this problem above in Section 3.1).

VGG. The next steps in deep CNN architectures were also associated with the ILSVRC challenge: for several years, top results in image classification were marked by new important ideas that later made their way into numerous other architectures as well. One of the most fruitful years was 2014, when the best ImageNet classification

---

[1]http://image-net.org/challenges/LSVRC/2012/results.html.

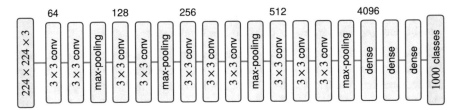

Fig. 3.8 VGG: decomposing large convolutions [802].

was achieved by the VGG network by Simonyan and Zisserman [802]; the name originates from the Visual Geometry Group in the University of Oxford. The main idea that defined the VGG family is that individual convolutions in a CNN virtually never need to be large: a 5×5 convolution can be expressed as a composition of two 3×3 convolutions without any pooling or nonlinearities in between, a 7×7 convolution is just three layers of 3×3 convolutions, and so on. Figure 3.8 shows the first successful network from the VGG family; note how max-pooling layers come after groups of two or three convolutional layers, thus decomposing larger convolutions. This results in much deeper networks with fewer weights, serving as additional regularization and at the same time making training and inference more efficient. Later architectures also experimented with expressing $n \times n$ two-dimensional convolutions as compositions of $1 \times n$ and $n \times 1$ one-dimensional convolutions, and this trick is also common in modern CNN architectures.

Inception and GoogLeNet. In the same year, Google presented *GoogLeNet*, a network by Szegedy et al. that won the object detection track of ILSVRC 2014 [836]. Together with a precursor work on "network-in-network" architectures by Lin et al. [522], it had three important ideas that have stayed with us ever since: Inception modules, 1×1 convolutions, and auxiliary classifiers.

First, "network-in-network" architectures take the basic idea of convolutional networks—applying the same simple transformation to all local windows over the input—and run with it a bit further than regular CNNs. Instead of just using a matrix of weights, they design special architectures for the "simple transformation" (not so simple anymore), so that a single layer is actually applying a whole neural network to each window of the input, hence the name. The architecture of these small networks from [836], called *Inception modules*, is shown in Fig. 3.9. Since then, there have been plenty of modifications, including Inception v2 and v3 [837] and later combinations of Inception with ResNet (see below).

Second, 1×1 convolutions play an important part in all variations of network-in-network modules. At first glance, it may appear that 1×1 convolutions are pointless. However, while they indeed do not collect any new features from neighboring pixels, they provide additional expressiveness by learning a (nonlinear, otherwise it is pointless indeed) transformation on the vector of features in a given pixel. In practice, this is usually needed to change the dimension of the feature vector, often reducing it before performing more computationally demanding transformations.

Fig. 3.9 Inception modules: (a) basic "naive" Inception v1 module [836]; (b) Inception v1 module with dimension reductions via 1×1 convolutions [836]; (c) sample Inception v2 module [837].

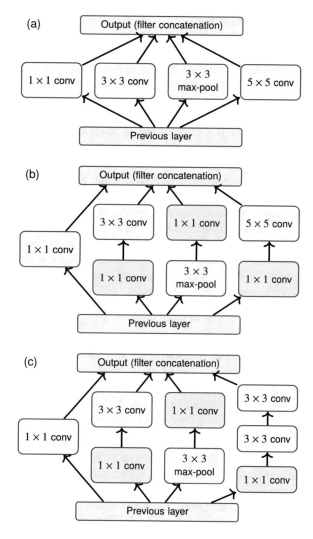

For example, a 3×3 convolution that maps a 512-dimensional vector into a 512-dimensional vector has $512 \times 3 \times 3 \times 512 = 9 \cdot 2^{18} \approx 2.36M$ weights. But if we first apply a 1×1 convolution to reduce the dimension to 64 and then map the result back into dimension 512, we add two convolutions with $512 \times 1 \times 1 \times 64 = 2^{15} = 32768$ weights each but reduce the 3×3 convolution to $64 \times 3 \times 3 \times 64 = 9 \cdot 2^{12}$ weights, for a total of $2 \cdot 2^{15} + 9 \cdot 2^{12} \approx 102K$ weights, a reduction by a factor of more than 20! The additional approximation that this kind of dimensionality reduction implies usually does not hurt and may even serve as additional structural regularization.

This idea has been widely used in architectures that try to minimize the memory footprint or latency of convolutional neural models. Figure 3.9b shows the Inception v1 module with 1×1 convolutions that perform these dimension reductions, and Figure 3.9c shows how Inception v2 has modified this module with the VGG basic idea of decomposing larger convolutions into compositions of 3×3 convolutions [837]. We do not show all variations of Inception modules here and refer to [836, 837] for more details.

Third, *GoogLeNet* is a deep network, it has 22 layers with trainable parameters, or 27 if you count pooling layers. When training by gradient descent, *GoogLeNet* faces problems that we discussed in Section 2.4 in relation to deep neural networks in general: error propagation is limited, and when top layers reach saturation it becomes very hard for bottom layers to train. To overcome this problem, Szegedy et al. [836] proposed to use auxiliary classifiers to help the loss gradients reach bottom layers. The *GoogLeNet* architecture (see Fig. 3.10) has two auxiliary classifiers that have separate classification heads (shallow networks ending in a classification layer). The loss functions are the same (binary cross-entropy classification loss), and they are simply added together with the main loss function to form the objective function for the whole network:

$$\mathcal{L}^{\text{GoogLeNet}} = \mathcal{L}^{\text{MainBCE}} + \alpha_1 \mathcal{L}^{\text{AuxBCE}_1} + \alpha_2 \mathcal{L}^{\text{AuxBCE}_2}.$$

The α coefficient was initialized to 0.3 and gradually reduced during training. This trick was intended to speed up the training of bottom layers on early stages of training and improve convergence, but Szegedy et al. report that in practice, convergence rate did not improve significantly, but the final performance of the network was better, so auxiliary classifiers served more like a regularizer.

ResNet. Despite these promising results, auxiliary classifiers are not widely used in modern architectures. The reason is that the main problem that they had been intended to solve, problems with error propagation after the saturation of top layers, was solved in a different way that proved to be much better. A *Microsoft Research* team led by Kaiming He developed and implemented the idea of *deep residual learning* [330] that was the main driving force behind the success of *ResNet* architectures that won ILSVRC 2015 in both classification (reducing the ImageNet Top-5 error rate to 3.5%) and object detection (with the Faster R-CNN detection architecture that we will discuss below in Section 3.3).

The basic structure of *ResNet* is simple: it is a composition of consecutive layers, and each of them is usually simply a convolutional layer, perhaps with batch normalization on top. The main difference is that in a residual unit, the layer that computes a function $F(\mathbf{x})$ for some input $\mathbf{x}$ (shown schematically in Fig. 3.11a) is supplemented with a direct residual connection that goes around the layer, so that the overall function that produces the kth layer output, denoted as $\mathbf{y}^{(k)}$, from the input vector $\mathbf{x}^{(k)}$ is computed as

$$\mathbf{y}^{(k)} = F(\mathbf{x}^{(k)}) + \mathbf{x}^{(k)},$$

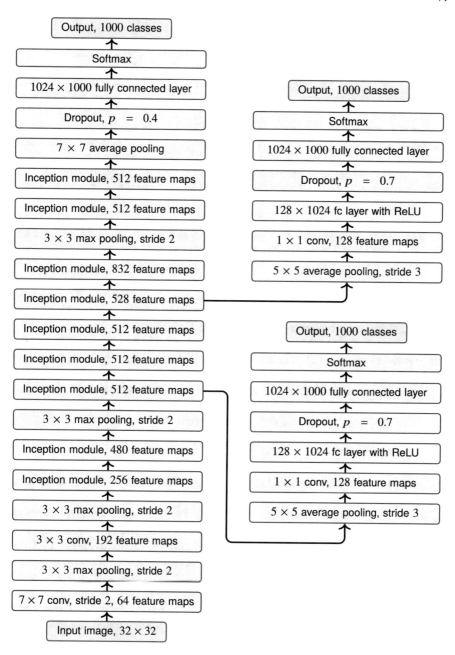

Fig. 3.10 The *GoogLeNet* general architecture.

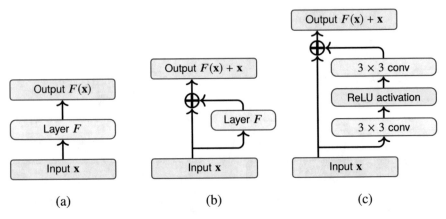

Fig. 3.11 Deep residual learning: (a) schematics of a simple layer; (b) schematics of a layer with a residual connection; (c) sample ResNet layer from [330].

where $\mathbf{x}^{(k)}$ is the input vector of the kth layer, $F(x)$ is the function that the layer computes, and $\mathbf{y}^{(k)}$ is the output of the residual unit that will later become $\mathbf{x}^{(k+1)}$ and will be fed to the next layer in the network (see Fig. 3.11b).

The name comes from the fact that if the layer as a whole is supposed to approximate some function $H(\mathbf{x})$, it means that the actual neural layer has to approximate the *residual*, $F(\mathbf{x}) \approx H(\mathbf{x}) - \mathbf{x}$; this does not complicate the problem for $F(\mathbf{x})$ much (if at all), but at the same time provides a direct way for the gradient to flow around $F(\mathbf{x})$. Now

$$\frac{\partial \mathbf{y}^{(k)}}{\partial \mathbf{x}^{(k)}} = 1 + \frac{\partial F(\mathbf{x}^{(k)})}{\partial \mathbf{x}^{(k)}},$$

and even if the layer F becomes completely saturated, its near-zero derivatives will not hinder training: the gradient will simply flow down to the previous layer unchanged.

Residual learning was not a new idea: it is the same *constant error carousel* idea that had been used in recurrent architectures for a long time, in particular in the famous long short-term memory (LSTM) architectures developed in the late 1990s by Hochreiter, Gers, and Schmidhuber [273, 350]. A recurrent network is, in essence, a very deep network by default (consider its computational graph when unrolled along the entire input sequence), and the same phenomenon leads to either exploding or vanishing gradients that effectively limit the propagation of information ("memory") in recurrent networks. The constant error carousel is precisely the idea of having a "direct path" for the gradient to flow.

However, He et al. were the first to apply this idea to "unrolled" deep convolutional networks, with great success. Note that a comparison of several residual architectures performed in [331] shows that the best results are achieved with the simplest possible residual connections: it is usually best to leave the direct path as free from any transformations (such as nonlinearities, batch normalizations, and the like) as possible. It even proved to be a bad idea to use control gates that could potentially learn to "open" and "close" the path around the layer $F(\mathbf{x})$, an idea that had been successful in LSTM-like recurrent architectures. Figure 3.11c shows a simple sample residual layer from [330], although, of course, many variations on this idea have been put forward both in the original paper and subsequent works.

Architecturally, this has led to the possibility of training very deep networks. Kaiming He coined the term "revolution of depth": VGG had 19 trainable layers, GoogLeNet had 22, but even the very first version of *ResNet* contained 152 layers. It is still a popular feature extraction backbone, usually referred to as *ResNet-152*, with a popular smaller variation *ResNet-101* with 101 layer (there is really neither space nor much sense in presenting the architectures of deep residual networks in the graphical form here). Theoretically, residual connections allow to train networks with hundreds and even thousands of layers, but experiments have shown that there is no or very little improvement in performance starting from about 200 layers.

Some of the best modern convolutional feature extractors result from a combination of the network-in-network idea coming from *Inception* and the idea of residual connections. In 2016, Szegedy et al. [835] presented *Inception-v4* and several versions of *Inception ResNet* architectures with new architectures for both small network units and the global network as a whole. The resulting architectures are still among the best feature extractors and often serve as backbones for object detection and segmentation architectures.

Striving for efficiency: MobileNet, SqueezeNet, and others. The very best results in basic problems such as image classification are achieved by heavy networks such as the *Inception ResNet* family. However, one often needs to make a trade-off between the final performance and available computational resources; even a desktop GPU may be insufficient for modern networks to run smoothly, and computer vision is often done on smartphones or embedded devices. Therefore, the need arises to develop architectures that save on the network size (memory, usually related to the number of weights) and its running time (usually depending on the number of layers) without losing much in terms of performance. Naturally, it would be great to have the best of both worlds: excellent performance and small networks. Below, we will not present the exact architectures (I believe that after giving one full example with *GoogLeNet*, a further presentation of complete architectures would only clutter the book with information that is easy to find and unnecessary to remember) but only survey the ideas that researchers have used in these architectures.

How can one save weights? We have discussed above that convolutions are a great structural regularizer: by applying the same weights across a large image, convolutions can extract features in an arbitrarily large input with a fixed and relatively small number of weights. But that's not all: convolutions themselves can also grow to be quite large.

Suppose, for instance, that you have a layer with 256 channels (a very reasonable number, on the low side even), and you want to apply a 5×5 convolution to get another 256 channels at the output. A straightforward four-dimensional convolution would have, as we discussed in Section 3.1,

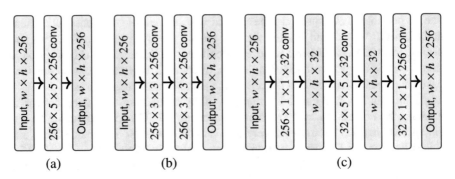

Fig. 3.12 Illustration for reducing the convolutions: (a) basic single convolution; (b) the VGG trick; (c) the bottleneck trick.

$$256 \times 5 \times 5 \times 256 = 1638400$$

weights (as shown in Fig. 3.12a). This is, of course, a big improvement compared to a feedforward layer that would have $256 \times$ width $\times$ height $\times 256$ weights, but it is often desirable to reduce these 1.6M weights further.

Let us briefly go through the main ideas used for this purpose in modern architectures. Note that all methods shown below, strictly speaking, are not equivalent to a single convolution, which is only natural: a network with 1.6M weights can be more expressive than a network with ten times fewer weights. Fortunately, it turns out that this added expressiveness usually does not improve performance and actually can deteriorate it due to overfitting or insufficient data to train so many weights.

First, we can use the VGG trick and represent a 5×5 convolution with a composition of two 3×3 convolutions (see Fig. 3.12b). This reduces the number of weights to $2 \times (256 \times 3 \times 3 \times 256) = 1179648$. It can be reduced even further if we represent 3×3 convolutions as compositions of 1×3 and 3×1, following [837].

Second, we can use the *bottleneck* trick that was first popularized by the *Inception* family of architectures. The 1.6 million weights in the layer above result from the fact that we have to multiply all dimensions of the convolution. But we can turn some of these multiplications into additions if we first compress the 256 channels down to a more manageable size with a 1×1 convolution, then do the spatial 5×5 convolution on the reduced tensor, again producing a tensor with a small number of channels (say 32 again), and only then expand it back with another 1×1 convolution. This method, illustrated in Fig. 3.12c, is somewhat akin to a low-rank approximation for the convolution tensor. Suppose that the bottleneck part has 32 channels; then the total number of weights in the three resulting convolutions will be

$$256 \times 1 \times 1 \times 32 + 32 \times 5 \times 5 \times 32 + 32 \times 1 \times 1 \times 256 = 41984,$$

with minor further reductions again available if we apply the VGG trick to the 5×5 convolution in the middle. At this point, we have already achieved a dramatic reduction in network size, reducing the total number of weights by a factor of more than 28.

The bottleneck idea was presented and successfully used in the *SqueezeNet* architecture that replaced *Inception* units with *Fire* modules that have a "squeeze-then-expand" structure: first use 1×1 convolutions to reduce the number of channels and then expand them back, applying a number of different convolutions and concatenating the outputs [382].

But even that's not all! Third, we can take the bottleneck approach even further by using *depthwise separable convolutions*. The idea is now to further decompose the tensor in the middle, a $32 \times 5 \times 5 \times 32$ convolution that still has all four factors present. This convolution mixes all channels together; but what if we leave the mixing for 1×1 convolutions (after all, that's exactly what they do) and concentrate only on the spatial part? Formally speaking, we replace a single convolution with 32 separate 5×5 convolutions, each applied only to a single channel. This definitely reduces the expressiveness of the convolution in the middle since now each channel in the result has access to only one channel in the input; but since the channels can freely exchange information in the 1×1 convolution, it usually does not lead to any significant loss of performance. In our running example, we could apply this idea to the bottleneck, shaving off one of the 32 factors and getting a total of

$$256 \times 1 \times 1 \times 32 + 32 \times 5 \times 5 + 32 \times 1 \times 1 \times 256 = 17184$$

weights. Alternatively, we could just forget about the whole bottleneck idea and do 256 depthwise separable convolutions instead of one of the 1×1 convolutions and the bottleneck, getting

$$256 \times 1 \times 1 \times 256 + 256 \times 5 \times 5 = 71936$$

weights. The second approach looks worse in this case, but, first, it depends on the actual dimensions, and second, compressing all features to an exceedingly small bottleneck does tend to lose information, so if we can achieve the same result without compressing the features it might result in better performance.

Depthwise separable convolutions were introduced by Francois Chollet in [152], where he noted that a basic *Inception* module can be represented as a depthwise separable convolution that mixes subsets of channels and presented the *Xception* modules that take this idea to its logical conclusion as we have just discussed. They also became the main tool in the construction of the *MobileNet* family of networks that were designed specifically to save on memory and still remain some of the best tools for the job [358].

Neural architecture search and EfficientNet. In the survey above, basic ideas such as compressing the channels with 1×1 convolutions are easy to understand, and we can see how researchers might come up with ideas like this. A more difficult question is how to come up with actual architectures. Who and how could establish that for *GoogLeNet* you need exactly two convolutional layers in the stem followed by nine basic *Inception* modules interspersed with max-pooling in just the right way?

The actual answer is simple: there is no theorem that shows which architecture is best; you just have to come up with a wide spectrum of different architectures that "make sense" in terms of dimensions, test a lot of them in practice, and choose the one that performs best.

This looks suspiciously like a machine learning problem: you have the building blocks (various convolutions, pooling layers, etc.) and a well-defined objective function (say, performance on the *ImageNet* test set after the training converges). Naturally, it was not long before researchers decided to automate this process. This problem is quite naturally formulated as a *reinforcement learning* problem: while we do not have a ready-to-use dataset, we can compute the objective function on any network. But computing the objective function is quite expensive (you need to train a large model to convergence). This approach to developing neural networks is known as *neural architecture search* (NAS); I will not go into more details about it and will refer to the main sources on NAS [532, 846, 930, 1031].

In convolutional architectures, neural architecture search yielded the *EfficientNet* family, proposed in 2019 by Tan and Le [847]. They introduced the *compound scaling* method, basically generalizing all of the above discussion into a single approach that scales network width, depth, and resolution according to a set of scaling coefficients. This approach by itself already allowed to improve existing architectures, but even more importantly, this generalization allowed the authors to formulate the problem of finding the best network in an efficient parameter space. The resulting networks outperformed all predecessors, setting a whole new Pareto frontier for the performance/efficiency trade-off.

To sum up, in this section we have seen the main ideas that constitute the state of the art in convolutional architectures. But note that everything that we have been talking about could be formulated in terms of "training on *ImageNet*", that is, all networks mentioned above solve the image classification problem. But this is only one problem in computer vision, and hardly even the most important one... how do we solve object detection, segmentation, and all the rest? Let's find out.

3.3 Case Study: Neural Architectures for Object Detection

In subsequent chapters, we will consider the use of synthetic data for computer vision. We have seen above which convolutional architectures are regarded as the best state-of-the-art feature extractors for images. However, computer vision encompasses many problems, and feature extraction is usually just the beginning. Indeed, even the basic setting of computer vision introduced in the 1960s—teaching a robot to look around itself and navigate the environment—involves much more than just image classification. When I am typing this text, I do not just recognize a "monitor" although it does take up most of my field of view: I can also see and distinguish the keyboard, my own hands, various objects on the screen all the way down to individual letters, and so on, all in the same image.

(a) (b)

Fig. 3.13 Sample training set annotations from the *OpenImages* dataset [61, 473, 489]: (a) object detection; (b) segmentation.

The real high-level problems in this basic computer vision setting are

- *object detection*, i.e., finding the location of an object in the image, usually formalized as a *bounding box* (rectangle defined by four numbers, usually the coordinates of two opposing angles), and classifying the object inside each bounding box;
- *instance segmentation*, i.e., finding the actual silhouette of every object in the image; this can be formalized as a separate classification problem for every pixel in the image: which object (or background) does this specific pixel belong to?

Figure 3.13 shows sample training set annotations from the *OpenImages* dataset, which is currently one of the largest available datasets of real data with object detection and segmentation labeling [61, 473, 489].

Note that in these new problems, the *output* of the network suddenly takes up a much higher dimension than ever before. In an ImageNet classification problem with 1000 classes, the output is a vector of probabilities assigned to these classes, so it has dimension 1000. In an object detection problem, the output has the same dimension 1000 plus four numbers defining the bounding box *for every object*, and as we will see shortly, it is usually even higher than that. In a classification problem, the output has, formally speaking, dimension 1000 *per pixel*, although in practice segmentation is rarely formalized in this straightforward way.

As much as I would like to, I have neither the space nor the willpower to make this chapter into a wide survey of the entire research field of computer vision. So in this section, I will attempt a more in-depth survey of one specific computer vision problem, namely *object detection*. This will showcase many important ideas in modern computer vision and will align with the case study in Section 6.4, where we will see how object detection architectures that we consider here benefit from the use of synthetic data.

Both object detection and segmentation have been around forever, at least since the 1960s. I will not dwell on classical approaches to these problems since, first, our focus is on deep learning, and second, most classical approaches have indeed been obsoleted by modern neural networks. I want to mention only one classical approach, the *selective search* algorithm developed in 2004 [234]. In brief, it represents the image as a graph where edge weights show similarities between small patches, starting from single pixels and gradually uniting them until the image is broken into a lot (usually several hundred) small patches. This is known as *pre-segmentation* or *sub-segmentation*, and the resulting patches are often called *superpixels*, i.e., the assumption is that the patches are so uniform that they definitely should belong to the same object. This may be a useful approach even today, and it is still used in some cases as preprocessing even for deep learning approaches [184], because after pre-segmentation the dimension of the problem is greatly reduced, from millions of pixels to hundreds or at most a few thousand of superpixels.

In 2012, selective search became the basis for a classical object detection algorithm [884] that worked as follows:

- use selective search to do pre-segmentation;
- greedily unite nearest neighbors in the resulting graph of patches; there can be lots of different proximity measures to try, based on color, texture, fill, size, and other properties;
- as a result, construct a large set of bounding boxes out of the superpixels; this is the set of candidates for object detection, but at this stage, it inevitably contains a lot of false positives;
- choose positive and negative examples, taking care to include hard negative examples that overlap with correct bounding boxes;
- train a classifier (SVM in this case) to distinguish between positive and negative examples; during inference, each candidate bounding box is run through the SVM to filter out false positives as best we can.

This pipeline will bring us to our first series of neural networks. But before we do that, we need to learn one more trick.

Convolutionalization and OverFeat. In the previous section, we have seen many wonderful properties of convolutional neural networks. But there is one more important advantage that we didn't mention there. Note how when we were counting the weights in a CNN, we never used the width and height of the input or output image, only the number of channels and the size of the convolution itself. That is because convolutions *don't care* about the size of their input: they are applied to all windows of a given size with shared weights, and it does not matter how many such windows the image contains. A network is called *fully convolutional* if it does not contain any densely connected layers with fixed topology and therefore can be applied to an input of arbitrary size.

But we can also turn regular convolutional neural networks, say *AlexNet* for image classification, into fully convolutional networks! In a process known as *convolutionalization*, we simply treat fully connected layers as 1×1 convolutions. The default *AlexNet* takes 224×224 images as input, so we can cover the input image by 224×224 windows and run every window through *AlexNet*; the fully connected layers at the end become 1×1 convolutions with the corresponding number of channels and have the same kind of computational efficiency. As a result of this process, we will transform the original image into a heatmap of various classes: every 224×224 window will become a vector of probabilities for the corresponding classes.

This procedure is very helpful; in particular, one of the first successful applications of modern deep neural networks to object detection, *OverFeat*, did exactly this, replacing the final classifier with a regression model that predicts bounding boxes and postprocessing the results of this network with a greedy algorithm that unites the proposed bounding boxes (naturally, such a procedure will lead to a lot of greatly overlapping candidates) [782]. This approach won the ILSVRC 2013 challenge in both object detection and object localization (a variant of object detection where it is known *a priori* that there is only one significant object in the picture, and the problem is to locate its bounding box).

Most modern architectures do not take this idea to its logical conclusion, i.e., do not produce vectors of class probabilities for input windows. But basically, all of them use convolutionalization to extract features, i.e., run the input image through the first layers of a CNN, which is often one of the CNNs that we discussed in the previous section. This CNN is called the *backbone* of an object detection or segmentation pipeline, and by using a fully convolutional counterpart of a backbone the pipelines can extract features from input images of arbitrary size.

Two-stage object detection: the R-CNN family. Let us now recall the object detection pipeline based on selective search from [884] and see how we can bring CNNs into the mix.

The first idea is to use a CNN to extract the features for object classification inside bounding boxes and perhaps also the final SVM that weeds out false positives. This was exactly the idea of R-CNN [276], a method that defined new state of the art for object detection around 2013–2014. The pipeline, illustrated in Figure 3.14a, runs as follows:

- run a selective search to produce candidate bounding boxes as above;
- run each region through a backbone CNN such as *AlexNet* (pretrained for image classification and possibly fine-tuned on the training set); on this stage, the original R-CNN actually warped each region to make its dimensions match the input of the CNN;
- train an SVM on the features produced by the CNN for classification to remove the false positives;
- train a separate bounding box regression on the same features used to refine the bounding boxes, i.e., shift their corners slightly to improve the localization of objects.

This approach was working very well but was very fragile in training (it had quite a few models that all had to be trained separately but needed to work well in combination) and hopelessly slow: it took about 45–50 seconds to run the *inference* of the R-CNN pipeline on a single image, even on a GPU! This was definitely

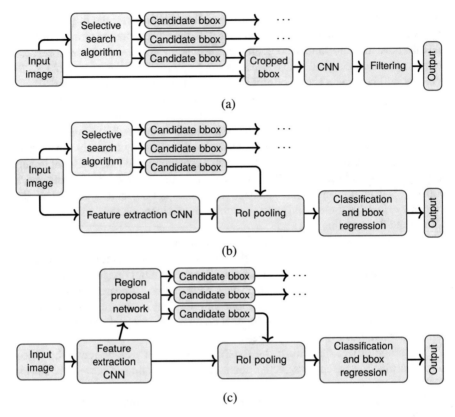

Fig. 3.14 The R-CNN family of architectures: (a) original R-CNN pipeline [276]; (b) Fast R-CNN [277]; (c) Faster R-CNN [718].

impractical, and further developments in this family of approaches tried to make R-CNN work faster.

The main reason for this excessive running time was that R-CNN needs to make a pass through the CNN for every region. Therefore, *Fast R-CNN* [277], illustrated in Fig. 3.14b, was designed so that it could use a single pass of the main backbone CNN for the whole image. The main idea of Fast R-CNN is to introduce a *region of interest (RoI) projection* layer that collects features from a region. The RoI projection layer does not have any weights; it simply maps a given bounding box to a given layer of features, translating the geometry of the original image to the (reduced) geometry in this layer of features. As a result, the tensors of features corresponding to different bounding boxes will have different dimensions.

To be able to put them through the same classifier, *Fast R-CNN* introduced the *RoI pooling* layer that performs max-pooling with dimensions needed to reduce all bounding boxes to the same size. As a result, for every bounding box we get a tensor of features with the same dimensions that can now be put through a network that performs object classification and bounding box regression (which means that it has four outputs for bounding boxes and C outputs for the classes). Only this last part of the network needs to be run for every bounding box, and the (much larger) part of the network that does feature extraction can be run once per image.

Fast R-CNN was two orders of magnitude faster than regular R-CNN at no loss of quality. But it was still relatively slow, and now the bottleneck was not in the neural network. The slowest part of the system was now the selective search algorithm that produced candidate bounding boxes.

The aptly named *Faster R-CNN* [718] removed this last bottleneck by producing candidate bounding boxes as part of the same neural network. In the *Faster R-CNN* architecture (now it is indeed a neural architecture rather than a pipeline of different models and algorithms), shown in Fig. 3.14c, the input image first goes through feature extraction and then the tensor of features is fed to a separate *region proposal network* (RPN). The RPN moves a sliding window of possible bounding boxes along the tensor of features, producing a score of how likely it is to have an object in this box and, at the same time, exact coordinates of this box. Top results from this network are used in the RoI projection layer, and then it works exactly as discussed above. Note that all of this processing now becomes part of the same computational graph, and the gradients flow through all new layers seamlessly: they are all at the end just differentiable functions.

To me, this is a perfect illustration of the expressive power of neural networks: if you need to do some additional processing along the way, you can usually just do it as part of the neural network and train the whole thing together, in an end-to-end fashion. Soon after *Faster R-CNN* appeared, it was further improved and sped up with R-FCN (region-based fully convolutional network), which introduced position-sensitive feature maps that encode information regarding a specific position in the bounding box ("left side of the object", "bottom right corner", etc.) [177]; we will not go into the details here. *Faster R-CNN* and R-FCN remain relevant object detection frameworks up to this day (they are considered to be slow but good), only the preferred backbones change from time to time.

One-stage object detection: YOLO and SSD. The R-CNN family of networks for object detection is known as *two-stage* object detection because even *Faster R-CNN* has two clearly distinguishable separate stages: one part of the network produces candidate bounding boxes, and the other part analyzes them, ranks their likelihood to be a true positive, and classifies the objects inside.

But one can also do object detection in a single pass, looking for both bounding boxes and the objects themselves at the same time. One of the first successful applications of this approach was the original YOLO ("you only look once") object detector by Redmon et al. [709]. This was, again, a single neural network, and it implemented the following pipeline:

- split the image into an $S \times S$ grid, where S is a fixed small constant (e.g., $S = 7$);
- in each cell, predict both bounding boxes and probabilities of classes inside them; this means that the network's output is a tensor of size

$$S \times S \times (5B + C),$$

where C is the number of classes (we do classification inside every cell separately, producing the probabilities $p(\text{class}_i \mid \text{obj})$ of various classes assuming that this cell does contain an object) and $5B$ means that each of B bounding boxes is defined by five numbers: four coordinates and the score of how certain the network is in that this box is correct;
- then the bounding boxes can be ranked simply by the overall probability

$$p(\text{class}_i \mid \text{obj})\,p(\text{obj})\text{IoU},$$

where $p(\text{obj})\text{IoU}$ is the target for the certainty score mentioned above: we want it to be zero if there is no object here and if there is, to reflect the similarity between the current bounding box and the ground truth bounding box, expressed as the intersection-over-union score (also known as the Jaccard similarity index).

All this could be trained end-to-end, with a single loss function that combined penalties for incorrectly predicted bounding boxes, incorrect placement of them, and wrong classes. Overall, YOLO minimizes

$$
\mathcal{L}(\theta) = \lambda_{\text{coord}} \sum_{i=1}^{S^2} \sum_{j=1}^{B} [\![\text{Obj}_{ij}]\!] \left(\left(x_i - \hat{x}_i(\theta)\right)^2 + \left(y_i - \hat{y}_i(\theta)\right)^2 \right)
$$

$$
+ \lambda_{\text{coord}} \sum_{i=1}^{S^2} \sum_{j=1}^{B} [\![\text{Obj}_{ij}]\!] \left(\left(\sqrt{w_i} - \sqrt{\hat{w}_i(\theta)}\right)^2 + \left(\sqrt{h_i} - \sqrt{\hat{h}_i(\theta)}\right)^2 \right)
$$

$$
+ \sum_{i=1}^{S^2} \sum_{j=1}^{B} [\![\text{Obj}_{ij}]\!] \left(C_i - \hat{C}_i(\theta)\right)^2 + \lambda_{\text{noobj}} \sum_{i=1}^{S^2} \sum_{j=1}^{B} [\![\neg\text{Obj}_{ij}]\!] \left(C_i - \hat{C}_i(\theta)\right)^2
$$

$$
+ \sum_{i=1}^{S^2} [\![\text{Obj}_{ij}]\!] \sum_{c} \left(p_i(c) - \hat{p}_i(c;\theta)\right)^2,
$$

where θ denotes the weights of the network, and network outputs are indicated as functions of θ.

Let us go through the original YOLO loss function in more detail as it provides an illustrative example of such loss functions in other object detectors as well:

- $[\![\text{Obj}_{ij}]\!]$ is the indicator of the event that the jth bounding box (out of B) in cell i (out of S^2) is "responsible" for an actual object appearing in this cell, that is, $[\![\text{Obj}_{ij}]\!] = 1$ if that is true and 0 otherwise;
- similarly, $[\![\text{Obj}_i]\!] = 1$ if and only if an object appears in cell i;
- the first two terms deal with the bounding box position and dimensions: if bounding box j in cell i is responsible for a real object, the bounding box should be correct, so we are bringing the coordinates of the lower left corner, width, and height of the predicted bounding box closer to the real one;

- the third and fourth terms are related to $\hat{C}_i(\theta)$, the network output that signifies the confidence that there is an object in this cell; it is, obviously, brought closer to the actual data C_i; note that since cells with and without objects are imbalanced (although this imbalance cannot hold a candle to the imbalance that we will see in SSDs), there is an additional weight λ_{coord} to account for this fact;
- the fifth term deals with classification: the internal summation runs over classes, and it brings the vector of probabilities $\hat{p}_i(c)$ that the network outputs for cell i closer to the actual class of the object, provided that there is an object in this cell;
- λ_{coord} and λ_{noobj} are hyperparameters, constants set in advance; the original YOLO used $\lambda_{\text{coord}} = 5$ and $\lambda_{\text{noobj}} = \frac{1}{2}$.

The original YOLO had a relatively simple feature extractor, and it could achieve results close to the best *Faster R-CNN* results in real time, with 40–50 frames per second while *Faster R-CNN* could do less than 10.

Further development of the idea to predict everything at once led to *single-shot detectors* (SSD) [540]. SSD uses a set of predefined *anchor boxes* that are used as default possibilities for each position in the feature map. It has a single network that predicts both class labels and the corresponding refined positions for the box angles for every possible anchor box. Applied to a single tensor of features, this scheme would obviously detect only objects of a given scale since anchor boxes would take up a given number of "pixels". Therefore, the original SSD architecture already applied this idea on several different scales, i.e., several different layers of features. The network is, again, trained in an end-to-end fashion with a single loss function.

Note that SSD has *a lot* of outputs: it has $M \times N \times (C + 4)$ outputs for an $M \times N$ feature map, which for the basic SSD architecture with a 300×300 input image came to 8732 outputs per class, that is, potentially millions of outputs per image. But this does not hinder performance significantly because all these outputs are computed in parallel, in a single sweep through the neural network. SSD worked better than the original YOLO, on par with or even exceeding the performance of *Faster R-CNN*, and again did it at a fraction of the computational costs.

Since the original YOLO, the YOLO family of one-stage object detectors has come a long way. I will not explain each in detail but will mention the main ideas incorporated by Joseph Redmon and his team into each successive iteration:

- YOLOv2 [710] tried to fix many localization errors and low recall of the original YOLO; they changed the architecture (added batch normalization layers and skip connections from earlier layers to increase the geometric resolution, predicted bounding box offsets rather than coordinates, etc.), pretrained their own high-resolution (448×448) classifier instead of using one pretrained on ImageNet (256×256), and added *multi-scale training*, i.e., trained on different image sizes;
- YOLO9000, presented in the same paper [710], generalized object detection to a large number of classes (9000, to be precise) by using the *hierarchical softmax* idea: instead of having a single softmax layer for 9000 classes, a hundred of which are various breeds of dog, let us first classify if the object is a living thing, then if it is an animal, get to a specific breed only after going down several layers in the decision tree;

- YOLOv3 [711] changed the feature extraction architecture and introduced a number of small improvements, in particular a multi-scale architecture similar to the feature pyramid networks that we will discuss below.

Another important addition to the family of one-stage object detectors was *RetinaNet* by Lin et al. [524]. The main novelty here is a modified loss function for object detection known as *focal loss*. One problem with one-stage object detection is that the output is wildly imbalanced: we have noted several times that one-stage detectors have a huge number of outputs that represent a wide variety of candidate bounding boxes, but how many of them can actually be correct?

The mathematical side of this problem is that even correctly classified examples (a bounding box in the background correctly classified as "no object") still contribute to the classification loss function: the usual cross-entropy loss equals $-\log p$ for an example where the correct class gets probability p. This is a small value when p is close to 1 (which it should be when the network is sure), but negative examples outweigh positive examples by a thousand to one, and these relatively small values add up. So focal loss downweighs the loss on well-classified examples, bringing it close to zero with an additional polynomial factor in the loss function: the focal loss for a correctly classified example is $-(1-p)^\gamma \log p$ for some $\gamma > 0$ instead of the usual cross-entropy loss $-\log p$. Focal loss has proved to be an extremely useful idea, used in a wide variety of deep learning models since the original work [524].

The YOLO family of networks and *RetinaNet* have defined state of the art in real-time object detection for a long time. In particular, YOLOv3 was the model of choice for at least 2 years, and this situation has begun to change only very recently. We will go back to the story of YOLO at the end of this section, but for now let us see the other main ideas in modern object detection.

Object detection at different scales: feature pyramid networks. To motivate the next set of ideas, let us analyze which specific problems have historically plagued object detection. If you look at the actual results in terms of how many objects are usually detected, you will see that the absolute numbers in object detection are pretty low: long after image classifiers beat *ImageNet* down to superhuman results of less than 5% top-5 test error (human level was estimated at about 5.1%) [329], the best object detectors were getting mean average precision of about 80% on the (relatively simple) PASCAL VOC dataset and struggled to exceed 35% mean average precision on the more realistic Microsoft COCO dataset. At the time of writing (summer of 2020), the best mAP on Microsoft COCO is about 55%, still very far from perfect [848, 900]. Why are the results so low?

One big problem lies in the different *scales* of objects that need to be recognized. Small objects are recognized very badly by most models discussed above, and there are plenty of them in real life and in realistic datasets such as *Microsoft COCO*. This is due to the so-called *effective stride*: by the time the region proposal network kicks in, the original geometry has already been compressed a lot by the initial convolutional layers. For example, the basic *Faster R-CNN* architecture has effective stride 16, i.e., a 16×16 object is only seen as a single pixel by the RPN. We could try to reduce effective stride mechanically, by doing convolutional layers without pooling

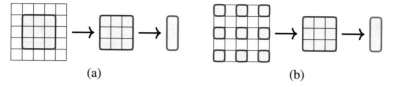

Fig. 3.15 Types of convolutions: (a) regular convolutions; (b) dilated convolutions.

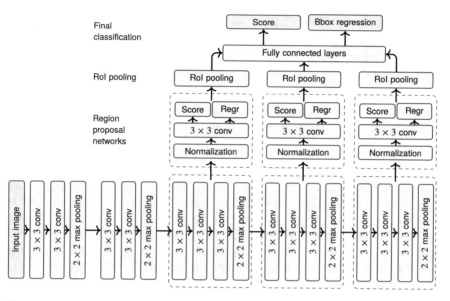

Fig. 3.16 Multi-scale object recognition: a sample architecture from [222] that does not have top-down connections.

and without reducing the geometry, but then the networks become huge and basically infeasible. On the other hand, there are also large objects: a car may be taking up either 80% of the photo or just a tiny 30×30 pixel spot somewhere; sometimes both situations happen on the same photo. What do we do?

One idea is to change how we do convolutions. Many object detection architectures use *dilated* (sometimes also called *atrous*) convolutions, i.e., convolutions whose input window is not a contiguous rectangle of pixels but a strided set of pixels from the previous layer, as shown in Fig. 3.15; see, e.g., [211] for a detailed explanation. With dilated convolutions, fewer layers are needed to achieve large receptive fields, thus saving on the network size. Dilated convolutions are used in Faster R-CNN, R-FCN, and other networks; see, e.g., a work by Yu and Koltun where dilated convolutions were successfully applied to semantic segmentation [980].

But this is just a trick that can improve the situation, not a solution. We still need to cope with multi-scale objects in the same image. The first natural idea in this direction is to gather proposals from several different layers in the backbone feature extraction

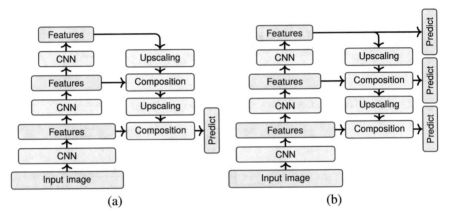

Fig. 3.17 Feature pyramid networks: (a) top-down architecture with predictions on the bottom level [108, 674]; (b) feature pyramid network [523].

network. In a pure form, this idea was presented by Eggert et al. [222], where the architecture has several (three, to be precise) different region proposal networks, each designed to recognize a given scale of objects, from small to large. Figure 3.16 shows an outline of their architecture; the exact details follow *Faster R-CNN* quite closely. The results showed that this architecture improved over basic *Faster R-CNN*, and almost the entire improvement came from far better recognition of small objects. A similar idea in a different form was implemented by Cai et al. [108], who have a single region proposal network with several branches with different scales of outputs (the branches look kind of like *GoogLeNet*).

But the breakthrough came when researchers realized that top layers can inform lower layers in order to make better proposals and better classify the objects in them. An early layer of a convolutional backbone can only look at a small part of the input image, and it may lack the semantic expressiveness necessary to produce good object proposals. Therefore, it would be beneficial to add *top-down* connections, i.e., use the semantically rich features produced on top layers of the CNN to help the early layers understand what objects they are looking at.

This is exactly the idea of *feature pyramid networks* (FPN), presented in 2016 by Lin et al. [523]. Actually, there are two ways to do that. One way is to use skip connections to get from the top down; the work [108] already contains an architecture that implements this idea, and it appeared in other previous works as well [674]; we illustrate it in Fig. 3.17a. The difference in the feature pyramid approach is that instead of sending down semantic information to the bottom layer, Lin et al. make RoI predictions independently on several layers along this top-down path, as shown in Fig. 3.17b; this lets the network better handle objects of different scales. Note that the upscaling and composition parts do not have to have complicated structure: in [523], upscaling is a single convolutional layer, and composition is done by using a single 1×1 convolution on the features and adding it to the result of upscaling.

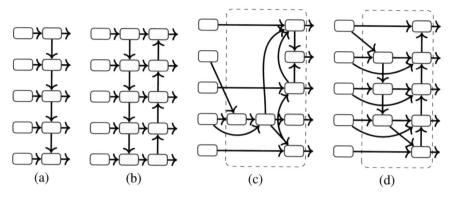

    (a)   (b)   (c)   (d)

Fig. 3.18 The top-down pathway in various pyramid architectures: (a) FPN [523]; (b) PANet [538]; (c) NAS-FPN [274]; (d) BiFPN from *EfficientDet* [848]. Dashed rectangles show repeated blocks.

The "predict" part can be complicated if needed, e.g., it can represent the RPN from *Faster R-CNN*.

Feature pyramids have become a staple of two-stage object detection. One important advance was Path Aggregation Network (PANet) [538], which added new pathways between layers of the pyramid and introduced adaptive feature pooling; specifically, they added another bottom-up pathway to the top-down one, as shown in Fig. 3.18b. Another advance was NAS-FPN [274], which used neural architecture search to find a better feature pyramid architecture. Its resulting architecture is shown in Fig. 3.18c; note that only one block is shown but it is repeated several times in the final architecture.

At present, research into feature pyramid architectures has culminated with *EfficientDet*, a network developed by *Google Brain* researchers Tan et al. [848]. Variations of this network currently show record performance, sometimes losing to solutions based on CSPNet, a new recently developed CNN backbone (we will not go into details on this one) [900]. *EfficientDet* introduced a new solution for the multi-scale feature fusion problem, an architecture called BiFPN (weighted bidirectional feature pyramid network); the outline of one of its repeated blocks is shown in Fig. 3.18d.

YOLOv4, YOLOv5, and current state of the art. I would like to finish this section with an interesting story that is unfolding right as I'm writing this book.

After YOLOv3, Joseph Redmon, the main author of all YOLO architectures that we have discussed above, announced that he stopped doing computer vision research[2]. For about 2 years, YOLOv3 basically defined the state of the art in object detection: detectors that could outperform it worked much slower. But in 2020, Alexey Bochkovskiy et al. released YOLOv4, a significantly improved (and again sped up) version of YOLO [78]. By using new methods of data augmentation, including even GAN-based style transfer (we will discuss such architectures in Chapter 10),

---

[2]He said that the main reason was that "the military applications and privacy concerns eventually became impossible to ignore". Indeed, modern computer vision can bring up certain ethical concerns, although it is far from obvious on which side the scales tip.

using the mixup procedure in training [993], and actually a large and diverse collection of other recently developed tricks (see Section 3.4), Bochkovskiy et al. managed to reach performance comparable to *EfficientDet* with performance comparable to YOLOv3.

But that's not the end of the story. In just two months, a different group of authors released an object detector that they called YOLOv5. At the time of writing (summer of 2020), there is still no research paper or preprint published about YOLOv5, but there is a code repository[3], and the blog posts by the authors claim that YOLOv5 achieves the same performance as YOLOv4 at a fraction of model size and with better latencies [622, 623]. This led to the usual controversy about naming, credit, and all that, but what is probably more important is that it is still not confirmed which of the detectors is better; comparisons are controversial, and third-party comparisons are also not really conclusive. Matters are not helped by the subsequent release of PP-YOLO [547], an improved reimplementation of YOLOv3 that also added a lot of new tricks and managed to outperform YOLOv4...

By the time you are reading this, the controversy has probably already been settled, and maybe you are already using YOLOv6 or later, but I think this snapshot of the current state of affairs is a great illustration of just how vigorous and alive modern deep learning is. Even in such a classical standard problem as object detection, a lot can happen very quickly!

3.4 Data Augmentations: The First Step to Synthetic Data

In the previous section, we have already mentioned *data augmentation* several times. Data augmentations are defined as transformations of the input data that change the target labels in predictable ways. This allows to significantly (often by several orders of magnitude) increase the amount of available data at zero additional labeling cost. In fact, I prefer to view data augmentation as the first step towards synthetic data: there is no synthetic data generation *per se*, but there is recombination and adaptation of existing real data, and the resulting images often look quite "synthetic".

The story of data augmentation for neural networks begins even before the deep learning revolution; for instance, Simard et al. [801] used distortions to augment the MNIST training set in 2003, and I am far from certain that this is the earliest reference. The MC-DNN network discussed in Section 3.2, arguably the first truly successful deep neural network in computer vision, also used similar augmentations even though it was a relatively small network trained to recognize relatively small images (traffic signs).

But let us begin in 2012, with *AlexNet* [477] that we have discussed in detail in Section 3.2. *AlexNet* was the network that made the deep learning revolution happen in computer vision... and even with a large dataset such as *ImageNet*, even back in

---

[3]https://github.com/ultralytics/yolov5.

2012 it would already be impossible without data augmentation! *AlexNet* used two kinds of augmentations:

- horizontal reflections (a vertical reflection would often fail to produce a plausible photo) and
- image translations; that is the reason why the *AlexNet* architecture, shown in Fig. 3.7, uses $224 \times 224 \times 3$ inputs while the actual *ImageNet* data points have 256 pixels by the side: the 224×224 image is a random crop from the larger 256×256 image.

With both transformations, we can safely assume that the classification label will not change. Even if we were talking about, say, object detection, it would be trivial to shift, crop, and/or reflect the bounding boxes together with the inputs—that is exactly what we mean by "changing in predictable ways". The resulting images are, of course, highly interdependent, but they still cover a wider variety of inputs than just the original dataset, reducing overfitting. In training *AlexNet*, Krizhevsky et al. estimated that they could produce 2048 different images from a single input training image. What is interesting here is that even though *ImageNet* is so large (*AlexNet* trained on a subset with 1.2 million training images labeled with 1000 classes), modern neural networks are even larger (*AlexNet* has 60 million parameters). Krizhevsky et al. had the following to say about their augmentations: "Without this scheme, our network suffers from substantial overfitting, which would have forced us to use much smaller networks" [477].

By now, this has become a staple in computer vision: while approaches may differ, it is hard to find a setting where data augmentation would not make sense at all.

To review what kind of augmentations are commonplace in computer vision, I will use the example of the *Albumentations* library developed by Buslaev et al. [103]; although the paper was only released in 2020, the library itself had been around for several years and by now has become the industry standard.

The first candidates are color transformations. Changing the color saturation, permuting color channels, or converting to grayscale definitely does not change bounding boxes or segmentation masks, as we see in Figure 3.19. This figure also shows two different kinds of blurring, jpeg compression, and various brightness and contrast transformations.

The next obvious category are simple geometric transformations. Again, there is no question about what to do with segmentation masks when the image is rotated or cropped: we can simply repeat the same transformation with the labeling. Figure 3.20 shows examples of both global geometric transformations such as flipping or rotation (Fig. 3.20a) and local distortions defined according to a grid (Fig. 3.20b). The same ideas can apply to other types of labeling; for instance, keypoints (facial landmarks, skeletal joints in pose estimation, etc.) can be treated as a special case of segmentation and also changed together with the input image.

All of these transformations can be chained and applied multiple times. Figure 3.21 shows a more involved example produced with the *Albumentations* library; it corresponds to the following chain of augmentations:

Fig. 3.19 Sample color transformations and blurring provided by the *Albumentations* library [103].

- take a random crop from a predefined range of sizes;
- shift, scale, and rotate the crop to match the original image dimension;
- apply a (randomized) color shift;
- add blur;
- add Gaussian noise;
- add a randomized elastic transformation for the image;
- perform *mask dropout*, removing a part of the segmentation masks and replacing them with black cutouts on the image.

That's quite a few operations! But how do we know that this is the best way to approach data augmentation for this particular problem? Can we find the best possible sequence of augmentations? Indeed, recent research suggests that we can look for a meta-strategy for augmentations that would take into account the specific problem setting; we will discuss these approaches in Section 12.2.

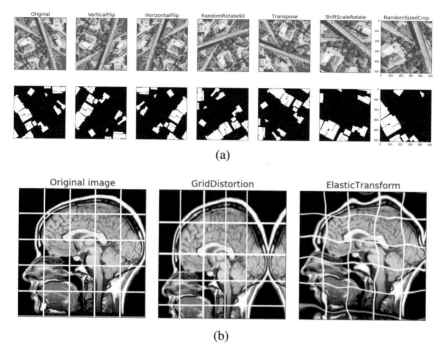

Fig. 3.20 Sample geometric transformations provided by the *Albumentations* library [103]: (a) global transformations; (b) local distortions.

But even that is not all! What if we take augmentation one step further and allow augmentations to produce more complex combinations of input data points? In 2017, this idea was put forward in the work titled "Smart Augmentation: Learning an Optimal Data Augmentation Strategy" by Lemley et al. [506]. Their basic idea is to have two networks, "Network A" that implements an augmentation strategy and "Network B" that actually trains on the resulting augmented data and solves the downstream task. The difference here is that "Network A" does not simply choose from a predefined set of strategies but operates as a generative network that can, for instance, blend two different training set examples into one in a smart way. I will not go into the full details of this approach, but Figure 3.22 provides two characteristic examples from [506].

This kind of "smart augmentation" borders on synthetic data generation: transformations are complex, and the resulting images may look nothing like the originals. But before we turn to actual synthetic data generation in subsequent chapters, let us discuss other interesting ideas one could apply even at the level of augmentation.

Mixup, a technique introduced by MIT and FAIR researchers Zhang et al. in 2018 [993], looks at the problem from the opposite side: what if we mix the labels together with the training samples? This is implemented in a very straightforward way: for two labeled input data points, Zhang et al. construct a convex combination

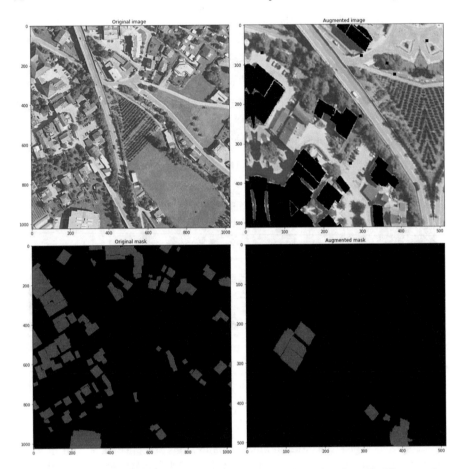

Fig. 3.21 An involved example of data augmentation by transformations from the *Albumentations* library [103].

Fig. 3.22 An example of "smart augmentations" by Lemley et al. [506]: the image on the left is produced as a blended combination of two images on the right.

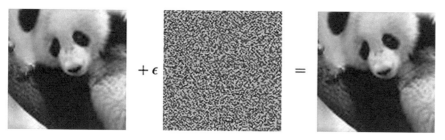

Fig. 3.23 The famous "panda-to-gibbon" adversarial example [292]: the image on the left is recognized by *AlexNet* as a panda with 57.7% confidence, but after adding small random-looking noise the network recognizes the image on the right as a gibbon with 99.3% confidence.

of both the inputs and the labels:

$$\tilde{\mathbf{x}} = \lambda \mathbf{x}_1 + (1 - \lambda)\mathbf{x}_2, \quad \text{where } \mathbf{x}_1, \mathbf{x}_2 \text{ are raw input vectors,}$$
$$\tilde{\mathbf{y}} = \lambda \mathbf{y}_1 + (1 - \lambda)\mathbf{y}_2, \quad \text{where } \mathbf{y}_1, \mathbf{y}_2 \text{ are one-hot label encodings.}$$

The blended label does not change either the network architecture or the training process: binary cross-entropy trivially generalizes to target discrete distributions instead of target one-hot vectors.

The resulting labeled data covers a much more robust and continuous distribution, and this helps the generalization power. Zhang et al. report especially significant improvements in training GANs. By now, the idea of mixup has become an important part of the deep learning toolbox: you can often see it as an augmentation strategy, especially in the training of modern GAN architectures.

The last idea that I want to discuss in this section is *self-adversarial training*, an augmentation technique that incorporates *adversarial examples* [292, 484] into the training process. Adversarial examples are a very interesting case that showcases certain structural and conceptual problems with modern deep neural networks. It turns out that for most existing artificial neural architectures, one can modify input images with small amounts of noise in such a way that the result looks to us humans completely indistinguishable from the originals but the network is very confident that it is something completely different. The most famous example from one of the first papers on adversarial examples by Goodfellow et al. [292], with a panda turning into a very confident gibbon, is reproduced in Fig. 3.23.

By now, adversarial examples and ways to defend against them have become a large field of study in modern deep learning; let me refer to [781, 963] for recent surveys on adversarial examples, attacks, and defenses.

In the simplest case, such adversarial examples are produced by the following procedure:

- suppose that we have a network and an input **x** that we want to make adversarial; let us say that we want to turn a panda into a gibbon;

- formally, it means that we want to increase the "gibbon" component of the network's output vector **y** (at the expense of the "panda" component);
- so we fix the weights of the network and start regular gradient ascent, but with respect to **x** rather than the weights!

This is the key idea for finding adversarial examples; it does not explain why they exist (it is not an easy question) but if they do, it is really not so hard to find them in the "white box" scenario, when the network and its weights are known to us, and therefore we can perform this kind of gradient ascent with respect to the input image.

So how do we turn this idea into an augmentation technique? Given an input instance, let us make it into an adversarial example by following this procedure for the current network that we are training. Then we train the network on this example. This may make the network more resistant to adversarial examples, but the important outcome is that it generally makes the network more stable and robust: now we are explicitly asking the network to work robustly in a small neighborhood of every input image. Note that the basic idea can again be described as "make the input data distribution cover more ground", but by now we have come quite a long way since horizontal reflections and random crops.

Note that unlike basic geometric augmentations, this may turn out to be a quite costly procedure. But the cost is entirely borne during training: yes, you might have to train the final model for two weeks instead of one, but the resulting model will, of course, work with exactly the same performance. The model architecture does not change, only the training process does.

One of the best recent examples for the combined power of various data augmentations is given by the architecture that we discussed in Section 3.3: YOLOv4 by Bochkovskiy et al. [78]. Similar to many other advances in the field of object detection, YOLOv4 is in essence a combination of many small improvements. Faced with the challenge of improving performance but not sacrificing inference speed, Bochkovskiy et al. systematize these improvements and divide them into two subsets:

- *bag of freebies* includes tricks and improvements that do not change the inference speed of the resulting network at all, modifying only the training process;
- *bag of specials* includes changes that do have a cost during inference but the cost is, hopefully, as small as possible.

The "bag of specials" includes all changes related to the architecture: new activation functions, bounding box postprocessing, a simple attention mechanism, and so on. But the majority of the overall improvement in YOLOv4 comes from the "bag of freebies"... which almost entirely consists of augmentations.

YOLOv4 uses all kinds of standard augmentations, self-adversarial training, and mixup that we have just discussed, and introduces new Mosaic and CutMix augmentations that amount to composing an input image out of pieces cropped out of other images (together with the objects and bounding boxes, of course). This is just one example but there is no doubt that data augmentations play a central role in modern computer vision: virtually every model is trained or pretrained with some kind of data augmentations.

3.5 Conclusion

In this chapter, we have seen a brief introduction to the world of deep learning for computer vision. We have seen the basic tool for all computer vision applications—convolutions and convolutional layers,—have discussed how convolutional layers are united together in deep feature extractors for images, and have seen how these feature extractors, in turn, can become key parts of object detection and segmentation pipelines.

There is really no hope to cover the entirety of computer vision in a single short chapter. In this chapter, I have decided to concentrate on giving at least a very brief overview of the most important ideas of convolutional architectures and review as an in-depth example of one basic high-level computer vision problem, namely object detection. Problems such as object detection and segmentation are very much aligned with the topic of this book, synthetic data. In synthetic images, both object detection and segmentation labeling come for free: since we control the entire scene, all objects in it, and the camera, we know precisely which object every pixel belongs to. On the other hand, hand-labeling for both problems is extremely tedious (try to estimate how long it would take you to label a photo like the one shown in Fig. 3.13!). Therefore, such basic high-level problems are key to the whole idea of synthetic data, and in Chapter 6 we will see how modern synthetic datasets help the networks that we have met in this chapter. In fact, I will devote a whole case study in Section 6.4 to just object detection with synthetic data.

In this chapter, we have also begun to discuss synthetic data, in its simplest form of data augmentations. We have seen how augmentations have become a virtually indispensable tool in computer vision these days, from simple geometric and color transformations to very complex procedures such as self-adversarial training. But still, even "smart" augmentations are just combinations and modifications of real data that someone has had to label in advance, so in Chapter 6 the time will come for us to move on to purely synthetic examples.

Before we do that, however, we need another introductory chapter. In the next chapter, we will consider generative models in deep learning, from a general introduction to the notion of generative models to modern GAN-based architectures and a case study of style transfer, which is especially relevant for synthetic-to-real domain adaptation.

Chapter 4
Generative Models in Deep Learning

So far, we have mostly discussed discriminative machine learning models that aim to solve a supervised problem, i.e., learn a conditional distribution of the target variable conditioned on the input. In this chapter, we consider *generative* models whose purpose is to learn the entire distribution of inputs and be able to sample new inputs from this distribution. We will go through a general introduction to generative models and then proceed to generative models in deep learning. First, we will discuss explicit density models that model distribution factors with deep neural networks and their important special case, normalizing flows, and explicit density models that approximate the distribution in question, represented by variational autoencoders. Then we will proceed to the main content, generative adversarial networks, discuss various adversarial architectures and loss functions, and give a case study of style transfer with GANs that is directly relevant to synthetic-to-real transfer.

4.1 Introduction to Generative Models

All the methods that we consider in this book lie in the domain of machine learning, and most of them use deep neural networks and thus fall into the deep learning category. In Chapter 2, we saw a very brief introduction to the most important paradigms of machine learning: probabilistic foundations of machine learning with the prior, posterior, and likelihood, supervised and unsupervised learning, and other basic concepts of the field.

In this chapter, it is time to discuss another important distinction that spans the entire machine learning: the distinction between *discriminative* and *generative* models. Intuitively, the difference is quite clear: if you need to distinguish between cats and dogs on a picture, you are training a discriminative model, and if you want to generate a new picture of a cat, you need a generative model.

© The Author(s), under exclusive license to Springer Nature Switzerland AG 2021
S. I. Nikolenko, *Synthetic Data for Deep Learning*, Springer Optimization
and Its Applications 174, https://doi.org/10.1007/978-3-030-75178-4_4

Formally, this difference can be expressed as follows:

- discriminative models learn a *conditional* distribution $p(y \mid \mathbf{x})$, where $\mathbf{x}$ is the input data point and y is the target variable;
- generative models learn a *joint* distribution $p(y, \mathbf{x})$ or $p(\mathbf{x})$, or at least learn some way to sample from this distribution (we will see the difference in this chapter).

It is also quite intuitively clear that learning a generative model is harder than learning a discriminative one: it should be harder to generate a new photo of a cat than just to distinguish between cats and dogs, right? This is also clear formally: if you have a generative model and can compute $p(y, \mathbf{x})$, to get $p(y \mid \mathbf{x})$ it suffices to compute $p(y, \mathbf{x})$ for all values of y (usually quite feasible in a classification problem) and normalize the result, but there is no clear way to go from $p(y \mid \mathbf{x})$ to $p(y, \mathbf{x})$.

In practice, this means that generative models either have to struggle with more difficult optimization problems (as we will see in the case of GANs) or have to make stronger assumptions about the structure of this distribution. Note that models for unsupervised learning are often generative because there is no clear target variable, but it is actually not necessary at all: e.g., we will discuss how a simple autoencoder, while being a perfectly fine model for unsupervised dimensionality reduction, does not quite work as a generative model yet.

Before proceeding further, let me begin with one important remark about generative models. Consider learning a generative model in the maximum likelihood framework:

$$\theta^* = \arg\max \prod_{\mathbf{x} \in \mathcal{D}} p(\mathbf{x}; \theta) = \arg\max \sum_{\mathbf{x} \in \mathcal{D}} \log p(\mathbf{x}; \theta),$$

where $p = p_{\text{model}}$ is the model probability density. Note that here D denotes the dataset, so if our model is sufficiently expressive to simply memorize the data, it will probably be the best strategy with respect to maximum likelihood; overfitting is a problem in generative models as well, and we have to alleviate it by introducing more restrictive assumptions and/or priors on $p(\mathbf{x}; \theta)$.

We can look at the same optimization problem from a different perspective. Consider the "data distribution" p_{data} that is defined as the uniform distribution concentrated at the points from the dataset D. In this case, the Kullback–Leibler divergence between p_{data} and p_{model} becomes

$$\mathrm{KL}\left(p_{\text{data}} \| p_{\text{model}}\right) = \sum_{\mathbf{x} \in D} \frac{1}{|D|} \log \frac{1}{|D| p_{\text{model}}(\mathbf{x})} d\mathbf{x} =$$

$$= -\frac{1}{|D|} \sum_{\mathbf{x} \in D} \log p_{\text{model}}(\mathbf{x}) d\mathbf{x} + \text{Const},$$

which means that minimizing the KL-divergence between p_{data} and p_{model} is exactly equivalent to maximizing the likelihood above. In other words, in the optimization problem the data points try to "pull up" the probability density p_{model}, while the

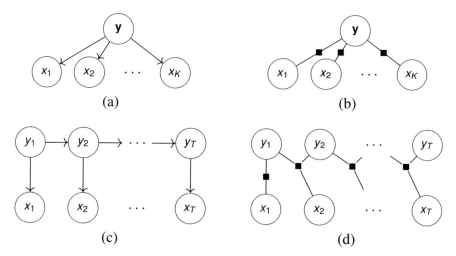

Fig. 4.1 Sample generative and discriminative models in machine learning: (a) directed graphical model for naive Bayes; (b) factor graph for logistic regression; (c) directed graphical model for an HMM; (d) factor graph for a linear chain CRF.

opposing constraints are our probabilistic assumptions and the fact that p_{model} has to sum to one.

In this section, let us consider a couple of examples that illustrate the difference and the inherent duality between generative and discriminative models.

For a simple yet very illustrative example, consider the naive Bayes classifier. It is a generative model that makes the assumption that features in $\mathbf{x} = (x_1 \dots x_K)$ are independent given the class label y:

$$p(y, \mathbf{x}) = p(y) \prod_{k=1}^{K} p(x_k \mid y).$$

Under such strong assumptions, even a generative model is rather straightforward to train: assuming that there are N classes, $y \in \{1, \dots, N\}$, and that each x_k is a discrete variable with M values, $x_k \in \{1, \dots, M\}$, the parameters of naive Bayes are

$$\theta_{kmn} = p(x_k = m \mid y = n), \qquad \theta_n = p(y = n),$$

and they can be learned independently by collecting dataset statistics. The same goes for other (e.g., continuous) distributions of x_k: naive Bayes assumptions let us learn them independently from subsets of the data. Once this model is trained, it becomes straightforward to sample from the joint distribution $p(y, \mathbf{x})$: first sample y and then sample x_k conditional on y. Figure 4.1a shows the directed graphical model corresponding to naive Bayes.

Let us rewrite the joint likelihood of the naive Bayes classifier in a different form, explicitly showing the dependence on parameters:

$$p\left(D \mid \theta\right) = \prod_{(\mathbf{x},y)\in D} p(y) \prod_{k=1}^{K} p\left(x_k \mid y\right) = \prod_{(\mathbf{x},y)\in D} \prod_{n=1}^{N} \left(\theta_n \prod_{m=1}^{M} \prod_{k=1}^{K} \theta_{kmn}^{[\![x_k=m]\!]} \right)^{[\![y=n]\!]} =$$

$$= \exp \left(\sum_{(\mathbf{x},y)\in D} \left(\sum_{n=1}^{N} [\![y=n]\!]\theta_n + \sum_{n=1}^{N}\sum_{m=1}^{M}\sum_{k=1}^{K} [\![y=n]\!][\![x_k=m]\!]\theta_{kmn} \right) \right),$$

where $[\![\cdot]\!]$ denotes 1 if the statement in the brackets holds and 0 if it does not.

In other words, the naive Bayes classifier makes the assumption that the log-likelihood

$$\log p\left(\mathbf{x}, y \mid \theta\right) = \sum_{n=1}^{N} [\![y=n]\!]\theta_n + \sum_{n=1}^{N}\sum_{m=1}^{M}\sum_{k=1}^{K} [\![y=n]\!][\![x_k=m]\!]\theta_{kmn}$$

is a linear function of trainable parameters with coefficients that represent certain features of the data point (indicators of various values for $\mathbf{x}$ and y).

Consider now a linear *discriminative* model that makes basically the same assumption: *logistic regression*. In logistic regression, for an input $\mathbf{x}$ and y we model

$$p\left(y = n \mid \mathbf{x}; \theta\right) = \frac{e^{\sum_{k=1}^{K} \theta_{nk}x_k}}{\sum_{n'=1}^{N} e^{\sum_{k=1}^{K} \theta_{n'k}x_k}} = \frac{1}{Z(\mathbf{x})} \exp \left(\sum_{k=1}^{K} \theta_{nk}x_k \right),$$

so

$$\log p\left(y = n \mid \mathbf{x}; \theta\right) = \sum_{k=1}^{K} \theta_{nk}x_k + \log Z(\mathbf{x}).$$

We now see that the expression for naive Bayes is a special case of this expression, for a specific (albeit quite general) form of the features. Logistic regression provides a simple decomposition for the conditional distribution $p\left(y \mid \mathbf{x}\right)$ similar to the naive Bayes decomposition, and its factor graph shown in Fig. 4.1b is very similar. Note, however, that now it is a factor graph rather than a directed graphical model: this is only a decomposition of $p\left(y \mid \mathbf{x}\right)$ rather than the full joint distribution $p(y, \mathbf{x})$.

Thus, as a discriminative model, logistic regression generalizes the naive Bayes model, providing greater flexibility for modeling the distribution $p\left(y \mid \mathbf{x}\right)$ without additional restrictions on the form of features imposed in the naive Bayes version, at no extra cost for training. On the other hand, the constant in the likelihood of logistic regression depends on $\mathbf{x}$. This means that for a general form of the features $\mathbf{x}$, it would be hard to compute, and logistic regression does not give us a convenient way to sample from the joint distribution $p(y, \mathbf{x})$. Naive Bayes provides precisely the special case when this constant is feasible and can be computed. Due to this duality, we say that naive Bayes and logistic regression form a *generative–discriminative pair*.

Another, slightly more involved example of a generative–discriminative pair is given by hidden Markov models (HMM) and conditional random fields (CRF). A *hidden Markov model* contains a sequence of observables $X = \{x_t\}_{t=1}^{T}$ and hidden states $Y = \{y_t\}_{t=1}^{T}$, with the following independence assumptions:

$$p(y \mid \mathbf{x}) = \prod_{t=1}^{T} p(y_t \mid y_{t-1}) p(x_t \mid y_t).$$

The corresponding directed graphical model is shown in Fig. 4.1c. Again, it is straightforward to sample from an HMM when the parameters are known: we can simply go along the direction of increasing time.

Similar to naive Bayes above, let us rewrite HMM in a different form:

$$p(y \mid \mathbf{x}) = \prod_{t=1}^{T} \exp\left(\sum_{i,j \in S} \theta_{ij} [\![y_t = i]\!] [\![y_{t-1} = j]\!] + \sum_{i \in S} \sum_{o \in O} \mu_{oi} [\![y_t = i]\!] [\![x_t = o]\!] \right),$$

where S is the set of hidden states, O is the set of values for the observables, and in terms of the hidden Markov model, we have

$$\theta_{ij} = \log p(y_t = i \mid y_{t-1} = j), \qquad \mu_{oi} = \log p(x = o \mid y = i).$$

Similar to logistic regression, we can introduce feature functions $f_k(y_t, y_{t-1}, x_t)$ corresponding to the indicators above; let us do it explicitly:

$$f_{ij}(y, y', x) = [\![y = i]\!] [\![y' = j]\!] \quad \text{for every transition,}$$
$$f_{io}(y, y', x) = [\![y = i]\!] [\![x = o]\!] \quad \text{for every observable.}$$

Note that each feature depends only on two adjacent hidden states and a single observation (actually even less than that, but let us generalize a little).

We can now do a trick similar to how logistic regression could be obtained from naive Bayes: we pass to a more general form of the features, expressing the conditional distribution $p(y \mid \mathbf{x})$ instead of the joint distribution $p(\mathbf{y}, \mathbf{x})$:

$$p(y \mid \mathbf{x}) = \frac{\prod_{t=1}^{T} \exp\left(\sum_{k=1}^{K} \theta_k f_k(y_t, y_{t-1}, x_t) \right)}{\sum_{y'} \prod_{t=1}^{T} \exp\left(\sum_{k=1}^{K} \theta_k f_k(y'_t, y'_{t-1}, x_t) \right)}.$$

This is exactly how a *linear chain conditional random field* (CRF) is defined:

$$p\left(\mathbf{y} \mid \mathbf{x}\right) = \frac{1}{Z(\mathbf{x})} \prod_{t=1}^{T} \exp\left(\sum_{k=1}^{K} \theta_k f_k(y_t, y_{t-1}, x_t)\right),$$

i.e., we simply allow for a more general form of the features and observables. The factor graph corresponding to a linear chain CRF is shown in Fig. 4.1d.

Note that since it is now truly infeasible to compute $Z(x)$, we have to resort to discriminative modeling, and again there is no easy way to sample from $p(x, y)$.

Conditional random fields represent a very interesting class of models. It is one of the few "simple" probabilistic models that still remain relevant for complex unstructured inputs in the world of deep learning: linear chain CRFs are commonly used "on top" of deep neural networks in sequence tagging problems that arise in natural language processing, semantic segmentation, or video analysis [29, 508]; for example, the BiLSTM-CNN-CRF architecture has become a staple in problems such as named entity recognition [144, 494, 569, 650, 938]. But I merely wanted to use CRFs as an example to illustrate the difference between generative and discriminative models; for further information on CRFs, I refer to [830, 978]; note also CRFs can also be thought of as recurrent neural networks and thus integrated into a neural architecture [1022].

To summarize this section, in these two examples we have seen a natural correspondence between generative and discriminative probabilistic models in machine learning. In general, discriminative models are usually more flexible and expressive than their generative counterparts when it comes to predicting the target variable; this is understandable as they are solving an easier problem. There are several reasons why generative models are important and can be preferable to their discriminative counterparts, but in this book it will be sufficient to say that sometimes we actually need to generate $\mathbf{x}$: draw a picture of a cat, refine a synthetic image to make it more realistic, or introduce additional variety in a dataset. Let us see what our options are for such a high-dimensional generation.

4.2   Taxonomy of Generative Models in Deep Learning and Tractable Density Models: FVBNs and Normalizing Flows

In the previous section, we have laid out the main foundations of generative models in machine learning and have seen two specific examples. In this part, we will discuss how generative models make their way into deep learning.

The general taxonomy of generative models in deep learning, shown in Fig. 4.2, follows Ian Goodfellow's famous NIPS 2016 tutorial [291] where he presented generative adversarial networks (GANs) and also discussed the place of generative models in modern machine learning and introduced a taxonomy similar to the one shown in Fig. 4.2. However, nodes in this taxonomy have received many new examples

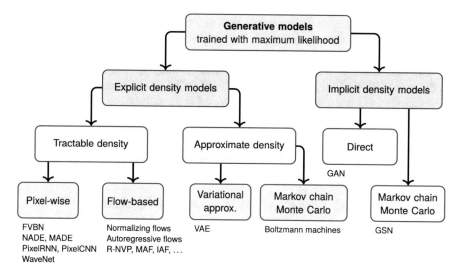

Fig. 4.2 Taxonomy of generative models in deep learning [291].

over the last years, so let us briefly go over these variations before proceeding to generative adversarial networks.

The first split in the taxonomy is the distinction between *explicit density* and *implicit density* models. The difference is that in the former, we make explicit probabilistic assumptions that allow us to write down the density of the resulting joint distribution as a product of smaller distributions. Both generative models that we saw in Section 4.1 are explicit density models: in fact, if you can draw a factor graph or directed graphical model for your set of assumptions, you are probably in the explicit density domain.

In reality, such explicit factorizations are almost always of the form characteristic for directed graphical models: to decompose a large joint distribution $p(\mathbf{x}) = p(x_1, \ldots, x_n)$, let us first of all note that

$$p(\mathbf{x}) = p(x_1, \ldots, x_n) = p(x_1)p(x_1 \mid x_2) \ldots p(x_n \mid x_1, x_2, \ldots, x_{n-1}).$$

This decomposition is, of course, always true. But now let us try to impose restrictions that would reduce the set of conditions in each of these factors. For instance, the naive Bayes classifier that we discussed above cuts each condition down to a single (target) variable:

$$p(\mathbf{x}, y) = p(y)p(x_1 \mid y) \ldots p(x_n \mid y).$$

The big divide in explicit density models happens depending on the fact whether we treat this simplified decomposition as the density itself, assuming it is tractable and learning its parameters, or as a class of approximations where we need to find the best one. Classical probabilistic models such as the ones we discussed in Section 4.1 belong to the class of tractable density models, but recent advances in deep learning have extended this class with a lot of new interesting cases.

The first class of tractable density models in deep learning are *fully visible belief nets* (FVBN), initially developed in Geoffrey Hinton's group back in the mid-1990s [242, 243] but currently being developed at a much larger scale. They make use of the decomposition above:

$$p_{\text{model}}(\mathbf{x}) = \prod_{i=1}^{n} p_{\text{model}}(x_i \mid x_1, x_2, \ldots, x_{i-1}),$$

and then use neural networks to model each of the probability distributions $p_{\text{model}}(x_i \mid x_1, x_2, \ldots, x_{i-1})$. There is no need to go into details as we will not use the models from this family too much, but let us briefly mention some notable examples.

Neural Autoregressive Distribution Estimation (NADE) [887] is an architecture that changes a standard autoencoder architecture, which maps $\mathbf{x}$ to $\hat{\mathbf{x}}$ with a loss function that makes $\hat{\mathbf{x}} \approx \mathbf{x}$, in such a way that the first component $\hat{x}_0$ does not depend on $\mathbf{x}$ at all, $\hat{x}_1$ depends only on x_0, $\hat{x}_2$ on $\mathbf{x}_0$ and $\mathbf{x}_1$, and so on. In this way, NADE changes a classical autoencoder into one that satisfies the decomposition above and can thus be used for generation. The next iteration of this model, *Masked Autoencoder for Distribution Estimation* (MADE) [272], achieved the same goal with clever use of masking and was able to average over different orderings of variables in $\mathbf{x}$, which improved things further.

In the domain of images, FVBNs were represented by PixelRNN and PixelCNN developed by DeepMind researchers Aäron van den Oord et al. [645]. The basic idea of the decomposition remains the same, and generation proceeds pixel by pixel, but now each factor $p_{\text{model}}(x_i \mid x_1, x_2, \ldots, x_{i-1})$ is modeled by a recurrent or convolutional architecture, respectively, that takes as input all previously generated pixels. The novelty here lies in how to organize the recurrent or convolutional network to combine information from a not-quite-rectangular set of pixels generated up to this point. Van den Oord et al. proposed special LSTM-based architectures that proceed by rows and columns for PixelRNN and special masked convolutions (the masks ensure that the network does not look ahead to pixels not yet generated) for PixelCNN. The PixelCNN model was later improved to PixelCNN++ by OpenAI researchers Tim Salimans et al. [755].

One important win for tractable density models was *WaveNet* by van den Oord et al. [642]. *WaveNet* made a breakthrough in text-to-speech, producing very lifelike speech samples and generally pushing the state of the art in audio generation. In essence, *WaveNet* is an FVBN model with one-dimensional dilated convolutions used to combine all previously generated inputs in the conditional part of the current distribution. WaveNet was later improved in [156] and made much more efficient in [140, 643]... but in order to add efficiency, DeepMind researchers had to push the latest version of WaveNet outside the domain of FVBNs.

The general problem with FVBNs is that by their very design, the generation process has to be sequential: in order to generate x_i, an FVBN has to know x_{i-1}, and in fact it is likely that x_{i-1} is one of the most important conditions for x_i due to its proximity: e.g., in audio generation, it stands to reason that the last tick is the most important, and in image generation neighboring pixels are probably important as well. So by design, there is no way to achieve parallel generation for FVBN-based architectures, which means that for high-dimensional problems such as image or audio generation such models will inevitably be quite slow: you have to run a neural network for each pixel in an image (millions of times) or for each value in an audio (tens of thousands times per second).

Fortunately, there is a different way to compose explicit density models that is rapidly gaining popularity in recent years. *Normalizing flows* provide a way to construct complex distributions as a composition of individual components represented by invertible transformations [462, 723]. We know that for $\mathbf{z} \in \mathbb{R}^d$, $\mathbf{z} \sim q(\mathbf{z})$, and an invertible function $f : \mathbb{R}^d \to \mathbb{R}^d$, the random variable $y = f(\mathbf{z})$ will have the density

$$q_y(\mathbf{z}) = q(\mathbf{z}) \left| \det \frac{\partial f^{-1}}{\partial \mathbf{z}} \right| = q(\mathbf{z}) \left| \det \frac{\partial f}{\partial \mathbf{z}} \right|^{-1}.$$

It is also easy to see what happens if we compose several such functions in a row:

$$\mathbf{z}_K = f_K \circ \ldots \circ f_1(\mathbf{z}_0), \quad \mathbf{z}_0 \sim q_0(\mathbf{z}_0),$$

$$\mathbf{z}_K \sim q_K(\mathbf{z}) = q_0(\mathbf{z}_0) \prod_{k=1}^{K} \left| \det \frac{\partial f_k}{\partial \mathbf{z}_{k-1}} \right|^{-1},$$

$$\log q_K(\mathbf{z}) = \log q_0(\mathbf{z}_0) - \sum_{k=1}^{K} \log \left| \det \frac{\partial f_k}{\partial \mathbf{z}_{k-1}} \right|.$$

This general idea is illustrated in Fig. 4.3a.

In this way, we can try to approximate a very complex distribution as a simple distribution (say, the standard Gaussian) transformed by a large composition of relatively simple transformations. Such transformations can include, for example, a *planar* (linear) flow that "cuts" the space with hyperplanes:

$$f(\mathbf{z}) = \mathbf{z} + \mathbf{u}h(\mathbf{w}^\top \mathbf{z} + b),$$

$$\left| \det \frac{\partial f}{\partial \mathbf{z}} \right| = \left| 1 + \mathbf{u}^\top \left(h'(\mathbf{w}^\top \mathbf{z} + b)\mathbf{w} \right) \right|,$$

or a *radial* flow that cuts it with spheres:

$$f(\mathbf{z}) = \mathbf{z} + \beta h(\alpha, r)(\mathbf{z} - \mathbf{z}_0), \quad \text{where}$$

$$r = \|\mathbf{z} - \mathbf{z}_0\|_2, \quad h(\alpha, r) = \frac{1}{\alpha + r}.$$

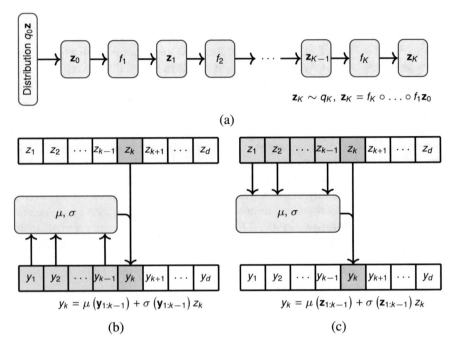

Fig. 4.3 Flow-based generative models: (a) the general idea of a flow as a composition of transformations; (b) masked autoregressive flow (MAF); (c) inverse autoregressive flow (IAF).

The only restriction here is that for every transformation $f(\mathbf{z})$, we need to have an efficient way to compute its Jacobian $\det \frac{\partial f}{\partial \mathbf{z}}$. At first glance, it does sound quite restrictive because computing determinants is hard. But note that if the Jacobian is triangular, its determinant is simply the product of the main diagonal! This is exactly the idea of *autoregressive flows* [456], where

$$y_i = f(z_1, \ldots, z_i), \quad \det \frac{\partial f}{\partial \mathbf{z}} = \prod_{i=1}^{d} \frac{\partial y_i}{\partial z_i}.$$

The simplest example of an autoregressive flow is given by *Real Non-Volume Preserving Flows* (RealNVP) [198] that are not very expressive but very efficient. Volume preservation here refers to the Jacobian: if $\det \frac{\partial f}{\partial \mathbf{z}} = 1$ then volume is preserved, and in RealNVP it is emphatically not so. For some functions $\mu, \sigma : \mathbb{R}^k \to \mathbb{R}^{d-k}$, we define

$$\mathbf{y}_{1:k} = \mathbf{z}_{1:k}, \quad \mathbf{y}_{k+1:d} = \mathbf{z}_{k+1:d} \cdot \sigma(\mathbf{z}_{1:k}) + \mu(\mathbf{z}_{1:k}),$$

i.e., we copy the first k dimensions and apply a linear transformation to the rest. As a result, the matrix of first derivatives that consists of a unit matrix follows by a lower triangular one, and the Jacobian can be computed as

$$\det \frac{\partial \mathbf{y}}{\partial \mathbf{z}} = \prod_{i=1}^{d-k} \sigma_i(\mathbf{z}_{1:k}).$$

Note that in this case, it is very easy to compute (completely in parallel!) and easy to invert even for non-invertible μ and σ.

There are two ways to generalize R-NVP to more expressive transformations. The first is *Masked Autoregressive Flows* (MAF) [652], illustrated in Fig. 4.3b:

$$y_1 = \mu_1 + \sigma_1 z_1, \quad y_i = \mu(\mathbf{y}_{1:i-1}) + \sigma(\mathbf{y}_{1:i-1})z_i.$$

This is a more expressive model whose Jacobian is still very simple, $\prod_i \sigma(\mathbf{y}_{1:i-1})$, and can be easily inverted. But now it cannot be computed in parallel (at least in general). Therefore, MAFs can be used efficiently for density estimation (i.e., training) but the result is slow to apply, just like FVBNs.

The other way is given by *Inverse Autoregressive Flows* (IAF) [456] that are shown in Fig. 4.3c and operate basically the other way around:

$$y_i = z_i \sigma(\mathbf{z}_{1:i-1}) + \mu(\mathbf{z}_{1:i-1}).$$

Now $\mathbf{y}$ depends only on $\mathbf{z}$, and it can be computed very efficiently (in parallel):

$$\mathbf{y} = \mathbf{z} \cdot \sigma(\mathbf{z}) + \mu(\mathbf{z}), \quad \det \frac{\partial \mathbf{y}}{\partial \mathbf{z}} = \prod_{i=1}^{d} \sigma_i(\mathbf{z}),$$

and if we consider a composition $f_K \circ \ldots \circ f_0(\mathbf{z})$, we get

$$\log q_K(\mathbf{z}_K) = \log q(\mathbf{z}) - \sum_{k=0}^{K} \sum_{i=1}^{d} \log \sigma_{k,i}.$$

Now it is very efficient to sample from this model, but it is harder to compute its density, although still possible analytically.

The deep learning part here is that we can parameterize μ and σ with neural networks. In particular, PixelRNN, MADE, and WaveNet can all be viewed as special cases of a normalizing flow. The next interesting step was taken in *Parallel WaveNet* [643] with the method of *parallel density distillation*.

WaveNet is the kind of model that is quick to learn (learning can be done in parallel) but slow to sample (sampling has to be sequential). It would be great to pass to an IAF to sample quickly:

$$\log p_X(\mathbf{x}) = \log p_Z(\mathbf{z}) - \log \left| \frac{\partial \mathbf{x}}{\partial \mathbf{z}} \right|, \quad \log \left| \frac{\partial \mathbf{x}}{\partial \mathbf{z}} \right| = \sum_t \log \frac{\partial f(\mathbf{z}_{\leq t})}{\partial z_t}.$$

Now sampling would work as $\mathbf{z} \sim q(\mathbf{z})$ (the logistic distribution in case of *WaveNet*), and then we get

$$x_t = z_t \cdot s(\mathbf{z}_{\leq t}, \theta) + \mu(\mathbf{z}_{\leq t}, \theta),$$

and for s and μ we can use the exact same kind of convolutional networks as in the basic *WaveNet*.

It is now easy to sample from this IAF, but entirely impractical to train it. The parallel density distillation trick is to train a student model $p_S(\mathbf{x})$ based on the teacher model $p_T(\mathbf{x})$ with loss function

$$\mathrm{KL}\,(P_S \| P_T) = H(P_S, P_T) - H(P_S).$$

Skipping the math, I will just say that the probabilistic assumptions and the form of the logistic distribution allow to easily compute the entropy and cross-entropy, and we can now generate $\mathbf{x}$ from the student, compute $p_T(x_t \mid \mathbf{x}_{<t})$ in parallel in the teacher, and then efficiently sample several x_t from $p_S(x_t \mid \mathbf{x}_{<t})$ for every t. Adding a few more auxiliary loss functions for the student network, we get the same kind of audio generation quality. But the original *WaveNet* could produce 172 samples per second, while the distilled *Parallel WaveNet* can use the same hardware to produce more than 500,000 samples per second...

The logical alternative is to generalize the kind of flows that can be efficient for both sampling and training (that is, RealNVP flows) as much as possible. I will not go into the details but note the GLOW model [455] that is able to generate high-resolution images and the WaveGLOW model [685] that combines GLOW and *WaveNet* to achieve even faster generation.

4.3 Approximate Explicit Density Models: VAE

In the previous section, we have considered tractable density models, where an explicit function is assumed to represent the probability distribution in question. Now we turn to approximate density models, where the explicit tractable family of distribution densities is used to find the best approximation to the data distribution p_{data} with a member of this family.

We will not consider in detail Boltzmann machines that can be viewed as Markov chain Monte Carlo approximations to a given density; see [348, 752, 753] for details. But let us consider, at least briefly, probably the currently most important representative of this family: *variational autoencoders* (VAE) [200, 457, 458].

To arrive at the variational autoencoder, let us start from a regular *autoencoder* architecture that consists of

- the *encoder* E that produces a latent code $\mathbf{z}$ from input $\mathbf{x}$ and
- the *decoder* D that tries to reconstruct $\mathbf{x}$ from the latent code $\mathbf{z}$.

In other words, a given image $\mathbf{x}$ is first compressed down to a latent representation $\mathbf{z} = E(\mathbf{x})$ and then "decompressed" back into

$$\hat{\mathbf{x}} = D(E(\mathbf{x}))$$

that is supposed to match the original $\mathbf{x}$. In this case, to train the composition of encoder E and decoder D, we can use various similarity metrics between $\hat{\mathbf{x}}$ and $\mathbf{x}$: in the simplest case, the L_2 or L_1-norm of the difference $\|\hat{\mathbf{x}} - \mathbf{x}\|$.

But this idea also proves to be insufficient to obtain a generative model. While a well-trained autoencoder can ensure that a real photo $\mathbf{x}$ will be reconstructed as such after $D(E(\mathbf{x}))$, this says nothing about decoding a random vector $\mathbf{z}$ of the same dimension. The latent vectors corresponding to real images (say, photos of cats) will most probably comprise a rather involved subset (subvariety) in the space of latent codes, and a random vector sampled from some standard distribution will have very little chance to fall into this subset, almost like a set of random pixels that have very little chance to comprise a photo-like image.

Thus, we need to "iron out the wrinkles" in this distribution. One way to do it would be to use strong assumptions for the latent code distribution, that is, use encoders and decoders that have a simple enough structure so that this subset of codes will necessarily have a simple structure as well.

In the extreme case, this reduces to *principal component analysis* (PCA), a method that can be thought of as an autoencoder with linear encoder and linear decoder. PCA minimizes the L_2-norm of the difference (Euclidean distance) between the points and their projections on the resulting subspace. The probabilistic interpretation of this model, known as *probabilistic PCA*, is formulated as a generative model for $\mathbf{x}$ with Gaussian noise:

$$\mathbf{x} \sim \mathcal{N}\left(\mathbf{x} \mid W\mathbf{z}, \sigma^2 \mathbf{I}\right), \qquad \mathbf{z} \sim \mathcal{N}\left(\mathbf{z} \mid \mathbf{0}, I\right).$$

But in deep learning for high-dimensional data, we would like to be able to train more expressive encoders and decoders. The variational autoencoder begins with the same idea as probabilistic PCA but substituting a nonlinear function instead of a linear projection:

$$p(\mathbf{x}, \mathbf{z} \mid \theta) = p(\mathbf{x} \mid \mathbf{z}, \theta) p(\mathbf{z} \mid \theta) = \mathcal{N}(\mathbf{z} \mid \mathbf{0}, \mathbf{I}) \prod_{j=1}^{D} \mathcal{N}(x_j \mid \mu_j(\mathbf{z}), \sigma_j(\mathbf{z})).$$

Now we can parameterize $\mu_j(\mathbf{z})$ and $\sigma_j(\mathbf{z})$ with a neural network with parameters θ; in VAEs, $\sigma_j(\mathbf{z})$ is usually assumed to be a constant σ.

Formally speaking, a VAE is a generative model that is trained to maximize the likelihood of every $\mathbf{x}$ in the dataset

$$p(D \mid \theta) = \prod_{\mathbf{x} \in X} \int p(\mathbf{x} \mid \mathbf{z}, \theta) p(\mathbf{z}) \mathrm{d}\mathbf{z} = \prod_{\mathbf{x} \in X} \int \mathcal{N}\left(\mathbf{x} \mid f(\mathbf{z}, \theta), \sigma^2 \mathbf{I}\right) p(\mathbf{z}) \mathrm{d}\mathbf{z}.$$

The prior distribution for $\mathbf{z}$ is usually taken to be the standard Gaussian $p(\mathbf{z}) = \mathcal{N}(\mathbf{z} \mid \mathbf{0}, \mathbf{I})$.

Note that this formalization does not have any encoders yet, only a decoder. To train this model, we can use stochastic gradient ascent: sample $\mathbf{z}_1, \ldots, \mathbf{z}_n$ from $p(\mathbf{z})$, approximate

$$p(\mathbf{x} \mid \theta) \approx \frac{1}{n} \sum_{i=1}^{n} \mathcal{N}\left(\mathbf{x} \mid f(\mathbf{z}_i, \theta), \sigma^2 \mathbf{I}\right) p(\mathbf{z}_i)$$

and maximize this approximation by gradient ascent along θ.

The problem with this approach is that $p(\mathbf{x} \mid \mathbf{z}, \theta)$ will be vanishingly small for almost all $\mathbf{z}$ except for $\mathbf{z}$ from a small region of $\mathbf{z}$'s that can produce an image similar to $\mathbf{x}$, so straightforward sampling would require exponentially (and completely impractically) many $\mathbf{z}_i$. This is where the *encoder* comes into the VAE framework: the encoder captures a distribution $q(\mathbf{z} \mid \mathbf{x})$ that is supposed to produce latent codes $\mathbf{z}$ from this neighborhood, i.e., $\mathbf{z}$'s with high values of $p(\mathbf{x} \mid \mathbf{z}, \theta)$.

To achieve this, we need $q(\mathbf{z})$ to serve as an approximation for $p(\mathbf{z} \mid \mathbf{x})$ for a given $\mathbf{x}$. Let us try to get this approximation in a relatively straightforward way, by minimizing the KL-divergence between the two distributions:

$$\begin{aligned}
\mathrm{KL}\left(q(\mathbf{z}) \| p(\mathbf{z} \mid \mathbf{x})\right) &= \mathbb{E}_{\mathbf{z} \sim q}\left[\log q(\mathbf{z}) - \log p(\mathbf{z} \mid \mathbf{x})\right] = \\
&= \mathbb{E}_{\mathbf{z} \sim q}\left[\log q(\mathbf{z}) - \log p(\mathbf{x} \mid \mathbf{z}) - \log p(\mathbf{z})\right] + \log p(\mathbf{x}) = \\
&= \mathbb{E}_{\mathbf{z} \sim q}\left[\log q(\mathbf{z}) - \log p(\mathbf{z})\right] - \mathbb{E}_{\mathbf{z} \sim q}\left[\log p(\mathbf{x} \mid \mathbf{z})\right] + \log p(\mathbf{x}) = \\
&= \mathrm{KL}\left(q(\mathbf{z}) \| p(\mathbf{z})\right) - \mathbb{E}_{\mathbf{z} \sim q}\left[\log p(\mathbf{x} \mid \mathbf{z})\right] + \log p(\mathbf{x}),
\end{aligned}$$

which means that

$$\log p(\mathbf{x}) - \mathrm{KL}\left(q(\mathbf{z}) \| p(\mathbf{z} \mid \mathbf{x})\right) = \mathbb{E}_{\mathbf{z} \sim q}\left[\log p(\mathbf{x} \mid \mathbf{z})\right] - \mathrm{KL}\left(q(\mathbf{z}) \| p(\mathbf{z})\right).$$

Now, since we are free to choose any distribution q, VAE makes $q(\mathbf{z})$ dependent on $\mathbf{x}$, turning it into $q(\mathbf{z} \mid \mathbf{x})$, and now the right-hand side of the equation above serves as the lower bound for the value $\log p(\mathbf{x})$, which we want to maximize. This is exactly the famous *variational lower bound* for our case, but since we will not use variational inference further I won't go into more details. Suffice it to say that the lower bound becomes exact if $q(\mathbf{z} \mid \mathbf{x})$ matches $p(\mathbf{z} \mid \mathbf{x})$ exactly, driving the KL-divergence down to zero, and if we use a sufficiently expressive model for $q(\mathbf{z} \mid \mathbf{x})$ we can hope that the lower bound will be sufficiently precise so that we can maximize the right-hand side. We achieve this, of course, by using a neural network to express $q(\mathbf{z} \mid \mathbf{x})$.

In other words, VAE consists of

- the encoder, a neural network that maps an input $\mathbf{x}$ into the parameters of a distribution $q(\mathbf{z} \mid \mathbf{x})$; let's assume (as VAEs usually do) that $q(\mathbf{z} \mid \mathbf{x})$ is a Gaussian whose parameters are produced by the encoder:

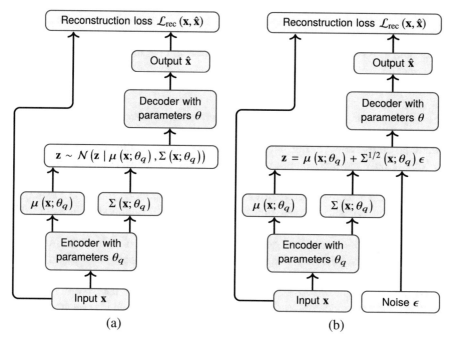

Fig. 4.4 Variational autoencoders: (a) basic idea with sampled **z**; (b) the reparametrization trick with noise ϵ sampled in advance.

$$q\left(\mathbf{z} \mid \mathbf{x}\right) = \mathcal{N}\left(\mathbf{z} \mid \mu(\mathbf{x}; \theta_q), \Sigma(\mathbf{x}; \theta_q)\right);$$

- the decoder, a neural network that maps a latent code **z** sampled from $q\left(\mathbf{z} \mid \mathbf{x}\right)$ into **x**.

How do we train a VAE? To train encoder parameters, we need to minimize

$$\mathbb{E}_{\mathbf{z}\sim q(\mathbf{z}|\mathbf{x})}\left[\log p\left(\mathbf{x} \mid \mathbf{z}\right)\right] - \mathrm{KL}\left(q\left(\mathbf{z} \mid \mathbf{x}\right) \| p\left(\mathbf{z} \mid \mathbf{x}\right)\right).$$

Since both p and q have a known standard form, say two Gaussians, the second term can be computed analytically as a function of the parameters θ_q. The first term cannot be computed exactly, and we need to approximate it by sampling. Actually, VAEs sample just a single value $\mathbf{z} \sim q\left(\mathbf{z} \mid \mathbf{x}\right)$, substitute it into the log-likelihood, and obtain a function of θ as a result; this is exactly what stochastic gradient descent does.

The resulting scheme is shown in Fig. 4.4a. Now it looks like we could simply use some standard reconstruction loss function such as the L_2-norm $\|\mathbf{x} - \hat{\mathbf{x}}\|^2$, but we still have one problem left: our autoencoder is sampling **z** between the encoder and decoder steps! It's no problem to sample from $\mathcal{N}\left(\mathbf{z} \mid \mu(\mathbf{x}; \theta_q), \Sigma(\mathbf{x}; \theta_q)\right)$ on the forward step, but it is not clear how to pass the gradients through the sampling process...

Fortunately, there is an option, called the *reparametrization trick*, that lets us sidestep this problem. The solution is simply to sample the noise ϵ from the standard Gaussian, $\epsilon \sim \mathcal{N}(\mathbf{0} \mid \mathbf{I})$, and rescale it with the results of the encoder:

$$\mathbf{z} = \mu(\mathbf{x}; \theta_q) + \Sigma^{1/2}(\mathbf{x}; \theta_q)\epsilon.$$

With the reparametrization trick, ϵ can be sampled in advance and treated as another input for the autoencoder; there is no need for sampling inside the network. Now the whole pipeline, shown in Fig. 4.4b, can be trained with gradient descent.

This is only the first, vanilla variation of VAE. There are plenty of extensions, and in recent years, variational autoencoders have become a very important class of models, starting to rival GANs in versatility and generation quality. For example, one can adapt VAEs for discrete objects; in particular in collaborative filtering [516, 789], dynamic VAEs are used to process sequential data [275], and so on, and so forth. In the high-dimensional generation of images, which is the most important application for us, VAEs are also rapidly approaching the quality of the best GANs. There are two main directions in this regard.

In *Vector Quantized VAE* developed by DeepMind researchers van den Oord et al. [644] and later extended by Razavi et al. [707], the latent code is quantized:

$$\text{Quant}(\mathbf{z}_e(\mathbf{x})) = \mathbf{e}_k, \qquad k = \arg\min_j \|\mathbf{z}_e(\mathbf{x}) - \mathbf{e}_j\|_2.$$

The gradient is simply copied over through the discrete layer (which would stop backpropagation otherwise). The embeddings are trained with the vector quantization algorithm, i.e., we simply bring $\mathbf{e}$ closer to the encoder outputs $\mathbf{z}_e(\mathbf{x})$, and the encoder is also brought closer to the embeddings. The resulting loss function is

$$\mathcal{L}^{\text{VQ-VAE}} = \log p\left(\mathbf{x} \mid \mathbf{z}_q(\mathbf{x})\right) + \|\text{sg}[\mathbf{z}_e(\mathbf{x})] - \mathbf{e}\|_2^2 + \beta \|\mathbf{z}_e(\mathbf{x}) - \text{sg}[\mathbf{e}]\|_2^2,$$

where sg (stopgradient) is the operator that stops gradient propagation: its forward pass computes the identity function, and the backward pass returns zero. The decoder is optimizing the first term in $\mathcal{L}^{\text{VQ-VAE}}$, the encoder deals with the first and third terms, and the embeddings themselves are trained with the second term.

The original VQ-VAE was based on PixelCNN and produced very reasonable images by 2017 standards, while VQ-VAE-2 has made the generation process hierarchical which led to high-resolution images with state-of-the-art generation quality.

But here too, NVIDIA researchers appear to stay ahead: *Nouveau VAE* (NVAE) by Vahdat and Kautz [888] is a VAE that is able to generate samples on par with the latest GANs. NVAE is also a hierarchical model:

$$p(\mathbf{z}) = \prod_l p\left(\mathbf{z}_l \mid \mathbf{z}_{<l}\right),$$

$$q\left(\mathbf{z} \mid \mathbf{x}\right) = \prod_l q\left(\mathbf{z}_l \mid \mathbf{z}_{<l}, \mathbf{x}\right),$$

$$\mathcal{L}_{\text{VAE}}(\mathbf{x}) = \mathbb{E}_q\left[\log p\left(\mathbf{x} \mid \mathbf{z}\right)\right] - \text{KL}\left(q\left(\mathbf{z}_1 \mid x\right) \| p(\mathbf{z}_1)\right)$$

$$- \sum_{l=2}^{L} \mathbb{E}_{q(\mathbf{z}_{<l}|\mathbf{x})}\left[\text{KL}\left(q\left(\mathbf{z}_l \mid \mathbf{x}, \mathbf{z}_{<l}\right) \| p\left(\mathbf{z}_l \mid \mathbf{z}_{<l}\right)\right)\right],$$

where $q\left(\mathbf{z}_{<l} \mid \mathbf{x}\right) = \prod_{i=1}^{l-1} q\left(\mathbf{z}_i \mid \mathbf{x}, \mathbf{z}_{<i}\right)$. It is trained similar to the basic VAE, via the reparametrization trick. The authors of NVAE have taken care to find the best architectures for the encoder and decoder and have used special tricks to stabilize training (an important problem for hierarchical models) and save memory. As a result, NVAE provides arguably some of the best generated samples for, say, high-definition faces, a common benchmark for generative models.

Variational autoencoders are starting to gain traction in synthetic data generation as well. As an interesting recent example, I would like to mention the work by Xiao et al. [953] who generate synthetic spatiotemporal aggregates, i.e., multi-scale images used for geospatial analysis and remote sensing, conditioned on both pixel-level and macroscopic feature-level conditions such as the road network. They introduce a novel deep conditional generative model (DCGM) architecture based on a VAE and demonstrate the usefulness of the resulting synthetic data for training models for downstream tasks.

But still, generative adversarial networks remain the most flexible and often the best class of modern generative models. Many examples of synthetic-to-real domain adaptation in the next chapter will make use of GANs. So starting from the next section, we will work through a brief review of generative adversarial networks, doing it in slightly more detail than the models we have discussed to this point.

4.4 Generative Adversarial Networks

From now on, our primary examples of generative models will come from the "Direct implicit density" class in Fig. 4.2 and will be represented by *generative adversarial networks* (GANs). In my opinion, the most clear motivation for GANs comes from the optimization problem associated with a black-box generative model, or, to be more precise, an evident lack of such a problem.

Indeed, suppose that you want a neural network to draw pictures of cats. It is no problem to design a convolutional architecture that accepts as input a vector of random numbers (we need some source of randomness, a neural network won't give us one by itself) and outputs a tensor that represents an image of a given dimension. It can even be a reasonably simple architecture... but what will the objective function be? We cannot write down a formal differentiable function that would capture the

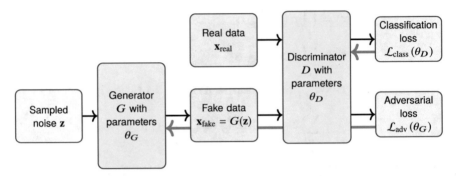

Fig. 4.5 The basic architecture of GANs. Thick green arrows show the flow of gradients: the discriminator is trained with a classification loss, and the generator is trained with an adversarial loss that is also computed with the discriminator's help.

"catness" of an image: that sounds suspiciously like exactly the problem that we are trying to solve. In Section 4.3, we have seen one way to approach this problem: get the basic inspiration for the loss function from autoencoders, but modify their architecture in such a way that the distribution of latent codes will be simple enough to sample from.

The main idea of generative adversarial networks is a different way to formalize this "catness" property. GANs do it via a separate network, the *discriminator*, that tries to distinguish between real objects from the p_{data} distribution and fake objects produced by the generator, from the p_g distribution. The discriminator (see Fig. 4.5 for a general illustration) is solving a binary classification problem, learning to output, say, 1 for real images and 0 for fake images. The generator, on the other hand, is trying to "fool" the discriminator into thinking that fake samples produced by the generator from random noise are in fact taken from the real dataset. This means that the generator's loss also depends on the current state of the discriminator; it is shown in Fig. 4.5 as the adversarial loss $\mathcal{L}_{adv}$. Note that in the most basic formulation, $\mathcal{L}_{adv}$ and $\mathcal{L}_{class}$ are the same function optimized in two different directions by the generator and the discriminator, but in our exposition (and in the history of GANs), they will become different almost immediately.

In an EM-like scheme, the generator G and discriminator D can be trained alternately, and in the ideal case the training would proceed as follows:

- at first the generator produces basically random noise, but the discriminator also cannot distinguish anything;
- so first we train the discriminator to differentiate real images from the random noise that G is producing;
- then we train the generator with an objective function of "fooling" the discriminator;
- but this will only be a generator that has learned to fool a very simple discriminator, so we continue this alternating training until convergence.

Formally speaking, the generator is a function

$$G = G(\mathbf{z}; \theta_g) : Z \to X,$$

while the discriminator is a function

$$D = D(\mathbf{x}; \theta_d) : X \to [0, 1].$$

The objective function for the discriminator is usually the binary classification error function, i.e., the binary cross-entropy

$$\mathbb{E}_{\mathbf{x} \sim p_{\text{data}}(\mathbf{x})} \left[\log D(\mathbf{x}) \right] + \mathbb{E}_{\mathbf{x} \sim p_g(\mathbf{x})} \left[\log(1 - D(\mathbf{x})) \right],$$

where $p_g(\mathbf{x}) = G_{\mathbf{z} \sim p_z}(\mathbf{z})$ is the distribution generated by G; in other words, during training the discriminator assigns label 0 to fake data produced by G and label 1 to real data.

The generator is learning to fool the discriminator, that is, in the simplest setting G is minimizing

$$\mathbb{E}_{\mathbf{x} \sim p_g(\mathbf{x})} \left[\log(1 - D(\mathbf{x})) \right] = \mathbb{E}_{\mathbf{z} \sim p_z(\mathbf{z})} \left[\log(1 - D(G(\mathbf{z}))) \right].$$

And now, combining the two, we get a typical minimax game:

$$\min_G \max_D V(D, G), \qquad \text{where}$$

$$V(D, G) = \mathbb{E}_{\mathbf{x} \sim p_{\text{data}}(\mathbf{x})} [\log D(\mathbf{x})] + \mathbb{E}_{z \sim p_z(z)} [\log(1 - D(G(z)))].$$

Note that this immediately makes training GANs into a much more difficult optimization problem than anything we have seen in this book before: generally speaking, optimization problems become much harder with every additional change of quantifiers. Much of what has been happening in the general theory and practice of GANs has been related to trying to simplify and streamline the training process.

The optimization problem above has some nice properties. One can show [290] that $\max_D V(D, G)$ is minimized exactly when $p_g = p_{\text{data}}$. It is also straightforward to show that for a fixed generator G, the optimal distribution for the discriminator D is the distribution

$$D_G^*(\mathbf{x}) = \frac{p_{\text{data}}(\mathbf{x})}{p_{\text{data}}(\mathbf{x}) + p_g(\mathbf{x})},$$

that is, simply the optimal Bayesian classifier between p_{data} and p_g.

The global minimum of the criterion is achieved if and only if $p_g = p_{\text{data}}$ almost everywhere; the criterion itself for optimal D is equivalent to minimizing

$$\text{KL} \left(p_{\text{data}} \left\| \frac{p_{\text{data}} + p_g}{2} \right. \right) + \text{KL} \left(p_g \left\| \frac{p_{\text{data}} + p_g}{2} \right. \right),$$

a symmetric similarity measure between two distributions known as the *Jensen–Shannon divergence*.

So we know that, at least in theory, a GAN should arrive at the correct answer and bring p_g as close to p_{data} as possible. But all these theoretical results hold for the generator objective function equal to $\mathbb{E}_{z\sim p_z(z)}[\log(1 - D(G(z)))]$, that is for the minimax problem with a single objective function for both G and D. Unfortunately, in practice it proves to be a very inconvenient objective function for the generator, leading to saturation and extremely slow convergence. Even the original work that introduced GANs [290] immediately suggested that instead of $\mathbb{E}_{z\sim p_z(z)}[\log(1 - D(G(z)))]$, one should minimize

$$-\mathbb{E}_{z\sim p_z(z)}[\log D(G(\mathbf{z}))].$$

Informally speaking, with this objective function we maximize the probability of D giving the wrong answer rather than minimize the probability of D giving the right answer; there is a difference!

Thus, GANs are commonly trained with an alternating EM-like scheme:

- fix the weights of G and update the weights of D according to minimizing the error function

$$\mathbb{E}_{\mathbf{x}\sim p_{data}(\mathbf{x})}\left[\log D(\mathbf{x})\right] + \mathbb{E}_{\mathbf{x}\sim p_g(\mathbf{x})}\left[\log(1 - D(\mathbf{x}))\right];$$

- fix the weights of D and update the weights of G according to minimizing the error function

$$-\mathbb{E}_{\mathbf{x}\sim p_{data}(\mathbf{x})}[\log D(\mathbf{x})].$$

The original GANs worked on toy examples such as MNIST, and one of the first truly successful architectures based on these principles was Deep Convolutional GAN (DCGAN) [696]. It used a fully convolutional architecture without max-pooling (using strided convolutions instead), added batch normalization layers, used the Adam optimizer that was new at the time, and added a few more tricks to improve the results. As a result, DCGAN learned to generate very reasonable interiors on the LSUN dataset, that is... 64×64 color images.

Since 2015, GANs have come a long way. Famous modern architectures include StyleGAN [438, 439] that is able to generate lifelike 1024×1024 images of human faces and BigGAN [90] able to generate 512×512 images from thousands of different categories when trained on ImageNet or JFT-300M [827]. Obviously, modern architectures are much larger and more involved than DCGAN, and the datasets are also orders of magnitude larger than LSUN. But there are also conceptually new ideas that have proven to be very useful for training high-quality GANs. In the rest of this chapter, we discuss these ideas from the general standpoint of GAN training, but delve into details only for those ideas that will be immediately useful for synthetic-to-real domain adaptation afterwards.

4.5 Loss Functions in GANs

In the previous section, we saw the basic idea of a GAN and saw their training process as optimization alternating between learning the parameters of G and D. As we have seen, even original GANs [290] did not use the same loss function for the generator and discriminator. Since 2015, many different loss functions have been proposed for the generator and discriminator in GANs. In this section, I will give a brief overview of these ideas and show the loss functions that are especially relevant for the GANs discussed in subsequent chapters. In the literature, these loss functions are usually called *adversarial* losses (recall Fig. 4.5).

One of the first but still commonly used ideas is *Least Squares GAN* (LSGAN) [580]. The problem with GANs shown above is that the error function (Jensen–Shannon divergence) is saturated when the generator distribution p_g is far from the correct answer: the discriminator D has a logistic sigmoid function at the end, and gradients for the generator have to come through this sigmoid first. Due to this saturation effect, training regular GANs is often slow and unstable.

The LSGAN idea flies in the face of everything you know about classification. LSGAN proposes to pass from a sigmoidal to a quadratic error function for the classification:

$$\min_D V_{\text{LSGAN}}(D) = \frac{1}{2}\mathbb{E}_{\mathbf{x}\sim p_{\text{data}}}\left[(D(\mathbf{x}) - b)^2\right] + \frac{1}{2}\mathbb{E}_{\mathbf{z}\sim p_{\mathbf{z}}}\left[(D(G(\mathbf{z})) - a)^2\right],$$

$$\min_G V_{\text{LSGAN}}(G) = \frac{1}{2}\mathbb{E}_{\mathbf{z}\sim p_{\mathbf{z}}}\left[(D(G(\mathbf{z})) - c)^2\right],$$

which means that the discriminator learns to output a for fake inputs and b for real inputs, while the generator tries to "convince" the discriminator to output c on fake inputs (G has no control over real inputs so that part always disappears from its objective function). This is highly counterintuitive for classification since trying to learn a classifier with least squares is usually a very bad idea: for example, the error begins to grow for inputs that are classified correctly and with high confidence!

However, there even exists a nice theoretical result that comes with LSGAN: if $b - c = 1$ and $b - a = 2$, that is, the generator is convincing the discriminator to output "I don't know" for fake data (c is exactly midway between a and b), the LSGAN optimization problem for the optimal discriminator

$$D^*_{\text{LSGAN}} = \frac{bp_{\text{data}}(\mathbf{x}) + ap_g(\mathbf{x})}{p_{\text{data}}(\mathbf{x}) + p_g(\mathbf{x})}$$

is equivalent to minimizing the Pearson χ^2 divergence between p_{data} and p_g. But in this case, practice again differs from theory: in practice, LSGAN is usually trained with $a = 0$ and $b = c = 1$. LSGAN has been shown to be more stable to changes in the architectures of G and D, easier to train, and less susceptible to mode collapse; in general, the quadratic adversarial loss function has become one of the staple methods in modern GAN-based architectures.

Another important adversarial loss function, and a very interesting one as well, comes from the *Wasserstein GAN* (WGAN) [27]. To explain what is going on here, we need to take a step back.

What does it mean to learn a probability distribution? It means that the learned distribution p_{model} should become similar to the given distribution p_{data}, that is, we would most probably like to minimize either $\text{KL}(p_{\text{data}} \| p_{\text{model}})$, $\text{KL}(p_{\text{model}} \| p_{\text{data}})$, or some other similarity measure from the same family such as the Jensen–Shannon divergence.

Suppose now that the two distributions, p_{data} and p_{model}, have disjoint supports. For example, suppose that p_{data} is the distribution of "color photos of cats of size 1024×1024"; this means that its support lies in the space of dimension $\mathbb{R}^{3 \cdot 2^{20}}$, which is a pretty big space! If we parameterize some model distribution p_{model} to have low-dimensional support (which also sounds very reasonable as we don't want to cover the entire space of dimension $3 \cdot 2^{20}$, we want to capture the cat photos), with overwhelming probability the intersection of their supports will be zero until p_{model} is already very similar to p_{data}.

Unfortunately, this throws a wrench into the usual similarity measures between distributions. The Kullback–Leibler divergence is

$$\text{KL}\left(p_{\text{data}} \| p_{\text{model}}\right) = \int p_{\text{data}}(\mathbf{x}) \log \frac{p_{\text{data}}(\mathbf{x})}{p_{\text{model}}(\mathbf{x})} d\mathbf{x},$$

so if their supports are disjoint the KL-divergence is infinite. The Jensen–Shannon divergence is not infinite, but it degenerates into a constant:

$$\text{JSD}\left(p_{\text{data}} \| p_{\text{model}}\right) =$$
$$= \frac{1}{2}\text{KL}\left(p_{\text{data}} \| \frac{p_{\text{data}} + p_{\text{model}}}{2}\right) + \frac{1}{2}\text{KL}\left(p_{\text{model}} \| \frac{p_{\text{data}} + p_{\text{model}}}{2}\right) = \log 2.$$

Alas, infinities and constants do not make for good objective functions: a small perturbation in p_{model} will not change either KL or JSD, so the gradients will be zero or nonexistent.

This sounds like a quite general critique, so why doesn't the entire machine learning fail in this way? The thing is, machine learning usually employs model distributions p_{model} that span the entire space. For example, we could add a full-dimensional Gaussian noise to the p_{model} distribution concentrated on a low-dimensional variety, thus extending its support to the entire space; this would solve the problem entirely (now it's no problem that p_{data} is concentrated on a low-dimensional subset) and would probably correspond to an L_2-norm somewhere in the error function.

This solution is fine if all you need is to find the maximum of p_{model} after training it, i.e., find the maximum likelihood or maximum *a posteriori* hypothesis. But for generative models, it may lead to problems: we don't really want to sample from the "blurred" noisy distribution. In explicit generative models, we can usually remove this noise after training the model, which solves the problem. But in GANs that would be a very difficult task because we do not really have the distribution density p_{model}, all we have is a black box that somehow manages to sample from it.

To solve this problem, *Wasserstein GAN* (WGAN) proposes to consider other similarity measures between p_{data} and p_{model}. I will not go into full mathematical details and refer to [27] for details. In brief, WGAN is based on the *Earth Mover distance*, also known as the *Wasserstein distance*:

$$W(p_{\text{data}}, p_{\text{model}}) = \inf_{\gamma \in \Pi(p_{\text{data}}, p_{\text{model}})} \mathbb{E}_{(\mathbf{x}, \mathbf{y}) \sim \gamma} \left[\|\mathbf{x} - \mathbf{y}\| \right],$$

where $\Pi(p_{\text{data}}, p_{\text{model}})$ is the set of joint distributions $\gamma(\mathbf{x}, \mathbf{y})$ whose marginals are p_{data} and p_{model}. In other words, $\gamma(\mathbf{x}, \mathbf{y})$ shows how much "earth" (probability mass) one has to move in order to change the "mound of earth" corresponding to p_{data} into the "mound" corresponding to p_{model} in an optimal way.

In the example above, if p_{data} and p_{model} look the same way and are concentrated on parallel straight lines at distance θ, the Earth Mover distance between them will be θ: you need to move total mass 1, moving each point over distance θ. Thus, its gradient will exist and gradient descent will actually bring the parallel lines closer together.

Wasserstein distance sounds exactly right for the task, but the functional that defines $W(p_{\text{data}}, p_{\text{model}})$ does not look like something that would be easy to compute or take gradients of. Fortunately, Kantorovich–Rubinstein duality says (again, let's skip the proofs) that the infimum

$$W(p_{\text{data}}, p_{\text{model}}) = \inf_{\gamma \in \Pi(p_{\text{data}}, p_{\text{model}})} \mathbb{E}_{(\mathbf{x}, \mathbf{y}) \sim \gamma} \left[\|\mathbf{x} - \mathbf{y}\| \right]$$

is equivalent to the supremum

$$W(p_{\text{data}}, p_{\text{model}}) = \sup_{\|f\|_L \leq 1} \left(\mathbb{E}_{\mathbf{x} \sim p_{\text{data}}} [f(\mathbf{x})] - \mathbb{E}_{\mathbf{x} \sim p_{\text{model}}} [f(\mathbf{x})] \right),$$

where the supremum is taken over by all functions with Lipschitz constant ≤ 1.

Since we want to train a generative model $p_{\text{model}} = g_\theta(\mathbf{z})$, it now simply remains to parameterize everything by neural networks. Let us introduce a network $f_{\mathbf{w}}$ for the function f and a network g_θ for g. Then training can again proceed in an alternating fashion, as follows:

- for a given g_θ update the weights of $f_{\mathbf{w}}$, maximizing

$$\mathbb{E}_{\mathbf{x} \sim p_{\text{data}}} [f(\mathbf{x})] - \mathbb{E}_{\mathbf{x} \sim p_{\text{model}}} [f(\mathbf{x})];$$

- for a given $f_{\mathbf{w}}$, compute

$$\nabla_\theta W(p_{\text{data}}, p_{\text{model}}) = \nabla_\theta \left(\mathbb{E}_{\mathbf{x} \sim p_{\text{data}}} [f(\mathbf{x})] - \mathbb{E}_{\mathbf{x} \sim p_{\text{model}}} [f(\mathbf{x})] \right) =$$
$$= -\mathbb{E}_{\mathbf{z} \sim Z} [\nabla_\theta f_{\mathbf{w}}(g_\theta(\mathbf{z}))].$$

It only remains to ensure that $f_{\mathbf{w}}$ is Lipshitz with constant $\leq l$. The original work does it in a very simple yet effective fashion: it clips the gradients to ensure that their norm does not exceed l. But it was soon found [303] that it is a much better idea to introduce a soft regularizer on the gradient, for example,

$$\lambda \mathbb{E}_{\hat{\mathbf{x}} \sim P_{\hat{\mathbf{x}}}} \left[\left(\|\nabla_{\hat{\mathbf{x}}} D(\hat{\mathbf{x}})\|_2 - 1 \right)^2 \right].$$

LSGAN and WGAN are the two most popular adversarial loss functions at the time of writing (late 2020). But there are other options that remain relevant as well.

In particular, *Energy-Based Generative Adversarial Network* (EBGAN) [1019] considers the discriminator as an energy function that assigns low energy values to regions near the data distribution and high energy values to the other regions. The generator in this setting is supposed to produce highly variable samples with minimal values of energy.

This approach allows to use as the discriminator basically any architecture, not necessarily a classifier that ends with a logistic sigmoid. The second idea from EBGAN [1019] is to use an autoencoder as the discriminator, outputting its reconstruction error, i.e.,

$$D(\mathbf{x}) = \|\text{Dec}(\text{Enc}(\mathbf{x})) - \mathbf{x}\|.$$

Now the low-energy regions are those that can be accurately reconstructed by this autoencoder, and high-energy regions are those that cannot. The idea is to use real images to train this autoencoder, under the assumption that fake images will not map nicely into the autoencoder's latent features and will not be reconstructed well. Figure 4.6 provides an illustration.

This is a fruitful idea but we also need to "help" the discriminator a little bit. Thus, the training loss for the discriminator includes both the reconstruction loss and a hinge loss that kicks in when $G(\mathbf{z})$ begins to produce reasonable images and asks D to differentiate between real and fake samples:

$$\mathcal{L}_D^{\text{EBGAN}}(\mathbf{x}, \mathbf{z}) = D(\mathbf{x}) + [m - D(G(\mathbf{z}))]_+,$$

where $[a]_+ = \max(0, a)$.

As for the generator, its adversarial loss in EBGAN is straightforward (there are variations, but let us skip them for now):

$$\mathcal{L}_G^{\text{EBGAN}}(\mathbf{z}) = D(G(\mathbf{z})).$$

Finally, the last adversarial loss that we will need comes from *Boundary Equilibrium Generative Adversarial Networks* (BEGAN) [66]. It follows the general idea of EBGAN (in fact, Fig. 4.6 is still perfectly relevant) but adds a little bit of Wasserstein

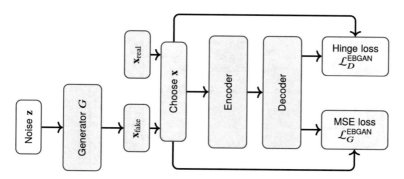

Fig. 4.6 The basic architecture of EBGAN [1019].

GAN's ideas. The idea of BEGAN is to keep the autoencoder structure but shift from optimizing the reconstruction loss in the discriminator directly to optimizing the distance between distributions of reconstruction losses from real and fake images. The authors of BEGAN argue that the Wasserstein distance between two distributions has a lower bound in the L_1-norm of the difference of their means, $|\mathbf{m}_1 - \mathbf{m}_2|$. We can capture this function on a given mini-batch of images, substituting instead of $\mathbf{m}_1$ the mean autoencoder loss $\mathcal{L}_{\mathrm{rec}}(\mathbf{x})$ for real images in the mini-batch and instead of $\mathbf{m}_2$ the mean autoencoder loss $\mathcal{L}_{\mathrm{rec}}(G(\mathbf{z}))$ for fake images in the mini-batch. Now the result should be maximized by the discriminator (which wants to pull these distributions apart) and, respectively, minimized by the generator (which wants to make the two distributions as similar as possible). We refer to [66] for further details and proofs.

At this point, we have seen several adversarial losses that can be substituted instead of the basic GAN loss that we discussed in Section 4.4. In the next section, we will consider some general GAN-based architectures that appear in the literature and in various applications very often and constitute the bulk of applications for generative adversarial networks.

4.6 GAN-Based Architectures

GANs as presented above are designed to learn to generate objects $\mathbf{x}$ from a domain defined by a dataset; the aim is to generate "fake" objects in such a way that these objects are indistinguishable from real ones. However, there are situations where the idea of adversarial learning still works great but the basic architecture shown in Fig. 4.5 needs certain modifications. In this section, we discuss three basic architecture ideas that have been used many times in very different applications: conditional GANs, adversarial autoencoders, and progressively growing GANs.

First, what if we want to generate objects of several different classes and control which class we are generating from now? Training several separate GANs would be

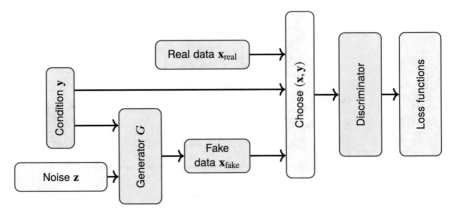

Fig. 4.7 The general conditional GAN architecture [602].

a waste of time and data: if you want to generate cats and dogs, it would be really helpful to join the dataset into one because both classes will have the same basic features almost up until the very end.

To train a single GAN for several classes, we can use a *conditional GAN*, first proposed almost immediately after the original GAN publication, in 2014 [602]. This is a straightforward extension: we supply the condition **y** to both generator and discriminator, as shown in Fig. 4.7. In our example, the generator would know whether it has to generate a cat or a dog, and the discriminator would know which animal the fake image is supposed to represent. A conditional GAN can utilize the same loss functions as a regular GAN, which we have discussed in Section 4.5.

The second important idea in this section deals with *adversarial autoencoders* invented by Makhzani et al. in 2016 [576]. In Section 4.3, we have discussed why autoencoders do not give rise to generative models by themselves: the distribution of latent codes may be quite complicated even in their low(er) dimensional space. In that section, we discussed variational autoencoders that provide one way to fix this problem by parameterizing the distribution of latent codes.

Adversarial autoencoders (AAE) represent a different way to fix this problem that makes use of the same basic idea of adversarial training, introducing a discriminator into the picture. But here the discriminator is not distinguishing between fake and real images, but rather between fake and real codes. The general AAE structure is shown in Fig. 4.8:

- the encoder Enc is trying to learn the distribution $q(\mathbf{z} \mid \mathbf{x})$, producing the latent code $\mathbf{z}_{\text{fake}}$;
- we call it $\mathbf{z}_{\text{fake}}$ because the discriminator D is trying to distinguish $\mathbf{z}_{\text{fake}}$ from $\mathbf{z}_{\text{real}}$ samples from a given distribution $p(\mathbf{z})$;
- at the same time, the decoder Dec is reconstructing the original **x** in the usual autoencoder fashion, producing a reconstruction $\hat{\mathbf{x}}$;

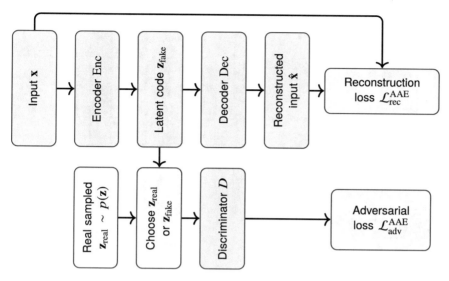

Fig. 4.8 Adversarial autoencoder [576].

- the loss function for this architecture is composed of the reconstruction loss $\mathcal{L}_{\text{rec}}^{\text{AAE}}$, which shows how similar $\hat{\mathbf{x}}$ and $\mathbf{x}$ are, and the adversarial loss $\mathcal{L}_{\text{adv}}^{\text{AAE}}$ for the discriminator, usually simply the binary cross-entropy.

Obviously, we choose $p(\mathbf{z})$ to be a simple standard distribution that we can sample from. This idea is similar in spirit to VAEs, but instead of minimizing the KL-divergence between $q(\mathbf{z})$ and a given prior, AAEs use an adversarial procedure.

Let me use the example of AAEs to illustrate the diversity of possible adversarial architectures, even when they are intended for the same problem. Suppose that we want to make a conditional AAE, say generate cats, dogs, and rabbits while being able to control which of these classes we are generating. The basic conditional AAE architecture can add the condition in the same way as a conditional GAN shown in Fig. 4.7, adding condition $\mathbf{y}$ as input to all three networks in the architecture: encoder, decoder, and discriminator, also choosing a different distribution for each class so that we can sample latent codes $\mathbf{z}$ separately from each class.

This, alas, is not the best idea (I haven't even drawn a figure about it) because this approach does not generalize to the semi-supervised setting: what if for some images we do not know their labels? It would still be useful to train the autoencoder, and we would even know which distribution to distinguish it from: let's simply take the mixture of all class distributions (uniform or with class priors if we know them).

But there still are two reasonable approaches to making a conditional AAE, each with its own properties. Figure 4.9 shows these two possible approaches:

- in Fig. 4.9a, the autoencoder does not care about class labels at all, and the label (with an extra option for unknown class) is fed only to the discriminator; this

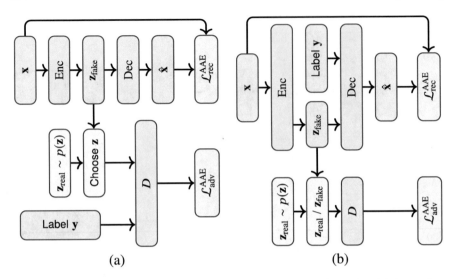

Fig. 4.9 Two versions of the conditional adversarial autoencoder [576]: (a) class labels are given only to the discriminator; (b) the decoder receives the class label.

means that the discriminator will try to associate each class with a separate mode of the distribution $p(\mathbf{z})$;

- in Fig. 4.9b, the class label is fed to the decoder; this means that the decoder now has the class information "for free", and the latent code is encoding only the *style* of the image; this architecture can lead to the *disentanglement* of style and content (in this case, the content is a class label); we will see more examples of such disentanglement in Section 4.7.

Both of these ideas appeared already in the original work [576], and since then AAEs have received many extensions and have been successfully used in many applications of generative models, including the generation of discrete objects such as molecular structures [419, 420, 679]. Adversarial autoencoders still remain a viable alternative to VAEs in many problems.

Our next stop is related to the problem of generating high-dimensional data. In 2014, GANs were doing a reasonably good job on 28×28 black-and-white images from the MNIST dataset but had no chance to handle reasonably sized photos. Improved architectures such as the above-mentioned DCGAN got generation up to 64×64 color images, but it was still a far cry from real-world applications. New ideas were needed.

Probably the most fruitful idea in this regard is *progressive growing* of GANs, which first appeared in the *ProGAN* model developed by NVIDIA researchers Karras et al. [435]. The basic idea is simple: suppose we want to generate high-resolution images (in reality, ProGAN reached 1024×1024 for human faces and 512×512 for more general datasets). It is not a big deal to train a regular GAN to generate 4×4 images. Then let's use this 4×4 image as an input (one could also say—as a

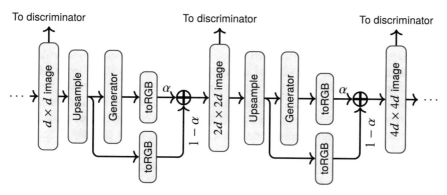

Fig. 4.10 Progressive growing of GANs: an excerpt from the generator structure of ProGAN [435].

condition) for the next GAN that performs basically superresolution, upsampling the 4×4 image to an 8×8 image, and so on, and so forth: each layer is only supposed to perform 2x superresolution, which is quite possible even for high resolutions.

Figure 4.10 illustrates this idea with an excerpt from the ProGAN generator's architecture. The idea is to gradually add new upsampling modules but keep training all layers in the deep architecture as the training progresses. To avoid sudden surprises that could upset previous layers as we shift to the next layer with untrained weights, ProGAN gradually "fades in" each new layer with a residual architecture shown in Fig. 4.10 and α gradually increasing from 0 to 1. The discriminator (not shown in Fig. 4.10) is also growing progressively together with the generator; we can downsample high-resolution images to get real images of any intermediate dimension.

ProGAN was an important breakthrough in GAN-based generation: suddenly GANs were able to produce high-resolution images, with high-quality latent space interpolations. It was widely publicized, and it gave rise to architectures such as BigGAN [89] and StyleGAN that we will discuss in the next section.

4.7 Case Study: GAN-Based Style Transfer

The primary use of GANs in this book is related to synthetic-to-real domain adaptation. As we will see in Chapter 10, one important approach to this problem is *refinement*, that is, trying to make synthetic data more realistic. This is usually done with GAN-based architectures.

Therefore, in this section let us consider the more general problem of *style transfer*, i.e., redrawing images from one style to another while preserving the content. This will allow us to discuss the main GAN-based architectures for style transfer that will be referenced a lot in Chapter 10.

We begin with an architecture that put artistic style transfer on the map back in 2015, *A Neural Algorithm of Artistic Style* by Gatys et al. [264]. They used a straightforward CNN and noted that high-level *content* information is preserved in features extracted by both lower and higher layers of the network, but the exact pixel-wise information is lost in higher layers. On the other hand, the *style* of an image is captured by correlations between extracted features, an idea that had been noted previously in [265] and has since become a staple in GANs in the form of the *texture loss*. In [264], correlations were formalized by Gram matrices of the corresponding features.

The key idea in [264] is that these representations are separable, and you can have an image that combines the content (feature activations) from one input image and style (feature correlations) from another. The basic idea is simple yet beautiful: let us fix a feature extractor CNN and perform gradient descent with respect to the *input image* $\mathbf{x}$ rather than with respect to the network weights. This idea is quite similar to the production of adversarial examples that we discussed in Section 3.4.

The loss function for the image will consist of similarities between feature activations of $\mathbf{x}$ and the content image $\mathbf{x}_c$ (the content loss $\mathcal{L}_{\text{content}}^{\text{GAT}}$), and similarities between Gram matrices of $\mathbf{x}$ and the style image $\mathbf{x}_s$ (the style loss $\mathcal{L}_{\text{style}}^{\text{GAT}}$):

$$\mathcal{L}^{\text{GAT}}(\mathbf{x}) = \alpha \mathcal{L}_{\text{content}}^{\text{GAT}}(\mathbf{x}, \mathbf{x}_c) + \beta \mathcal{L}_{\text{style}}^{\text{GAT}}(\mathbf{x}, \mathbf{x}_s), \quad \text{where}$$

$$\mathcal{L}_{\text{content}}^{\text{GAT}}(\mathbf{x}, \mathbf{x}_c) = \sum_{l=1}^{L} w_{\text{content}}^{(l)} \cdot \frac{1}{2} \sum_{i,j} \left(F_{i,j}^{(l)}(\mathbf{x}) - F_{i,j}^{(l)}(\mathbf{x}_c) \right)^2,$$

$$\mathcal{L}_{\text{style}}^{\text{GAT}}(\mathbf{x}, \mathbf{x}_s) = \sum_{l=1}^{L} w_{\text{style}}^{(l)} \cdot \frac{1}{4 W_l^2 H_l^2} \sum_{i,j} \left(G_{i,j}^{(l)}(\mathbf{x}) - G_{i,j}^{(l)}(\mathbf{x}_s) \right)^2,$$

where $F^{(l)}(\mathbf{x})$ denotes the features extracted by the CNN at layer l from $\mathbf{x}$, $G^{(l)}(\mathbf{x})$ is the Gram matrix of these features, $G_{i,j}^{(l)}(\mathbf{x}) = \sum_k F_{i,k}^{(l)}(\mathbf{x}) F_{j,k}^{(l)}(\mathbf{x})$, W_l and H_l are the dimensions of the feature tensor at layer l, $w_{\text{content}}^{(l)}$ and $w_{\text{style}}^{(l)}$ are constant weights with which different layers occur in the content and style loss, respectively, and α and β are constants. This architecture is illustrated in Fig. 4.11 with a sample convolutional architecture with five layers that produce the necessary features.

The method of Gatys et al. worked very well for artistic style transfer, and it was the first style transfer method to be widely publicized around 2015, when pictures with "photos made into Picasso paintings" briefly flooded the Web. This approach, however, has an important drawback: to perform style transfer, you need to actually perform gradient descent on the image pixels, which for a high-resolution image is basically equivalent to training a neural network with several million weights to convergence. This process takes a long time, may have trouble converging, may get stuck in local minima, and so on. So this is where GANs came into style transfer.

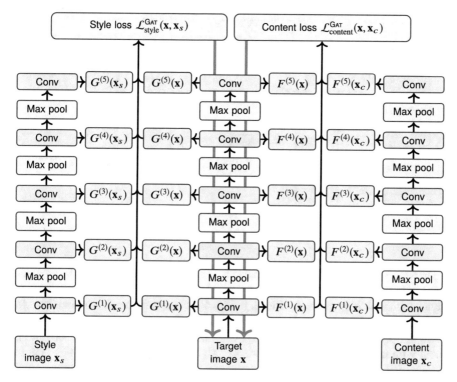

Fig. 4.11 Artistic style transfer by Gatys et al. [264]. Green arrows denote gradient flow: note that optimization here is done with respect to the target image **x**, and the neural network weights are fixed.

The most direct GAN-based approach to style transfer is provided by the *pix2pix* model developed by Isola et al. [389]. This is a straightforward conditional GAN, illustrated in Fig. 4.12:

- the generator receives as input (condition) an image in one style and is supposed to produce the same image in a different (target) style;
- the discriminator receives two images (the fake or real image and the condition) and tries to distinguish real images in the target style and fake images produced by the generator.

Later, *pix2pix* was further improved in the *pix2pixHD* model [909] that uses enhanced architectures for encoder and decoder and scales the original *pix2pix* up to high-definition images. In particular, the generator in *pix2pixHD* consists of two parts with the second part enhancing the result of the first. The "styles" in this idea do not even have to be actual image styles. One could argue that, for instance, StackGAN [994], a model for text-to-photo image synthesis, is also a style transfer model very similar to *pix2pix*, but the source style here is not an image but rather a textual description of what the image should look like.

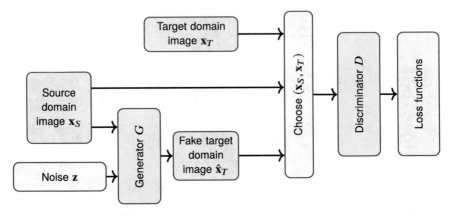

Fig. 4.12 The *pix2pix* model [389].

However, the conditional GAN that serves as the foundation for *pix2pix* models has a principled limitation: it requires a *paired* dataset that contains images of two styles with matching content. Indeed, in Fig. 4.12 inputs to the discriminator in two different styles have to match, and without it, nothing would prevent the generator from simply memorizing a few images in the target style and always outputting them with no regard to its input. For some style transfer applications, this is not a problem: for instance, if we want to create a realistic photo from a segmentation map, we need a dataset of segmentation maps, and where else could they come from if not from segmenting real images? But in other applications, we are not as lucky: for instance, we would be hard-pressed to find photographs that perfectly match the content of Monet paintings. Synthetic-to-real style transfer also falls into the latter category: usually, we produce synthetic data by CGI rendering, and there is no perfectly matching scene that we could photograph in the real world.

How can we avoid this problem? We are again facing a task similar to how at the beginning of this chapter we wanted to capture the "catness" of an image. Now the problem is how to capture the *content* of a given image with no regard to its style. And again, an interesting solution comes by introducing more networks into the architecture. The idea of CycleGAN [1025], illustrated in Fig. 4.13, is to train not one but *two* style transfers at the same time. Given a source domain X and a target domain Y, we train

- a generator $F : X \rightarrow Y$ that translates images from the source domain to the target domain; this is the function that we actually want to train as the result of the whole process;
- another generator $G : Y \rightarrow X$ that translates images from the target domain to the source domain;
- a discriminator D_X that learns to distinguish between real and fake images in the source domain X;
- another discriminator D_Y that learns to distinguish between real and fake images in the target domain Y.

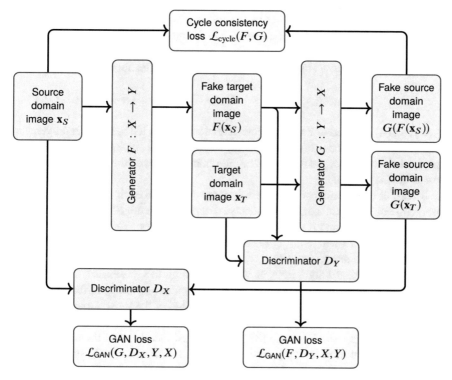

Fig. 4.13 The general architecture of CycleGAN [389] (compared to previous diagrams, nodes with random noise and random choice of discriminator inputs are omitted for clarity).

CycleGAN has the usual loss functions for these two GANs: classification loss functions for D_X and D_Y and adversarial loss functions for G and F; the original CycleGAN model [1025] used the basic GAN loss that we discussed in Section 4.4,

$$\mathcal{L}_{\text{GAN}}(F, D_Y, X, Y) = \mathbb{E}_{\mathbf{x} \sim p_{\text{data}}^X} \left[\log \left(1 - D_Y(F(\mathbf{x})) \right) \right],$$
$$\mathcal{L}_{\text{GAN}}(G, D_X, Y, X) = \mathbb{E}_{\mathbf{y} \sim p_{\text{data}}^Y} \left[\log \left(1 - D_X(G(\mathbf{y})) \right) \right],$$

but one can use any modern adversarial loss that we discussed in Section 4.5 for a similar architecture. The novelty here is the idea that once we translate an image $\mathbf{x}$ from the source domain to the target domain *and back*, the resulting image $G(F(\mathbf{x}))$ should match $\mathbf{x}$ exactly, and we can use a simple pixel-wise loss to capture this fact. In the original CycleGAN [1025], this *cycle consistency loss* was defined as the L_1-norm of the difference:

$$\mathcal{L}_{\text{cycle}}(F, G) = \mathbb{E}_{\mathbf{x} \sim p_{\text{data}}^X} \left[\| G(F(\mathbf{x})) - \mathbf{x} \|_1 \right] + \mathbb{E}_{\mathbf{y} \sim p_{\text{data}}^Y} \left[\| F(G(\mathbf{y})) - \mathbf{y} \|_1 \right],$$

where the expectations mean that we sample $\mathbf{x}$ from the dataset of source domain images and $\mathbf{y}$ from the dataset of target domain images (but they do not have to match each other's content any more).

The final loss is now simply a linear combination of the above:

$$\mathcal{L}^{\text{CYCLE}} = \mathcal{L}_{\text{GAN}}(F, D_Y, X, Y) + \mathcal{L}_{\text{GAN}}(G, D_X, Y, X) + \lambda \mathcal{L}_{\text{cycle}}(F, G).$$

CycleGAN proved to be a very powerful idea that has led to many improvements and derivative models. It has become an important tool for image generation and manipulation models, and we will encounter many such cycles and cycle consistency losses in Chapter 10.

But all of these models need relatively large datasets in each style. Often it's not a problem, but, for instance, for artistic style transfer there are only that many Monet paintings available, and if you have a few hundred different styles or want to let the user specify their own style, problems arise as well. To perform a few-shot style transfer, we need one more idea.

While GAN-based style transfer was progressing to CycleGAN, the basic idea of Gatys et al., that is, disentangling style and content by using different statistics of features, also received further developments. One of the most important recently developed tools in style transfer was the idea of *adaptive instance normalization* (AdaIN) by Huang and Belongie [371]. Let us first recall one of the basic building blocks of modern deep learning models: *batch normalization* [387]. In a batch normalization (BN) layer, mini-batch statistics are used to reduce internal covariate shift in the previous layer's outputs:

$$\text{BN}(\mathbf{x}) = \gamma \left(\frac{\mathbf{x} - \mu(X)}{\sigma(X)} \right) + \beta,$$

where X denotes a mini-batch of inputs, and statistics are computed according to this mini-batch, while γ and β are BN layer parameters learned from data.

Statistics used in a BN layer are a natural candidate for the same kind of statistics that define a style. But, of course, averaging over a mini-batch of completely different images will not help style transfer, so for style transfer we move to *instance normalization* $\text{IN}(\mathbf{x})$: the exact same transformation as $\text{BN}(\mathbf{x})$ but with statistics computed over every channel of a single image separately.

The first application of this idea to style transfer was in *conditional instance normalization* [210], a method where γ and β are trained separately for every style:

$$\text{CIN}(\mathbf{x}; s) = \gamma_s \left(\frac{\mathbf{x} - \mu(\mathbf{x})}{\sigma(\mathbf{x})} \right) + \beta_s.$$

But this still required running a separate training procedure for every style and could not extend to large numbers of styles. The next logical step, taken in *adaptive instance normalization* (AdaIN), is to use the same transformation for the content image $\mathbf{x}_c$ but with parameters γ and β learned as statistics of the style image $\mathbf{x}_s$:

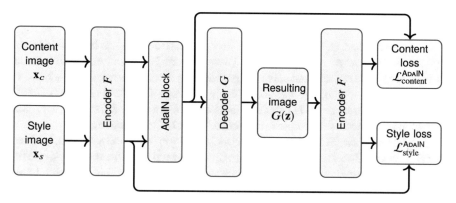

Fig. 4.14 The basic AdaIN model from [371].

$$\text{AdaIN}(\mathbf{x}_c, \mathbf{x}_s) = \sigma(\mathbf{x}_s)\left(\frac{\mathbf{x}_c - \mu(\mathbf{x}_c)}{\sigma(\mathbf{x}_c)}\right) + \mu(\mathbf{x}_s).$$

Huang and Belongie [371] showed that AdaIN can work for style transfer even in a very simple architecture, illustrated in Fig. 4.14. They applied it to artistic style transfer, where the need for such methods is especially dire because we cannot rely on having a large dataset of images in a given style. In this architecture, the AdaIN layer is applied to the result of a convolutional encoder F (specifically, a VGG encoder in this case). Then the result is decoded back with a decoder G, and the whole architecture is trained with two losses:

$$\mathcal{L}^{\text{ADAIN}} = \mathcal{L}^{\text{ADAIN}}_{\text{content}} + \lambda \mathcal{L}^{\text{ADAIN}}_{\text{style}},$$

where

- the content loss $\mathcal{L}^{\text{ADAIN}}_{\text{content}}$ is basically the autoencoder loss, checking that if we encode back the result of G by F, we get the same tensor of features that we started with:
$$\mathcal{L}^{\text{ADAIN}}_{\text{content}} = \|F(G(\mathbf{z})) - \mathbf{z}\|_2,$$
where $\mathbf{z} = \text{AdaIN}(F(\mathbf{x}_c), F(\mathbf{x}_s))$;
- the style loss $\mathcal{L}^{\text{ADAIN}}_{\text{style}}$ compares the statistics (that are supposed to encode the style) of the encoding of the style image $\mathbf{x}_s$ and the resulting image with supposedly swapped styles:

$$\mathcal{L}^{\text{ADAIN}}_{\text{style}} = \sum_{l=1}^{L} \left\|\mu\left(F^{(l)}(G(\mathbf{z}))\right) - \mu\left(F^{(l)}(\mathbf{x}_s)\right)\right\|_2 +$$

$$+ \sum_{l=1}^{L} \left\|\sigma\left(F^{(l)}(G(\mathbf{z}))\right) - \sigma\left(F^{(l)}(\mathbf{x}_s)\right)\right\|_2,$$

where $F^{(l)}$ denotes the output of layer l in the VGG encoder F.

The most important feature of AdaIN layers is that the style can be encoded from style features or even a single image. Therefore, models based on AdaIN do not require large datasets and can solve the problem we faced a few paragraphs back. In particular, the *Few-Shot Unsupervised Image-to-Image Translation* model (FUNIT) developed by Liu et al. [537] is able to transfer between many different styles, each of which is defined by a few pictures (hence "few-shot" in the title).

The basic idea in FUNIT includes a conditional generator and a multitask adversarial discriminator. The discriminator is solving several classification tasks, one for each class, deciding whether the input image is a real image of this class or a translation output produced by the generator. FUNIT training is done by the optimization problem

$$\min_{D} \max_{G} \mathcal{L}^{\text{FUNIT}}(D, G),$$
$$\mathcal{L}^{\text{FUNIT}}(D, G) = \mathcal{L}_{\text{GAN}}^{\text{FUNIT}}(D, G) + \lambda_1 \mathcal{L}_{\text{rec}}^{\text{FUNIT}}(G) + \lambda_2 \mathcal{L}_{\text{feat}}^{\text{FUNIT}}(G),$$

where

- $\mathcal{L}_{\text{GAN}}^{\text{FUNIT}}(D, G)$ is the usual conditional GAN loss,

- $\mathcal{L}_{\text{rec}}^{\text{FUNIT}}(G)$ is the reconstruction loss that says that if we use the same image for both style and content, we should get the same image as the output:

$$\mathcal{L}_{\text{rec}}^{\text{FUNIT}}(G) = \mathbb{E}_{\mathbf{x}} \left[\| \mathbf{x} - G(\mathbf{x}, \mathbf{x}) \|_1 \right];$$

- $\mathcal{L}_{\text{feat}}^{\text{FUNIT}}(G)$ is the feature matching loss that tries to match the discriminator features from class images $\{\mathbf{y}_1, \ldots, \mathbf{y}_K\}$ and translation output $\hat{\mathbf{x}}$:

$$\mathcal{L}_{\text{feat}}^{\text{FUNIT}}(G) = \mathbb{E}_{\mathbf{x}, \{\mathbf{y}_1, \ldots, \mathbf{y}_K\}} \left[\left\| D_f(\hat{\mathbf{x}}) - \frac{1}{K} \sum_{k=1}^{K} D_f(\mathbf{y}_k) \right\|_1 \right],$$

where $D_f(\mathbf{x})$ are the features taken from the penultimate layer of the discriminator; in other words, the feature matching loss is directly asking the generator to fool the discriminator into confusing $\hat{\mathbf{x}}$ and $\{\mathbf{y}_1, \ldots, \mathbf{y}_K\}$, serving as a kind of regularizer.

The interesting part of FUNIT is the generator architecture, shown in Fig. 4.15. It explicitly distinguishes between content and style encoders: the former is applied to the content image $\mathbf{x}$, and the latter to each of the input class images $\{\mathbf{y}_1, \ldots, \mathbf{y}_K\}$ that together define the style. Then the features produced from style images are averaged to obtain the style code $\mathbf{z}_{\mathbf{y}}$, and this code, after a few more fully convolutional layers, is used as the style input (coefficients μ and σ) for the AdaIN blocks in the decoder.

The next step along these lines was taken in the *Multimodal Unsupervised Image-to-Image Translation* (MUNIT) model by Huang et al. [374] that combines the ideas of FUNIT and CycleGAN. The autoencoder used in MUNIT is very similar to the

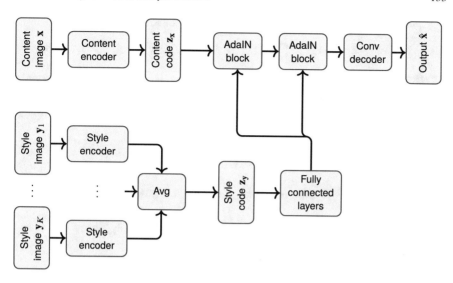

Fig. 4.15 FUNIT generator architecture [537].

Fig. 4.16 Sample MUNIT results on synthetic-to-real and real-to-synthetic transfer [374]: (a) SYNTHIA to *Cityscapes*; (b) *Cityscapes* to SYNTHIA.

FUNIT architecture shown in Fig. 4.15: from an input image $\mathbf{x}$, one encoder F^c produces the content code $\mathbf{c}$ and another encoder E^s produces the style code $\mathbf{s}$, which is then used in the decoder (generator) G via AdaIN blocks. Similar to FUNIT, this explicit disentanglement of style and content allows to perform style transfer by combining the style code from one image and content code from another in the decoder; Fig. 4.16 shows sample MUNIT results in synthetic-to-real and real-to-synthetic transfer.

The difference lies in the general structure of how these autoencoders are used, as shown in Figure 4.17. The overall loss function in MUNIT is

$$\mathcal{L}^{\text{MUNIT}} = \mathcal{L}^{\text{MUNIT}}_{\text{GAN},1} + \mathcal{L}^{\text{MUNIT}}_{\text{GAN},2} + \lambda_1 (\mathcal{L}^{\text{MUNIT}}_{\text{rec},x1} + \mathcal{L}^{\text{MUNIT}}_{\text{rec},x2}) +$$
$$+ \lambda_2 (\mathcal{L}^{\text{MUNIT}}_{\text{rec},c1} + \mathcal{L}^{\text{MUNIT}}_{\text{rec},c2}) + \lambda_3 (\mathcal{L}^{\text{MUNIT}}_{\text{rec},s1} + \mathcal{L}^{\text{MUNIT}}_{\text{rec},s2}), \quad \text{where}$$

- $\mathcal{L}^{\text{MUNIT}}_{\text{GAN},1}$ is the adversarial loss in the domain of $\mathbf{x}_1$ and $\mathcal{L}^{\text{MUNIT}}_{\text{GAN},2}$, in the domain of $\mathbf{x}_2$;
- $\mathcal{L}^{\text{MUNIT}}_{\text{rec},x1}$ and $\mathcal{L}^{\text{MUNIT}}_{\text{rec},x2}$ are reconstruction losses for autoencoders applied to images $\mathbf{x}_1$ and $\mathbf{x}_2$, respectively, (not shown in Fig. 4.17):

$$\mathcal{L}^{\text{MUNIT}}_{\text{rec},x1} = \mathbb{E}_{\mathbf{x}_1} \left[\left\| G_1(E_1^c(\mathbf{x}_1), E_1^s(\mathbf{x}_1)) - \mathbf{x}_1 \right\|_1 \right];$$

- $\mathcal{L}^{\text{MUNIT}}_{\text{rec},c1}$ and $\mathcal{L}^{\text{MUNIT}}_{\text{rec},c2}$ are content reconstruction losses for the two respective domains:

$$\mathcal{L}^{\text{MUNIT}}_{\text{rec},c1} = \mathbb{E}_{\mathbf{c}_1, \mathbf{s}_2' \sim q} \left[\left\| E_2^c(G_2(\mathbf{c}_1, \mathbf{s}_2')) - \mathbf{c}_1 \right\|_1 \right],$$

where $q(\mathbf{s}) = \mathcal{N}(\mathbf{s} \mid \mathbf{0}, \mathbf{I})$ is the standard Gaussian prior for style codes;
- $\mathcal{L}^{\text{MUNIT}}_{\text{rec},s1}$ and $\mathcal{L}^{\text{MUNIT}}_{\text{rec},s2}$ are style reconstruction losses for the two respective domains:

$$\mathcal{L}^{\text{MUNIT}}_{\text{rec},s1} = \mathbb{E}_{\mathbf{c}_2, \mathbf{s}_1' \sim q} \left[\left\| E_1^s(G_1(\mathbf{c}_2, \mathbf{s}_1')) - \mathbf{s}_1' \right\|_1 \right].$$

With this architecture, MUNIT can learn to perform style transfer in a fully unsupervised fashion, with unpaired data and domains encoded by sets of images.

Note that style transfer models can, in principle, be used directly for synthetic-to-real style transfer, making synthetic images more realistic similar to how artistic style transfer models would turn a Monet landscape into a photo. Many models mentioned in this section provide synthetic-to-real sample results; in particular, Figure 4.16 reproduces a few examples of the synthetic-to-real style transfer for outdoor scenes produced by MUNIT [374]. We will speak in much more detail about synthetic-to-real refinement in Chapter 10.

All of these ideas have come together in the most recent style transfer models. One famous architecture is *StyleGAN* by NVIDIA researchers Karras et al. [437] whose main use case is to generate human faces with conditions (styles) taken from other faces at different "granularity levels". The basic idea is to use the styles extracted from style images as inputs on different layers of the generator, and the styles are defined and used in the AdaIN fashion. On the other hand, the levels themselves follow the progressive growing idea that we have discussed in Section 4.6.

The StyleGAN architecture is illustrated in Fig. 4.18. I will not go into the full details of this (admittedly quite complicated) architecture, but let me briefly explain the main data flow in Fig. 4.18:

- StyleGAN uses a progressively growing architecture; Fig. 4.18 shows the first two blocks with 4×4 and 8×8 outputs, and subsequent blocks have the same structure as the 8×8 block in Fig. 4.18;
- StyleGAN does not have a latent noise input at the beginning; it begins with a constant $4 \times 4 \times 512$ tensor;

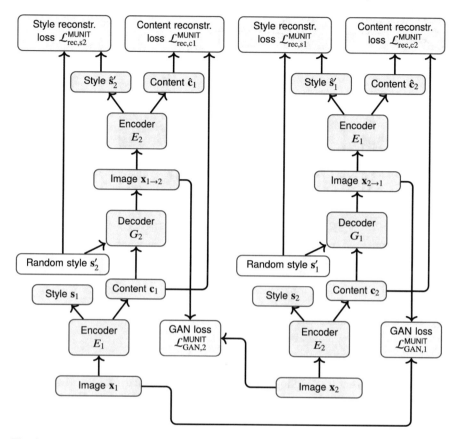

Fig. 4.17 MUNIT general architecture [374].

- the noise is injected on every progressive level after convolutional blocks; it is uncorrelated Gaussian noise that undergoes scaling with learned per-feature factors (denoted as B in Fig. 4.18);
- the style is added in AdaIN layers in the usual AdaIN structure that we have discussed above; the style vectors are produced from a given latent vector $\mathbf{z}$, which is mapped to a weight vector $\mathbf{w}$ by a fully connected mapping network, and then the vector $\mathbf{w}$ is specialized into styles (γ, β) for each AdaIN block with learned affine transformations denoted as A in Fig. 4.18.

With this architecture, StyleGAN is able to have very fine-grained control over the resulting image because styles and noise at different levels of progressive growing produce changes of different "scale" in the final image. In particular, for human faces low resolutions (4×4, 8×8) control the pose, general hairstyle, and large details such as sunglasses; facial features, specific details of the hairstyle, and facial expressions are governed by middle resolutions (16×16, 32×32); final layers (up to 1024×1024) control the color scheme and very fine details. This means that

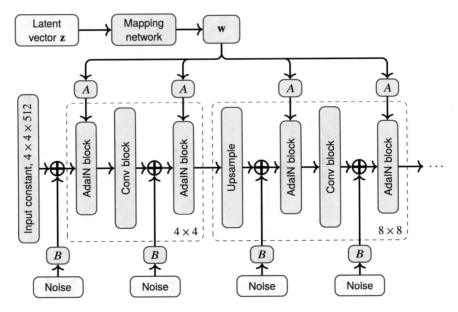

Fig. 4.18 StyleGAN architecture [437].

by using several different style vectors **w** for different resolutions, we can combine several faces into one, controlling even what exactly we want to take from each input. A similar idea has also been used in the *Face Swapping GAN* (FSGAN) by Nirkin et al. [632].

The next steps were taken in StyleGAN2, the next iteration by the same NVIDIA team of Karras et al. [440]. There have been two further developments after that: a semi-supervised version of StyleGAN that performs disentanglement learning in high resolution [629] and a variation of StyleGAN training with additional augmentations that further improves the resulting quality on limited size datasets [436]. Style transfer and related tasks remain a very active and interesting field of study, and I definitely expect many new models and improvements in the nearest future.

4.8 Conclusion

This chapter has been an attempt to provide a very brief introduction into the world of generative deep learning models. Naturally, in this short chapter we have not been able to give full justice to this wide topic. But I hope that the basic ideas we have outlined here will suffice to give the reader a first acquaintance, sufficient to understand Chapter 10, and the references provided in this chapter will help learn more about generative models in deep learning.

The last section with the style transfer case study was far from random: style transfer is the underlying task for synthetic-to-real refinement that we will discuss in Section 10.1 and Section 10.4. As we have already noted, new models for style transfer can be almost directly applied to make synthetic data look more realistic. But the next chapter will paint a wider picture: it will survey many different ideas for synthetic-to-real domain adaptation. Still, many of them are based on generative models, especially generative adversarial networks, so expect a lot more GAN-based architectures in Chapter 10.

But before we proceed to domain adaptation, in the next chapters we will consider more straightforward applications of synthetic data. In the next chapter, we begin with the history of early works in computer vision that seem to be inseparable from synthetic data in one form or another.

Chapter 5
The Early Days of Synthetic Data

It may appear that synthetic data has become instrumental only very recently, with the rise of modern computer graphics that allows for near-photorealistic imagery. But in fact, synthetic data has been used throughout the history of computer vision, starting from its very inception in the 1960s. In this chapter, we begin with the early days of synthetic data, show some of the earliest models and applications of computer vision, and discuss aspects of computer vision that have always been very hard or even impossible to do without synthetic data.

5.1 Line Drawings: The First Steps of Computer Vision

In this chapter, we will see how synthetic data started its journey and what its early days looked like. We will see how synthetic data has permeated computer vision starting from its very first steps; then in subsequent chapters, we will proceed to the modern state of the art and how synthetic data can help with it as well. But the journey of synthetic data begins much earlier than the rise of deep learning...

Computer vision in its modern form began almost as soon as the field of artificial intelligence started in earnest, and right off the bat, it was a very ambitious endeavour. A famous anecdote saying that Marvin Minsky asked his undergraduate student Gerald Sussman to solve computer vision as a summer project is not entirely true; it appears that the real task was to "connect a camera to a computer and do something with it", Minsky did not expect much.

However, this anecdote definitely does reflect the spirit of the times. In 1966, the *Summer Vision Project* by Seymour Papert was intended for a group of about 10 summer interns and asked them to construct a segmentation system: "The summer vision project is an attempt to use our summer workers effectively in the construction of a significant part of a visual system... The primary goal of the project is to construct a system of programs which will divide a vidisector picture into regions such as likely

© The Author(s), under exclusive license to Springer Nature Switzerland AG 2021
S. I. Nikolenko, *Synthetic Data for Deep Learning*, Springer Optimization
and Its Applications 174, https://doi.org/10.1007/978-3-030-75178-4_5

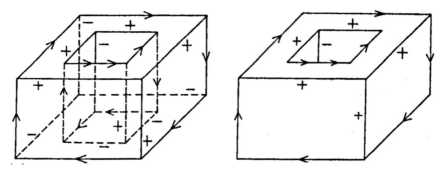

Fig. 5.1 Sample line drawings from [379]: an "X-ray" graph with labeled edges (left) and a picture graph with labeled edges (right).

objects, likely background areas, and chaos... The final goal is OBJECT IDENTIFI-CATION which will actually name objects by matching them with a vocabulary of known objects" [654].

This kind of optimism was, of course, unwarranted. Back in the 1960s and 1970s, usable segmentation on real photographs was very hard to achieve, even as a purely academic exercise. As a result, computer vision researchers turned to... synthetic data. In this case, the main purpose of early synthetic data was to provide easier test problems for computer vision algorithms, back when many computer vision problems were widely believed to be amenable to an algorithm rather than a large trainable model. Synthetic images in the form of artificial drawings are all over early computer vision.

Consider, for example, the *line labeling* problem: given a picture with clearly depicted lines, mark which of the lines represent concave edges and which are convex. Figure 5.1 shows a sample picture with this kind of labeling for two embedded parallelepipeds, taken from the classic paper by David Huffman, *Impossible Objects as Nonsense Sentences* [379]. The figure on the left shows a fully labeled "X-ray" image with all invisible edges also clearly labeled, and the figure on the right shows only the visible part, but still fully labeled. The problem is, given a graph such as the one on the right but without labels, to produce the labeling, marking convex edges with "+" and concave edges with "−".

A different but very similar kind of problem was posed by Maxwell Clowes in his seminal paper *On Seeing Things* [162]. A sample illustration of this problem is shown in Fig. 5.2a. Here, Clowes distinguishes different types of corners based on how many edges form them and what kind of edges (concave or convex) they are. Basically, the problem is the same: find out the shape of a polyhedron defined as a collection of edges projected on a plane, i.e., shown as a line drawing.

Note how both Huffman and Clowes abstract out the problem of actually con-structing this graph (in modern computer vision, it would be known as *edge detection*) and start operations assuming that the line graph has been constructed. This is exactly why they need synthetic data: real-world edge detection in the 1970s would present too many problems. On the other hand, abstracting away real photographs misses

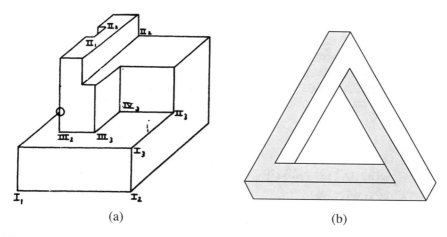

(a) (b)

Fig. 5.2 Line drawings and impossible objects: (a) an example from [162]; (b) the Penrose triangle (image taken from *Wikipedia*).

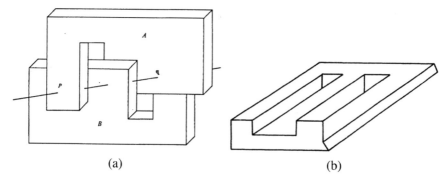

(a) (b)

Fig. 5.3 Hard-to-detect impossible objects: (a) an example from [379]; (b) an example from [162].

important information on the textures and lighting that could be used to deduce which edges are convex and which are concave. Modern single-image 3D scene reconstruction models try to make good use of such information.

Both researchers worked (as far as I know, independently) towards the goal of developing algorithms for this problem. They addressed it primarily as a topological problem, used the language of graph theory, and developed conditions under which a 2D line drawing can correspond to a feasible 3D object. It appears that they both were fascinated with impossible objects, 2D drawings that look realistic and satisfy a lot of reasonable necessary conditions for realistic scenes but at the same time still cannot be truly realized in 3D space. Think of M.C. Escher's drawings or the famous Penrose triangle (shown in Fig. 5.2b). Figure 5.3 shows two of the more complex polyhedral shapes, the one on the left (Fig. 5.3a) taken from Huffman's work and on the right (Fig. 5.3b), from Clowes. They are both impossible but it is not so easy to spot the impossibility by examining local properties of their graphs of lines.

Algorithms in both papers did not need training sets but they were tested entirely on artificially produced line drawings, one of the first and simplest examples of synthetic data in computer vision. While by modern standards this sounds like a very artificial example and the problem seems to belong more to the field of computational geometry than computer vision, it is very characteristic for early work on vision, relying more on the algorithmic approach than bottom-up feature extraction and learning.

5.2   Synthetic Data as a Testbed for Quantitative Comparisons

As we saw in the last section, in the early days of computer vision, many problems were tackled not by machine learning but by direct attempts to develop algorithms for specific tasks. Let us take as the running example for this section the problem of measuring *optical flow*. This is one of the basic low-level computer vision problems: estimate the distribution of apparent velocities of movement for the pixels of an image.

Optical flow can help find moving objects on an image or estimate (and probably subtract) the movement of the camera itself, but its most common use is for *stereo matching*: by computing optical flow between images taken at the same time from two cameras, one can construct a stereo view of the scene, do depth estimation, and generally go a long way towards true 3D scene understanding. Mathematically, optical flow is a field of vectors that estimate the movement of every pixel between a pair of images.

Algorithms for optical flow estimation fall into two categories: *dense* optical flow estimation aims to calculate the difference between entire images while *sparse* algorithms track only a small subset of given points. Figure 5.4 shows sample optical flow estimation on a Shibuya crossing video with two representative algorithms: Farnebäck's algorithm for dense optical flow estimation [232] and the Lucas–Kanade algorithm for the sparse version [561], as implemented in the *opencv* library [390].

Typical for low-level computer vision problems and early work on computer vision in general, both Farnebäck's and Lucas–Kanade methods are fixed algorithms, there are no parameters to be learned from data and no training sets to be labeled. This is a natural approach for this problem since manual labeling of optical flow is a gargantuan task, and there still do not exist large-scale datasets of real images with ground truth labeling for optical flow estimation; we will discuss contemporary synthetic datasets in Section 6.2, but none of them has any large-scale component composed of real scenes.

However, another question arises: while there is no need to have a training set for classical optical flow estimation algorithms, what about a *test set*? How can we find out which algorithm works better if there are no datasets with ground truth answers where we could compare them?

Fig. 5.4 Sample optical flow estimation, example code taken from [521]: (a) first original frame; (b) second original frame; (c) dense optical flow between the frames computed by Farnebäck's algorithm [232]; (d) sparse optical flow for several keypoints estimated by the Lucas–Kanade algorithm [561] over a longer video sequence.

Early works in computer vision often overlooked this question entirely. For example, in their seminal 1981 paper [561], Lucas and Kanade do not provide any quantitative estimates for how well their algorithm works, they just give several examples based on a single real-world stereo pair, shown in Figure 5.5. This is entirely typical: all early works on optical flow estimation develop and present new algorithms but do not provide any means for a principled comparison beyond testing them on a few real examples.

When enough algorithms had been accumulated; however, the need for an honest and representative comparison became really pressing. For an example of such a comparison, we jump forward to 1994, to the paper by Barron et al. called *Performance of Optical Flow Techniques* [42]. They present a taxonomy and a survey of optical flow estimation algorithms, including a wide variety of differential (such as Lucas–Kanade), region-based, and energy-based methods. But to get a fair experimental comparison, they needed a test set with known correct answers. And, of course, they turned to synthetic data for this.

Barron et al. concisely sum up the pluses and minuses of synthetic data that remain true up to this day: "The main advantages of synthetic inputs are that the 2D motion fields and scene properties can be controlled and tested in a methodical fashion. In particular, we have access to the true 2D motion field and can therefore quantify performance. Conversely, it must be remembered that such inputs are usually

Fig. 5.5 Real-world stereo pair used for testing in [561].

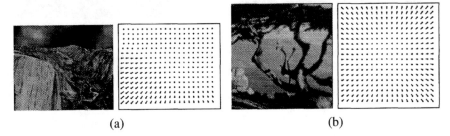

(a)   (b)

Fig. 5.6 Sample synthetic optical flow fields from [42]: (a) a translation sequence, the camera moves horizontally along a larger image; (b) a diverging sequence, the camera moves along its line of sight perpendicular to the image.

clean signals (involving no occlusion, specularity, shadowing, transparency, etc.) and therefore this measure of performance should be taken as an optimistic bound on the expected errors with real image sequences" [42].

Specifically, for simple examples in their comparison, they used sinusoidal inputs and a moving dark square over a bright background. For more complex tasks, they used synthetic camera motion. For example, if you take a still real photograph and start cropping out a moving window over this image, as a result, you basically get a camera moving perpendicular to its line of sight along the image, and the optical flow is fully known because you control the speed of this motion (see Fig. 5.6a). And if you take a window and start expanding it, you get a camera moving outwards, along its line of sight and perpendicular to the image plane (see Fig. 5.6b). These synthetic data generation techniques were simple and may seem naive by contemporary standards, but they did allow for a reasonable quantitative comparison: it turned out that even on such seemingly simple inputs, classical optical flow estimation algorithms gave very different answers, with widely varying accuracy.

This example illustrates a more general point: while classical algorithms for low-level computer vision problems almost never involved any learning of parameters, to make a fair comparison, one needs a test set with known correct answers, and this is exactly where synthetic data can step up. In problems where large-scale real datasets remain unavailable, such as optical flow estimation, synthetic test sets are widely used to this day.

5.3   ALVINN: A Self-Driving Neural Network in 1989

The idea of a self-driving vehicle is very old. A magic flying carpet appears in *The Book of the Thousand Nights and One Night*, where it is attributed to the court of King Solomon. Leonardo da Vinci left a sketch of a clockwork cart that would go through a preprogrammed route. For a comprehensive historical overview, I refer to an article by Marc Weber that shows plenty of drawings and photographs of early attempts at self-driving cars [922]. To borrow just one example, back in 1939, *General Motors* had a large pavillion at the New York World's Fair; in an aptly named *Futurama* ride (did Matt Groening take inspiration from *General Motors*?), they showed their vision of cities of the future. These cities included smart roads that would inform smart vehicles, and they would drive autonomously with no supervision from humans.

Reality still has not quite caught up with this vision, and the path to self-driving vehicles has been challenging. But successful experiments with self-driving cars started as early as the 1970s. In particular, in 1977, the Tsukuba Mechanical Engineering Laboratory in Japan developed a computer that would automatically steer a car based on visually distinguishable markers that city streets would have to be fitted with.

In the 1980s, a project funded by DARPA produced a car that could actually travel along real roads using computer vision. This was the famous ALV (Autonomous Land Vehicle) project, and its vision system was able to locate roads on images from cameras, solve the inverse geometry problem, and send the resulting three-dimensional road center points to the navigation system. The vision system, called VITS (for Vision Task Sequencer) and summarized in a paper by Turk et al. [879] (Figures 5.7 and 5.8 are taken from there), did not use machine learning in the

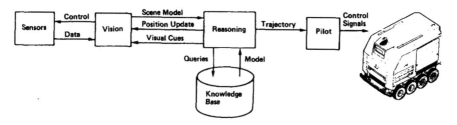

Fig. 5.7   General architecture of the ALV self-driving car control system [879].

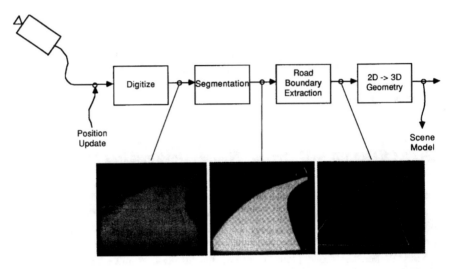

Fig. 5.8 Architecture of the VITS vision system designed for the ALV self-driving car [879].

contemporary meaning of the word. Similar to other computer vision systems of those days, as we discussed above, it relied on specially developed segmentation, road boundary extraction, and inverse geometry algorithms. Figure 5.7 shows the general architecture of ALV as shown in [879], and Figure 5.8 shows the pipeline of the vision system.

The paper reports that ALV "could travel a distance of 4.2 km at speeds up to 10 km/h, handle variations in road surface, and navigate a sharp, almost hairpin, curve". After obstacle avoidance had been added to the navigation system, *Alvin* (that was the nickname of the actual self-driving car produced by the ALV project) could steer clear of the obstacles on a road and speed up to 20 km/h on a clear road. These were really impressive results for the time, and they were achieved purely by custom-made algorithms: *Alvin* did not learn, did not need any data, and therefore had no use for synthetic datasets.

But this "no learning" stance changed very soon. In 1989, Dean A. Pomerleau published a paper on NIPS (this was the second NIPS, a very different kind of conference than what it has blossomed into now) called *ALVINN: An Autonomous Land Vehicle In a Neural Network* [680]. This was one of the first attempts to produce computer vision systems for self-driving cars based on machine learning. What's even more interesting for a contemporary reader is it was based on a neural network!

The basic architecture of ALVINN is shown in Fig. 5.9. The neural network had two inputs of different nature: 30×32 videos supplemented with 8×32 range finder data. The neural architecture was a classic two-layer feedforward network: input pixels go through a hidden layer and then on to the output units that try to recognize what is the curvature of the turn ahead, so that the vehicle could stay on the road. There is also a special road intensity feedback unit that simply tells whether the road was lighter or darker than the background on the previous input image.

Fig. 5.9 Architecture of the ALVINN neural network [680].

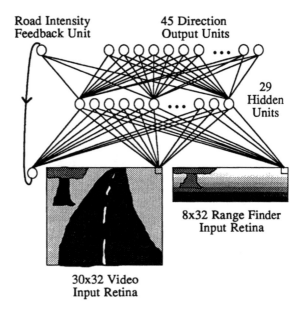

The next step was to train ALVINN. To learn, the neural network needed a training set with ground truth labels. Dean Pomerleau writes the following about training data collection: "Training on actual road images is logistically difficult, because in order to develop a general representation, the network must be presented with a large number of training exemplars depicting roads under a wide variety of conditions. Collection of such a data set would be difficult, and changes in parameters such as camera orientation would require collecting an entirely new set of road images..." This is exactly the kind of problem with labeling and dataset representativeness that we discussed in Chapter 1 for modern machine learning, only Pomerleau was talking about these problems in 1989, in the context of 30×32 pixel videos.

And what is his solution? Let us read on: "...To avoid these difficulties, we have developed a simulated road generator which creates road images to be used as training exemplars for the network". As soon as researchers needed to solve a real-world computer vision problem with a neural network, synthetic data appeared. This was one of the earliest examples of automatically generated synthetic datasets used to train a real computer vision system. Actually, the low resolution of the sensors in those days made it even easier to use synthetic data. Figure 5.10 shows two images used by ALVINN, a real one on the left and a simulated one on the right. It does not take too much effort to achieve photorealism in synthetic data when the real camera works like this.

The results were positive. On synthetic test sets, ALVINN could correctly (within two units) predict turn curvature approximately 90% of the time, on par with the best handcrafted algorithms of the time. In real-world testing, ALVINN could drive a car along a 400 meter path at a speed of half a meter per second. Interestingly, the limiting

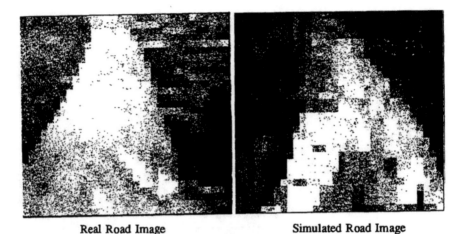

Real Road Image   **Simulated Road Image**

Fig. 5.10 Sample frames from real and synthetic video inputs used by ALVINN [680].

factor here was the speed of processing for the computer systems. Pomerleau reports that ALV's handcrafted algorithms could achieve a speed of 1 meter per second, but only on a much faster *Warp* computer, while ALVINN was working on the onboard Sun computer, and he expected "dramatic speedups" after switching to *Warp*.

This section probably features too many quotes from the ALVINN paper [680], but this work is such an excellent early snapshot of the positive features of synthetic data that I cannot but quote again from Pomerleau's conclusions: "Once a realistic artificial road generator was developed, backpropagation produced in half an hour a relatively successful road following system. It took many months of algorithm development and parameter tuning by the vision and autonomous navigation groups at CMU to reach a similar level of performance using traditional image processing and pattern recognition techniques."

Pomerleau says this more as a praise for machine learning over handcrafted algorithm development, but it also highlights the advantages of synthetic data over manual labeling: it would take much more than half an hour if Pomerleau's group had to label every single image from a real camera by hand. And to train for different driving conditions, they would need to collect a whole new dataset in the field, with a real car, rather than just tweak the parameters of the synthetic data generator.

By now, computer vision systems for self-driving cars are hard to imagine without synthetic datasets; we will discuss them in much more detail in Section 7.2. But modern approaches to synthetic data for autonomous vehicles often concentrate on not just creating a collection of labeled synthetic images and video but rather on providing an interactive simulation, a virtual city that could be used to train the driving algorithm as well. This brings us to our next topic.

5.4 Early Simulation Environments: Robots and the Critique of Simulation

Robotics is not quite as old as artificial intelligence: the challenge of building an actual physical entity that could operate in the real world was too big a hurdle for the first few years of AI. However, robotics was recognized as one of the major problems in AI very early on, and as soon as it became possible, people started to build real-world robots. One of the earliest attempts at a robot equipped with a vision system, the *Stanford Cart* built in the 1970s, is shown in Figure 5.11 (the pictures are taken from a later review paper by Hans Moravec [613]).

The Cart had an onboard TV system, and a computer program tried to drive the Cart through obstacle courses based on the images broadcast by this system. Based on several images taken from different camera positions (a kind of "super-stereo" vision), its vision algorithm tried to find interest points (features), detect obstacles, and avoid or go around them. It was extremely successful for such an early system, although performance was less than stellar: the Cart moved in short lurches, about 1 meter, every 10–15 minutes. Still, in these lurches, the Cart could successfully avoid real-life obstacles.

As we have already discussed, before the 1990s, computer vision was very seldom based on learning of any kind: researchers tried to devise algorithms, and data was only needed to test and compare them. This is also true for robotics: early robots such as the Cart had hardcoded algorithms for vision, pathfinding, and everything else. However, experiments with the Cart and similar robots taught researchers that it is far too costly and often entirely impossible to validate their ideas in the real world. Most researchers decided that they want to first test the algorithms in computer simulations and only then proceed to the real world. There are two main reasons for this:

- it is, of course, far easier, faster, and cheaper to test new algorithms in a simulated world than embed them into real robots and test in reality;
- simulations can abstract away many problems that a real-world robot has to face, such as unpredictable sources of noise in sensor readings, imperfections in the hardware, accumulating errors, and so on; it is important to be able to distinguish whether your algorithm does not work because it is a bad idea in general or because the sensor noise is too large in this particular case.

Fig. 5.11 The Stanford Cart [613].

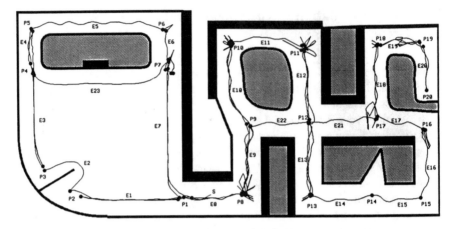

Fig. 5.12 Sample exploration results with systematic and 10% random error from [479].

Hence, robotics moved to the "simulate first, build second" principle which it abides by this day.

For example, Benjamin Kuipers and Yung-Tai Byun developed an approach to robot exploration and mapping based on a semantic hierarchy of spatial representations [479, 481]. This means that their robot is supposed to gradually work its way up from the *control level*, where it finds distinctive places and paths, through the *topological level*, where it creates a topological network description of the environment, and finally to the *geometric level*, where the topology is converted to a geometric map by incorporating local information about the distances and global metric relationships between the places. Figure 5.12 shows a sample trajectory of this navigation algorithm that explores a maze under noisy sensor readings.

The method itself was a seminal work, but it is not our subject right now and I will not go into any more details about it. I would like to note, however, their approach to implementing and testing the method: Kuipers and Byun programmed (in Common Lisp, by the way) a two-dimensional simulated environment called the *NX Robot Simulator*. The virtual robot in this environment has access to sixteen sonar-type distance sensors and a compass and moves by two tractor-type chains. Another interesting part of this simulation is that Kuipers and Byun took special care to implement error models for the sonars that actually reflect real-life errors.

Figure 5.13a shows a sample picture from their simulation; on the left, you can see a robot shooting sonar rays in 16 directions, and the histogram on the right shows sensor readings (with vertical lines) and true distances (with X and O markers). Note how the O markers represent a systematic error due to specular reflection, much more serious than the deviations of X markers that result from normal random error. The authors also made it into a software product with a GUI interface, which was much harder to do in the 1980s than it is now; a sample screenshot is shown in Fig. 5.13b.

The algorithms worked fine in a simulation, and the simulation was so realistic that it actually allowed to transfer the results to the real world. In a later work [480],

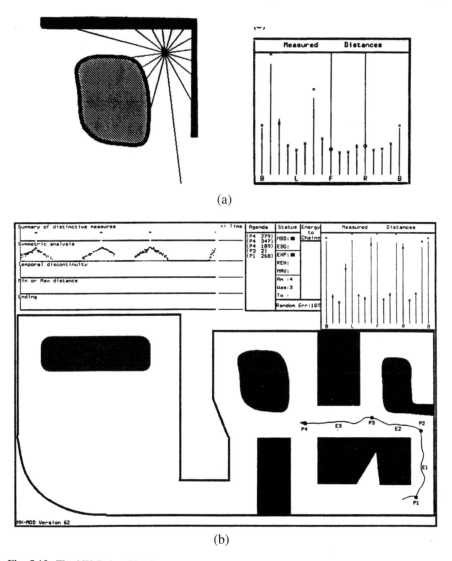

(a)

(b)

Fig. 5.13 The NX Robot Simulator: (a) sample image from the simulation [479]; (b) graphical interface of the simulation program [481].

Kuipers et al. report on their experiments with two physical mobile robots, Spot and Rover, that quite successfully implemented their algorithms on two different sensorimotor systems.

Note that synthetic data was much harder to do back in the 1980s than today. As another early example, I can refer to the work by Raczkowsky and Mittenbuehler [694] who discussed camera simulations in robotics. It is mostly devoted

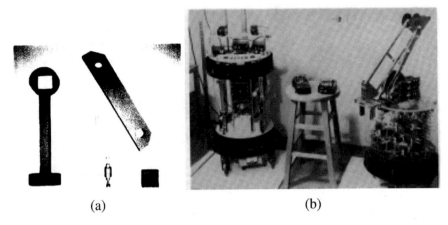

(a)   (b)

Fig. 5.14 Early robotics: (a) synthetic data from [694]; (b) Allen, Herbert, Tom, and Jerry from the MIT AI Lab [93].

to the construction of a 3D scene, and back in 1989, you had to do it all yourself, so the paper covers:

- surfaces, contours, and vertices that define a 3D object;
- optical surface properties including Fresnel reflection, diffuse reflectance, flux density, and more;
- light source models complete with radiance, wavelengths, and so on;
- and finally a camera model that simulates a lens system and electronic hardware of the robot's camera.

In those days, only after working through all that could you produce such marvelous photorealistic images as the 200×200 synthetic photo of some kind of workpieces shown in Fig. 5.14a.

Fortunately, by now, most of these problems have already been worked out in modern 3D modeling software or gaming engines. However, camera models are still relevant in modern synthetic data applications. For instance, an important use case for, e.g., a smartphone manufacturer might be to retrain or transfer its computer vision models when the camera changes, and you need a good model for both old and new cameras in order to capture this change and perform this transition.

Despite these successes, not everybody believed in computer simulations for robotics. In the same book as [480], another chapter by Rodney Brooks and Maja Mataric, aptly titled *Real Robots, Real Learning Problems* [93], had an entire section devoted to warning researchers in robotics from relying on simulations too much. Brooks and Mataric put it as follows: "Simulations are doomed to succeed. Even despite best intentions there is a temptation to fix problems by tweaking the details of a simulation rather than the control program or the learning algorithm. Another common pitfall is the use of global information that could not possibly be available to a real robot. Further, it is often difficult to separate the issues which are intrinsic to the simulation environment from those which are based in the actual learning problem."

Basically, they warned that computer vision had not been solved yet, and while a simulation might provide the robot with information such as "there is food ahead", in reality, such high-level information would never be available. This, of course, remains true to this day, and modern robotic vision systems make use of all modern advances in object detection, segmentation, and other high-level computer vision tasks (where synthetic data also helps a lot, by the way, as we see in the rest of this book).

This sounds like some very basic points that are undoubtedly true, and they sound more as a part of the problem setting than true criticism. However, Brooks also presents a much more interesting criticism which is not so much against synthetic data and simulations as against the entire computer vision program for robotics; while this is an aside for this book, this is an interesting aside, and I want to elaborate on it.

Brooks' ideas were summarized in two of his seminal papers, *Intelligence Without Representation* [92] and *Intelligence Without Reason* [91]. In the former, Brooks argues that abstract representation of the real world, which was a key feature of contemporary AI solutions, is a dangerous weapon that can lead to self-delusion. He says that real-life intelligence has not evolved as a machine for solving well-defined abstract problems such as chess playing or theorem proving: intelligence in animals and humans is inseparable from perception and mobility. This was mostly a criticism of early approaches to AI that indeed concentrated on abstractions such as block worlds or knowledge engineering.

In *Intelligence Without Reason*, Brooks goes further to argue that abstraction and knowledge are basically unavailable to systems that have to operate in the real world, that is, to robotic systems. For example, he mentions vision algorithms based on line drawings that we discussed in Section 5.1 and admits that although some early successes in line detection had dated back to the 1960s, even in the early 1990s, we did not have a reliable way to convert real-life images to line drawings. "Try it! You'll be amazed at how bad it is," Brooks comments, and this comment is not very far from the truth even today.

Brooks presents four key ideas that he believes to be crucial for AI:

- *situatedness*, i.e., placing AI agents in the real world "with continuity, surprises, or ongoing history"; Brooks agrees that such a world would be hard to accurately reflect in a simulation and concludes that "the world is its own best model";
- *embodiment*, i.e., physical grounding of a robot in the real world; this is important precisely in order to avoid self-delusional pitfalls that abstract simulations may lead to; apart from new problems, the embodiment may also present solutions to abstract problems by grounding the reasoning and conclusions in the real world;
- *intelligence*, which Brooks proposes to model after animals simpler than humans, concentrating at first on perception and mobility and only then moving to abstract problem solving, just like we humans did in the process of evolution;
- *emergence*, where Brooks makes the distinction between traditional AI systems whose components are functional (e.g., a vision system, a pathfinding system, and so on) and behaviour-based systems where each functional unit is responsible for end-to-end processing needed to form a given behaviour (e.g., obstacle avoidance, gaze control, etc.).

As for simulations, Brooks concludes that they are examples of precisely the kind of abstractions that may lead to overly optimistic interpretations of results and argues for complete integrated intelligent mobile robots.

Interestingly, this resonates with the words of Hans Moravec that he wrote in his 1990 paper about the *Stanford Cart* robot that I referenced above [613]: "My conclusion is that solving the day to day problems of developing a mobile organism steers one in the direction of general intelligence, while working on the problems of a fixed entity is more likely to result in very specialized solutions."

Brooks put his ideas into practice, leading a long-term effort to create mobile autonomous robots in the MIT AI lab. The robots developed there—Allen, Herbert, Tom, and Jerry, shown in Fig. 5.14b—were designed to interact with the world rather than plan and carry out plans. This work soon ran into technological obstacles: the hardware was just not up to the task in the late 1980s. But Brooks' ideas live on: *Intelligence Without Representation* has more than 2000 citations and is still being cited to this day, in fields ranging from robotics to cognitive sciences, nanotechnology, and even law (AI-related legislation is a very interesting topic, by the way).

So are simulations useful for robotics? Of course, and increasingly so! While I believe that there is a lot of truth to the criticism shown above, in my opinion in most applications, it boils down to the following: when your robot works in a simulation, it does not yet mean that it will work in real life. This is, of course, true.

On the other hand, if your robot does not work even in a simulation, it is definitely too early to start building real systems. Moreover, modern developments in robotics such as the success of reinforcement learning seem to have a strange relationship with Brooks' ideas. On the one hand, this is definitely a step in the direction of creating end-to-end systems that are behaviour-oriented rather than composed of clear-cut predesigned functional units. On the other hand, in the modern state of reinforcement learning, it is entirely hopeless to suggest that complex systems such as robots could be trained in real life: they absolutely need simulations because they require millions, if not billions, of training episodes. We will see examples of such systems in Chapter 7.

But while researchers are training robots in simulation up to this day, accurate 3D modeling is not quite enough to create a simulation that would be useful for robotics. Let us conclude this chapter with another early example in robotics that illustrates this point.

5.5 Case Study: MOBOT and The Problems of Simulation

In the previous section, we have talked about robotic simulations in general: what they are and why they are inevitable for robotics based on machine learning. We even touched upon some of the more philosophical implications of simulations in robotics, discussing early concerns on whether simulations are indeed useful or may become a dead end for the field. Here, let us consider in more detail the example of MOBOT, an important project about a robot navigating indoor environments.

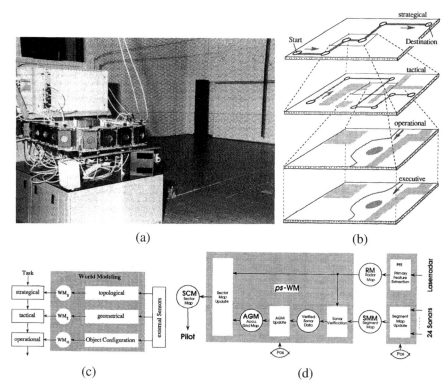

Fig. 5.15 Illustrations for the MOBOT project [97, 219, 356, 414, 872, 1030]: (a) the robot itself; (b) hierarchy of MOBOT's abstraction layers; (c) world modeling in MOBOT; (d) the ps-WM architecture.

MOBOT (**Mo**bile **Ro**bot, and that is one of the most straightforward acronyms you will see in this section) was developed in the first half of the 1990s by a group led by Ewald von Puttkamer in the University of Kaiserslautern [97, 219, 356, 414, 872, 1030].

The MOBOT (specifically, the MOBOT-IV version) is shown in Fig. 5.15a. Note the black boxes that form a 360 degree belt around the robot: these are sonar sensors, and we will come back to them later. The main problem that MOBOT developers were solving was navigation in the environment, that is, constructing the map of the environment and understanding how to go where the robot needed to go. There was a nice hierarchy of abstraction layers that gradually grounded the decisions down to the most minute details, illustrated in Fig. 5.15b. And there were three different layers of the world modeling, too; the MOBOT viewed the world differently depending on the level of abstraction, as shown in Fig. 5.15c.

In essence, this came down to the same old problem: figure out the sensor readings and map them to all these nice abstract layers where the robot could run pathfinding algorithms such as the evergreen A* [323]. Apart from the sonars, the robot also had a laser radar, and the overall scheme of the ps-WM (**p**ilot-specific **W**orld **M**odeling; I told you the acronyms would only get weirder) project [414], shown in Fig. 5.15d, looks quite involved. Note that there are several different kinds of maps that need updating.

But in this book, we are especially interested in the MOBOT project because it contained one of the earliest examples of a full-scale 3D simulation environment for robotics. It is the *3d7 Simulation Environment* [872]; the obscure name does not refer to a nonexistent seven-sided die but is again an acronym for "**3D S**imulation **E**nvironment". The 3d7 environment was developed for MOBOT-IV, an autonomous mobile robot that was supposed to navigate indoor environments; it had general-purpose ambitions rather than simply being, say, a robot vacuum cleaner, because its scene understanding was inherently three-dimensional, while for many specific tasks, a 2D floor map would be quite sufficient.

The overall structure of 3d7 is shown in Figure 5.16a. It is quite straightforward: the software simulates a 3D environment, robot sensors, and robot locomotion, which lets the developers to model various situations, choose the best algorithms for sensory data processing and action control, and so on, just like we have discussed above.

The main point I wanted to make with this example is this: creating realistic simulations is very hard. Usually, when we talk about synthetic data, we are concentrating on computer vision, and we are emphasizing the work it takes to create a realistic 3D environment. It is indeed a lot, but just creating a realistic 3D scene is not the end of the story for robotics. For 3D environment modeling, 3d7 contained an environment editor that lets you place primitive 3D objects such as cubes, spheres, or cylinders and also more complex objects such as chairs or tables. It produced a scene complete with the labels of semantic objects and geometric primitives that make up these objects, like the one shown in Fig. 5.16b.

But then the fun part began. MOBOT-IV contained two optical devices: a laser radar sensor and a brand new addition compared to MOBOT-III, an infrared range scanning device. This means that in order to make a useful simulation, the 3d7 environment had to simulate these two devices.

It turns out that both these simulations represent interesting projects. LARS, the Laser Radar Simulator, was designed to model the real laser radar sensor of MOBOT-III and the new infrared range scanner of MOBOT-IV. It produced a simulation illustrated in Figure 5.16c. As for sonar range sensors, the corresponding USS2D simulator (**U**ltrasonic **S**ensor **S**imulation **2D**) was even more interesting. It was based on the work [478] that takes about thirty pages of in-depth acoustic modeling; I will not go into the details but trust me, there are a lot of details there. The end result was a set of sonar range readings corresponding to the reflections from nearest walls; a sample set of readings is shown in Fig. 5.16d.

But wait, that's not all! After all of this is done, you only have simulated the sensor readings! To actually test your algorithm, you also need to model the actions your robot can take and how the environment will respond to these actions. In the case of 3d7, this means a separate locomotion simulation model for robot movement called SKy (**S**imulation of **K**inematics and **D**ynamics of wheeled mobile robots),

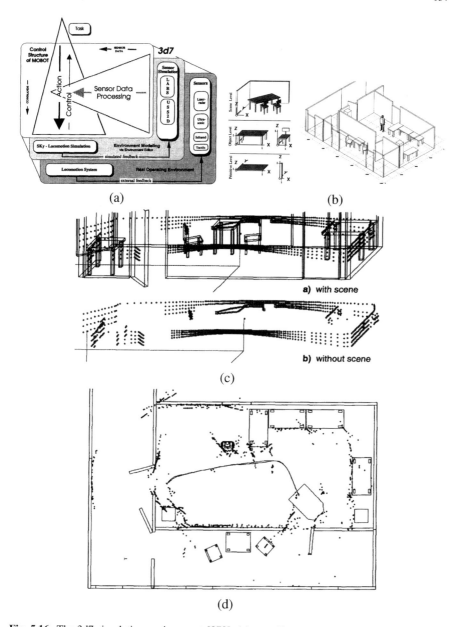

Fig. 5.16 The 3d7 simulation environment [872]: (a) overall structure; (b) sample 3D scene; (c) LARS laser simulation; (d) USS2D ultrasonic sensor simulation.

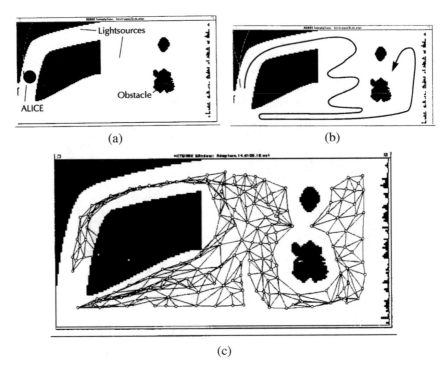

Fig. 5.17 Learning in the MOBOT project [1030]: (a) sample test environment; (b) test trajectory; (c) topogical representation produced by SOMs.

which also merited its own paper but which we definitely will not go into. All of these pieces can be found in Fig. 5.16a.

The MOBOT project did not contain many machine learning models, it was mostly operated by fixed algorithms designed to work with sensor readings as shown above. Even the 3d7 simulation environment was mostly designed to help test various data processing algorithms (world modeling) and control algorithms (e.g., path planning or collision avoidance), a synthetic data application similar to the early computer vision we talked about before.

But at some point, MOBOT designers did try out some machine learning. The work [1030] was titled *Realtime-learning on an Autonomous Mobile Robot with Neural Networks*, which sounds quite relevant to this book. These are not, however, the neural networks that we are used to: in fact, Zimmer and von Puttkamer used *self-organizing maps* (SOM), sometimes called Kohonen maps in honor of their creator [465], to cluster sensor readings.

The problem setting is as follows: as the robot moves around its surroundings, it collects sensor information. The basic problem is to construct a topological map of the floor with all the obstacles. To do that, the robot needs to be able to recognize places where it has already been, i.e., to cluster the entire set of sensor readings into "places" that can serve as nodes for the topological representation. Due to the imprecise nature of robotic movements, we cannot rely on the kinematic model of where we tried to go: small errors tend to accumulate. Instead, the authors propose to cluster sensor readings: if the current vector of readings is similar to what we have already seen before, we are probably in approximately the same place.

And again we see the exact same effect: while Zimmer and von Puttkamer do present experiments with a real robot, most experiments for SOM training were done with synthetic data. This data was collected in a test environment that looked as shown in Fig. 5.17a with test trajectories such as the one shown in Fig. 5.17b. And indeed, when the virtual robot had covered this trajectory, SOMs clustered nicely and allowed to construct a graph, a topological representation of the territory, like the one in Fig. 5.17c.

5.6 Conclusion

In this chapter, we have seen how synthetic data became a staple of machine learning starting from its very early years, mostly concentrated in computer vision, of course. Synthetic data was the only way to go for early computer vision algorithms that could not really handle real data. It has been invaluable in creating test sets for all sorts of computer vision algorithms in problems that would be impossibly hard to label manually (such as optical flow estimation).

We have also seen the main components of a robotic simulation system, discussed the many different aspects that need to be simulated, and seen how this all came together in one early robotic project, the MOBOT from the University of Kaiserslautern. Simulations were important for robotics in the early days, and by now, they are becoming indispensable.

But these are only the early steps. Starting from the next chapter, we will be getting close to the current state of the art of using synthetic data to train deep learning models. Our next topic is to see how synthetic data can help models that solve basic computer vision problems such as image classification, object detection, segmentation, or the same optical flow estimation that we have touched upon in this chapter. We will also review the most important synthetic datasets that have been used as testbeds for this research.

Chapter 6
Synthetic Data for Basic Computer Vision Problems

It is time to put the pedal to the metal: starting from this chapter, we will discuss the current state of the art in various aspects of synthetic data. This chapter is devoted to basic computer vision problems: we begin with low-level problems such as optical flow estimation and stereo image matching, proceed to datasets of basic objects that can be used to train computer vision models, discuss in detail the case study of synthetic data for object detection, and finish with several different use cases such as synthetic datasets of humans, OCR, and visual reasoning.

6.1 Introduction

In this chapter, we present an overview of several directions for using synthetic data in computer vision, surveying both popular synthetic datasets that have been widely used in recent studies and the studies themselves. We organize this chapter by classifying datasets and models with respect to use cases, from generic object detection and segmentation problems to specific domains such as face recognition. All of these domains benefit highly from pixel-perfect labeling available by default in synthetic data, both in the form of classical computer vision labeling—bounding boxes for objects and segmentation masks—and labeling types that would be very hard or impossible to do by hand: depth estimation, stereo image matching, 3D labeling in voxel space, and others.

In Section 6.2, we begin with low-level computer vision problems such as optical flow or stereo disparity estimation. Next we proceed to basic high-level computer vision problems, including recognition of basic objects (Section 6.3), a case study of improving object detection with synthetic data (Section 6.4), and solving other high-level computer vision problems (Section 6.5). We will also discuss several more specialized directions: human-related computer vision problems such as face recognition or crowd counting in Section 6.6 and more narrow vision-related problems such as character and text recognition and visual reasoning in Section 6.7. I refer to Table 6.1 for a brief overview of major synthetic datasets considered in this chapter.

© The Author(s), under exclusive license to Springer Nature Switzerland AG 2021
S. I. Nikolenko, *Synthetic Data for Deep Learning*, Springer Optimization
and Its Applications 174, https://doi.org/10.1007/978-3-030-75178-4_6

Table 6.1   An overview of synthetic datasets discussed in this chapter

Name	Year	Ref	Size / comments
Low-level computer vision			
Tsukuba Stereo	2012	[671]	1800 high-res stereo image pairs
MPI-Sintel	2012	[105]	Optical flow from an animated movie
Middlebury 2014	2014	[764]	33 high-res stereo datasets
Flying Chairs	2015	[202]	22K frame pairs with ground truth flow
Flying Chairs 3D	2015	[586]	22K stereo frames
Monkaa	2015	[586]	8591 stereo frames
Driving	2015	[586]	4392 stereo frames
UnrealStereo	2016	[1008]	Data generation software
Underwater	2018	[641]	Underwater synthetic stereo pairs generator
Datasets of basic objects			
YCB	2015	[690]	77 objects in 5 categories
ShapeNet	2015	[122]	>3M models, 3135 categories, rich annotations
ShapeNetCore	2017	[973]	51K manually verified models, 55 categories
UnrealCV	2017	[690]	Plugin for UE4 to generate synthetic data
VANDAL	2017	[115]	4.1M depth images, 9K objects, 319 categories
SceneNet	2015	[320]	Automated indoor synthetic data generator
SceneNet RGB-D	2017	[588]	5M RGB-D images from 16K 3D trajectories
DepthSynth	2017	[677]	Realistic simulation of depth sensors
PartNet	2018	[608]	26671 models, 573535 annotated part instances
Falling Things	2018	[868]	61.5K images of YCB objects in virtual envs
ADORESet	2019	[47]	Hybrid dataset for object recognition testing
Datasets of synthetic people			
ViHASi	2008	[699]	Silhouette-based action recognition
Agoraset	2014	[171]	Crowd scenes generator
LCrowdV	2016	[145]	1M videos, 20M frames with crowds
PHAV	2017	[815]	40K action recognition videos, 35 categories
SURREAL	2017	[889]	145 subjects, 2.6K sequences, 6.5M frames
SyRI	2018	[34]	Virtual humans in UE4 with realistic lighting
GCC	2019	[907]	15K images with 7.6M subjects
CLOTH3D	2019	[67]	Virtual humans in various clothing

6.2 Low-Level Computer Vision

Low-level computer vision problems include, in particular, *optical flow estimation*, i.e., estimating the distribution of apparent velocities of movement along the image, *stereo image matching*, i.e., finding the correspondence between the points of two images of the same scene from different viewpoints, *background subtraction*, and so on. Algorithms for solving these problems can serve as the foundation for computer vision systems; for example, optical flow is important for motion estimation and video compression. Low-level problems can usually be approached with methods that do not require modern large-scale datasets or much learning at all, e.g., classical differential methods for optical flow estimation. However, at the same time, ground truth datasets are very hard to label manually, and hardware sensors that would provide direct measurements of optical flow or stereo image correspondence are difficult to construct (e.g., commodity optical flow sensors simply run the same estimation algorithms).

All of these reasons make low-level computer vision one of the oldest problems where synthetic data was successfully used, originally mostly for evaluation (see also Section 5.2); we illustrate the datasets considered in this section in Fig. 6.1. Works as far back as late 1980s [529] and early 1990s [43] presented and used synthetic datasets to evaluate different optical flow estimation algorithms. In 1999, Freeman et al. [240] presented a synthetically generated world of images, with labeling derived from the corresponding 3D scenes, designed to train and evaluate low-level computer vision algorithms.

A modern dataset for low-level vision is *Middlebury* presented by Baker et al. [36]; in addition to ground truth real-life measurements taken with specially constructed computer-controlled lighting, they provide realistic synthetic imagery as part of the dataset and include it in a large-scale evaluation of optical flow and stereo correspondence estimation algorithms that they undertake. The *Middlebury* dataset played an important role in the development of low-level computer vision algorithms [765, 775], but its main emphasis was still on real imagery, as evidenced by its next version, *Middlebury 2014* [764].

Peris et al. [671] presented the *Tsukuba CG Stereo Dataset* with synthetic data and ground truth disparity maps and showed improvements in disparity classification quality. Butler et al. [105] presented a synthetic optical flow dataset *MPI-Sintel* derived from the short animated movie *Sintel*[1] produced as part of the *Durian Open Movie Project*. The main characteristic feature of *MPI-Sintel* is that it contains the same scenes with different render settings, varying quality and complexity; this approach can provide a deeper understanding of where various optical flow algorithms break down. This is an interesting idea that has not yet found its way into synthetic data for deep learning-based computer vision but might be worthwhile to investigate.

An interesting study by Meister and Kondermann [596] shows that while real and synthetic data (synthetic being very high-quality, produced with ray tracing) yield

---

[1]http://www.sintel.org/

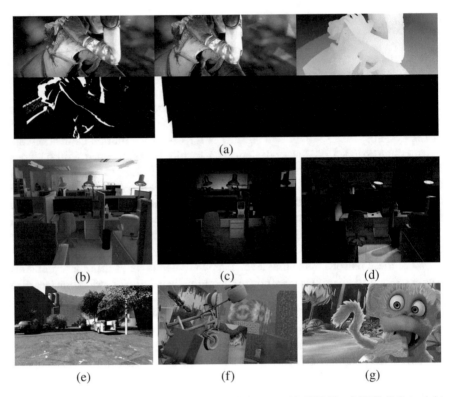

Fig. 6.1 Sample images from synthetic low-level datasets: (a) *MPI-Sintel* [105] (left to right: left view, right view, disparities; bottom row shows occluded and out-of-frame pixels); (b–d) *Tsukuba CG Stereo Dataset* [671] with different illumination conditions: (b) daylight, (c) flashlight, (d) lamps; (e) *Driving* [586]; (f) *FlyingThings3D* [586]; (g) *Monkaa* [586].

approximately the same results for optical flow detection in terms of mean endpoint error, the spatial distributions of errors are different, so synthetic data in this case may supplement real data in unexpected ways.

As the field moved from classical unsupervised approaches to deep learning, state-of-the-art models began to require large datasets that could not be produced in real life, and after the transition to deep learning synthetic datasets started to dominate. Dosovitsky et al. [202] presented a large-scale synthetic dataset called *Flying Chairs* from a public database of 3D chair models, adding them on top of real backgrounds to train a CNN-based optical flow estimation model. Mayer et al. [586] extended this work from optical flow to disparity and scene flow estimation, presenting three synthetic datasets produced in *Blender* (similar to *Sintel*):

- *FlyingThings3D* with everyday objects flying along randomized trajectories;
- *Monkaa*, made from its namesake animated short film, with soft nonrigid motion and complex details such as fur;
- *Driving* with naturalistic dynamic outdoor scenes from the viewpoint of a driving car (for more outdoor datasets see Section 7.2).

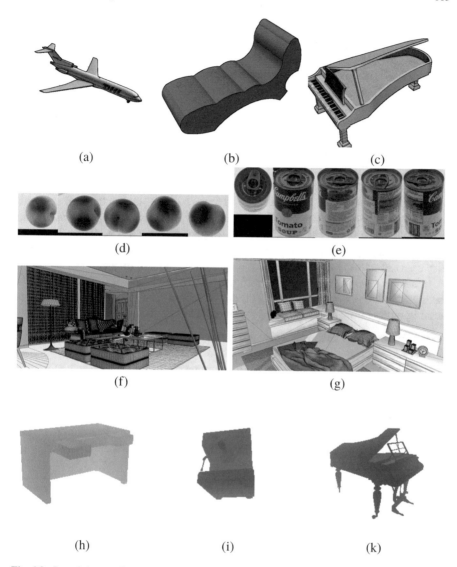

Fig. 6.2 Sample images from synthetic datasets of basic objects: (a–c) shapes from *ShapeNet* [122]: (a) airplane, (b) chair, (c) grand piano); (d-e) from YCB Object and Model set [110]: (d) peach, (e) tomato soup can; (f-g) 3D scenes from *SceneNet* [319, 320]: (f) living room, (g) bedroom; (h–k) depth images from VANDAL [115]: (h) desk, (i) coffee maker, (k) grand piano.

The *Flying Chairs* dataset was also later extended with additional modalities to *ChairsSDHom* [384] with optical flow ground truth and *Flying Chairs 2* [385] with occlusion weights and motion boundaries.

The *UnrealStereo* dataset by Zhang et al. [1008] is a data generation framework for stereo scene analysis based on *Unreal Engine 4*, designed to evaluate the robustness of stereo vision algorithms to changes in material and other scene parameters. Many datasets that we describe below for high-level problems, such as SceneNet RGB-D [588] or SYNTHIA [731], also contain labeling for optical flow and have been used to train the corresponding models.

Olson et al. [641] consider an unusual special case for this problem: *underwater* disparity estimation. Their work is also interesting in the way how they produce synthetic data: Olson et al. project real underwater images on randomized synthetic surfaces produced in Blender, and then use rendering tools developed to mimic the underwater sensors and characteristic underwater effects such as fast light decay and backscattering. They produce synthetic stereo image pairs and use the dataset to train disparity estimation models, with successful transfer to real images.

In a recent work, Mayer et al. [585] provide an overview of different synthetic datasets for low-level computer vision and compare them from the standpoint of training optical flow models. They come to interesting conclusions:

- first, for low-level vision synthetic data does not have to be realistic, *Flying Chairs* works just fine;
- second, it is best to combine different synthetic datasets and train in a variety of situations and domains; this ties into the domain randomization idea that we will discuss in Section 9.1;
- third, while realism itself is not needed, it does help to simulate the flaws of a specific real camera; Mayer et al. show that simulating, e.g., lens distortion and blur or Bayer interpolation artifacts in synthetic data improves the results on a real test set afterwards.

The question of realism remains open for synthetic data, and we will touch upon it many times in this book. While it does seem plausible that for low-level problems such as optical flow estimation "low-level realism" (simulating camera idiosyncrasies) is much more important than high-level scene realism, the answer may be different for other problems.

6.3   Datasets of Basic Objects

Basic high-level computer vision problems, such as object detection or segmentation, fully enjoy the benefits of perfect labeling provided by synthetic data, and there is plenty of effort devoted to making synthetic data work for these problems. Since making synthetic data requires the development of 3D models, datasets usually also feature 3D-related labeling such as the depth map, labeled 3D parts of a shape, volumetric 3D data, and so on. There are many applications of these problems, including object detection for everyday objects and retail items (where a high number of classes and frequently appearing new classes make using real data impractical), counting and detection of small objects, basically all applications of semantic and instance segmentation (where manual labeling is especially hard to obtain), and more.

Many works apply synthetic data to recognizing everyday objects such as retail items, food, or furniture, and most of them draw upon the same database for 3D models. Developed by Chang et al. [122], *ShapeNet*[2] indexes more than three million models, with 220,000 of them classified into 3,135 categories that match *WordNet* synsets. Apart from class labels, *ShapeNet* also includes geometric, functional, and physical annotations, including planes of symmetry, part hierarchies, weight and materials, and more; we show samples from ShapeNet and other datasets of basic objects in Fig. 6.2. Researchers often use the clean and manually verified *ShapeNet-Core* subset that covers 55 common object categories with about 51,000 unique 3D models [973].

ShapeNet has become the basis for further efforts devoted to improving labelings. In particular, region annotation (e.g., breaking an airplane into wings, body, and tail) is a manual process even in a synthetic dataset, while shape segmentation models increasingly rely on synthetic data [971]; this also relates to the 3D mesh segmentation problem [787]. Based on *ShapeNet*, Yi et al. [972] developed a framework for scalable region annotation in 3D models based on active learning, and Chen et al. [137] released a benchmark dataset for 3D mesh segmentation.

A recent important effort related to *ShapeNet* is the release of *PartNet* [608], a large-scale dataset of 3D objects annotated with fine-grained, instance-level, and hierarchical 3D part information; it contains 573,585 part instances across 26,671 3D models from 24 object categories. *PartNet* was mostly intended to serve as a benchmark for 3D object and scene understanding, but the corresponding 3D models will no doubt be widely used to generate synthetic data.

One common approach to generating synthetic data is to reuse the work of 3D artists that went into creating the virtual environments of video games. For example, Richter et al. [724, 725] captured datasets from the *Grand Theft Auto V* video game (see also Section 7.2). They concentrated on semantic segmentation; note that getting pixel-wise labels for segmentation still required manual labor, but the authors claim that by capturing the communication between the game and the graphics hardware, they have been able to cut the labeling costs (annotation time) by orders of magnitude. Once the annotator has worked through the first frame, the same combinations of meshes, textures, and shaders reused on subsequent frames can be automatically recognized and labeled, and the annotators are only asked to label new combinations. In essence, the game engine provides perfect superpixels that are persistent across frames (we discussed superpixels in Section 3.3).

As *Grand Theft Auto V* and other games became popular for collecting synthetic datasets (see also Section 7.2), more specialized solutions began to appear. One such solution is *UnrealCV* developed by Qiu et al. [689, 690], an open-source plugin for the popular game engine *Unreal Engine 4* that provides commands that allow to get

---

[2]https://www.shapenet.org/

and set camera location and field of view, get the set of objects in a scene together with their positions, set lighting parameters, modify object properties such as material, and capture from the engine the image and depth ground truth for the current camera and lighting parameters. This allows to create synthetic image datasets from realistic virtual worlds.

Robotics has motivated the appearance of synthetic datasets with objects that might be subject for manipulation, usually with fairly accurate models of their physical properties. Computer vision objectives in these datasets usually relate to robotic perception and include segmentation, depth estimation, object pose estimation, and object tracking. In particular, Choi et al. [148] present a dataset of 3D models of household objects for their tracking filter, while Hodan et al. [351] provide a real dataset of textureless objects supplemented with 3D models of these objects that provide the 6D ground truth poses. Lee et al. [502] test existing tracking methods with simulated video sequences with occlusion effects. Papon and Schoeler [655] consider the problem of object pose and depth estimation in indoor scenes. They have developed a synthetic data generator and trained on 7000 randomly generated scenes with ≈60K instances of 2842 pose-aligned models from the *ModelNet10* dataset [944], showing excellent results in transfer to real test data.

The *Yale-CMU-Berkeley* (YCB) Object and Model set presented by Calli et al. [110] contains a set of 3D models of objects commonly used for robotic grasping together with a database of real RGB-D scans and physical properties of the objects, which makes it possible to use them in simulations.

Once we have these basic objects, the next step is to put them into context; it is no coincidence that one of the main object detection datasets is called *Common Objects in Context*. For instance, in real-world object detection and segmentation datasets the problem becomes much harder if the same picture contains objects on different scales (small and large in terms of the proportion of the picture) and if the objects are embedded into a rich context. The backgrounds become especially hard if they contain other objects that do not need to be recognized but that might confuse the model; they are usually called *distractors*. Naturally, if you have a synthetic chair centered on a white background, like in the images above, the corresponding object detection or segmentation problem will be easy, and a network trained on this kind of dataset will not get you very far in real object detection.

So what can we do about it? On the surface, it looks like we might have to bite the bullet and start developing complex backgrounds that capture realistic 3D scenes. Researchers actually do it in, say, creating simulations and datasets for training self-driving cars that we will consider in Section 7.2, and it is an entirely reasonable investment of time and effort. But in high-level object detection problems such as object detection or segmentation, sometimes even much simpler things can work well. Some of the hardest cases in these problems stem from complex interactions between objects: partial occlusions, different scales caused by different distances to the camera, and so on. So why don't we use a more or less generic scene and just put the objects there at random, striving to achieve a cluttered and complicated scene but with little regard to physical plausibility?

This plays into the narrative of *domain randomization*, a general term that means randomizing the parameters of synthetic scenes in order to capture as wide a variety of synthetic data as possible. The idea is that if the network learns to do its job on an extremely wide and varied distribution of data, it will hopefully do the job well on real data as well, even if individual samples of this synthetic data are very far from realistic. Domain randomization is instrumental in synthetic data research, and we will discuss it in more detail in Section 9.1, but even in its simplest forms it is useful for producing computer vision datasets.

When you put this idea into practice, you get datasets like *Flying Chairs* that we have considered above and *Falling Things*. The *Falling Things* (FAT) dataset by NVIDIA researchers Tremblay et al. [868] contains about 61500 images of 21 household objects taken from the YCB dataset and placed into virtual environments under a wide variety of lighting conditions, with 3D poses, pixel-perfect segmentation, depth images, and 2D/3D bounding box coordinates for each object. Virtual environments are realistic enough, but the scenes are purely random, with a lot of

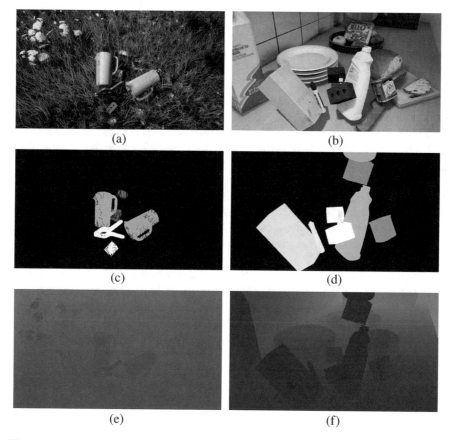

Fig. 6.3 Sample images from the *Falling Things* dataset [868]: (a–b) RGB images, (c–d) ground truth segmentation maps; (e–f) depth maps.

occlusions and objects just flying in the air in all directions; we show sample images from FAT in Fig. 6.3.

The *Falling Things* dataset also demonstrates one more trend in synthetic data: it can be downloaded as a standalone dataset, but even mere 21 objects result in a dataset which is 42GB in size. This is a common theme: as synthetic datasets grow in scale, it becomes less and less practical to render them completely in advance and transfer the pictures over a network. Procedural generation is increasingly used to avoid this and render images only on a per-need basis; we will return to this idea in Section 12.1.

Recent works begin to use synthetic datasets of everyday objects in more complex ways, in particular by placing them in real surroundings. Abu Alhaija et al. [5] and Georgakis et al. [270] propose procedures to augment real backgrounds with synthetic objects (see also Section 9.3 where we will discuss placing real objects on real backgrounds). In [5], the backgrounds come from the KITTI dataset of outdoor scenes and the objects are synthetic models of cars, while in [270] the authors place synthetic objects into indoor scenes with an eye towards home service robots.

Synthetic objects have been used on real backgrounds many times before, but the main distinguishing feature of [5] and [270] is that they are able to paste synthetic objects on real surfaces in a way consistent with the rest of the background scene. Abu Alhaija et al. developed a pipeline for automated analysis that recognized road surfaces on 360° panoramic images, but at the same time they conclude that the best (with respect to the quality of the resulting segmentation model) way to insert the cars was to do it manually, and almost all their experiments used manual car placement.

These experiments showed that state of the art models for instance segmentation and object detection yield better results on real validation tests when trained on scenes augmented with synthetic cars. Georgakis et al. use the algorithm from [852] to extract supporting surfaces from an image and place synthetic objects on such surfaces with proper scale; they show significant improvements by training on hybrid real+synthetic datasets. One of the latest and currently most advanced pipelines in this direction for autonomous driving is AADS [513] that we will discuss in Section 7.2.

By this time, the reader might wonder just how much effort has to go into creating a synthetic dataset of one's own. If you need a truly large-scale dataset with photorealistic quality of rendering, it may be a lot, and so far there is no way to save on the actual design of 3D models. But, as it often happens in the machine learning industry, people are working hard to commoditize the things that all these projects have in common, in this case the randomization of 3D scenes and backgrounds, object placement, lighting modifications, and other parameters, as well as procedural generation of these randomized scenes.

One recent example is NVIDIA's Dataset Synthesizer (NDDS) [859], a plugin for *Unreal Engine 4* that allows computer vision researchers to easily turn 3D models and textures into ready-to-use synthetic datasets. NDDS can produce RGB images, segmentation maps, depth maps, and bounding boxes, and if the 3D models contain keypoints for the objects then these keypoints and object poses can be exported too. What is even more important, NDDS has automated tools for scene randomization: you can randomize lighting conditions, camera location, poses, textures, and more.

Basically, NDDS makes it easy to create your own dataset similar to, say, the *Falling Things* dataset. The result can look precisely like the examples in Fig. 6.3.

NVIDIA researchers are already using NDDS to produce synthetic datasets for computer vision. For example, SIDOD (Synthetic Image Dataset for 3D Object Pose Recognition with Distractors) by Jalal et al. [395] is a synthetic dataset which is relatively small by today's standards, only 144K stereo image pairs, but it is one of the first datasets to combine all types of outputs with flying distractors.

In general, by now researchers have relatively easy access to large datasets of 3D models of everyday objects to generate synthetic environments (we will see more of this in Sections 7.2 and 7.3), add synthetic objects as distractors to real images, place synthetic objects on real backgrounds in smarter ways, and so on. Although RGB-D datasets with real scans are also increasingly available as the corresponding hardware becomes available (see, e.g., the survey [236] and [151]), they cannot compete with synthetic data in terms of the quality of labeling and diversity of environments (see also Section 9.1). Over the next sections, we will see how this progress helps to solve the basic computer problems: after all, recognizing *synthetic* objects is never the end goal.

6.4 Case Study: Object Detection With Synthetic Data

Even a book might not have margins large enough to fit a thorough discussion of each and every computer vision problem. Therefore, I will intersperse this book, which is primarily a high-level survey, with several sections that represent more detailed case studies, digging into more detail on specific problems or applications. The first such case study deals with synthetic data for object detection. In Section 3.3, we have already given an overview of the main architectures used for object detection, and here we will see some of the key works that have applied synthetic data to object detection and presented insights that remain relevant for synthetic data today. We will review seven papers in roughly chronological order.

We are concentrating on training neural networks with synthetic data, so our story of synthetic data for object detection cannot begin earlier than the first deep learning models for this problem. But it does not begin much later either: our first paper in this review is by Peng et al. [664]; it came out on ICLR 2015, and the preprint is dated 2014. Back in 2014, the deep learning revolution in computer vision was still in early stages, so in terms of image classification architectures that could serve as backbones for object detection researchers only had *AlexNet*, VGG, and *GoogLeNet* (the first in the Inception line; see Section 3.2 for more details). But at the time, there was little talk about "backbones": the state of the art in object detection, reporting a huge improvement over the ILSVRC2013 detection track winner OverFeat [783] (31.4% mIoU vs. 24.3% for OverFeat), was R-CNN [276], the most straightforward two-stage object detection architecture that we discussed in Section 3.3.

In 2014, researchers were not sure if synthetic data was helpful at all. Moreover, the synthetic data they had was far from photorealistic, it was more like the *ShapeNet*

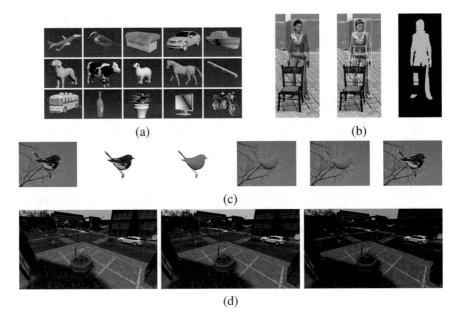

(a)   (b)

(c)

(d)

Fig. 6.4 Sample synthetic images for object detection: (a) samples from the dataset used in [664]; (b) sample 3D bounding box from [77]; (c) different variations of synthetic data compared in [664]; (d) varied lighting conditions from [77].

dataset we discussed above. The work by Peng et al. was in many ways intended to study this very question: can you improve object detection or, say, learn to recognize new categories with synthetic data that looks like the sample shown in Fig. 6.4a. Thus, the main question for Peng et al. was to separate different "visual cues", i.e., different components of an object. Simplistic synthetic data does pretty well in terms of shape, but poorly in terms of texture or realistic varied poses, and the background will have to be inserted separately so it probably will not match too well. Given this discrepancy in quality, what can we expect from object detection models?

To study this question, Peng et al. propose an object detection pipeline that looks like R-CNN but is actually even simpler than that. They used *AlexNet* pretrained on *ImageNet* as a feature extractor, and trained classifiers on features extracted from region proposals. Then they started testing for robustness to various cues, producing different synthetic datasets and testing object detection performance on a real test set after training on these datasets. Fig. 6.4c shows six different versions of training data that they compared: RGB and uniform gray objects superimposed against real, grayscale, and pure white backgrounds. With a thorough experimental study, Peng et al. obtained very interesting results that do not follow the standard intuition that the more details you have, the better the results will be. The simplest synthetic dataset, with uniform gray objects on white backgrounds, yields very reasonable results and significantly outperforms gray objects on more complex backgrounds. On the other hand, experiments in [664] show that adding a more varied set of views (front, side, and others) for a given object always helps, sometimes significantly.

The actual results in [664] did not really represent state of the art in object detection even in 2015, and are definitely not relevant today; neither are the object detection approaches and architectures. But conclusions and comparisons show an important trend that goes through many early results on synthetic data for computer vision: for many models, the details and textures do not matter as much since the models are looking for shapes and object boundaries. If that is the case for your model, then it is much more important to have a variety of shapes and poses, and textures can be left as an afterthought. Looking back from the present time, I would probably generalize this lesson: *different cues may be of different importance to different models*. So unless you are willing to invest serious resources into making an effort to achieve photorealism across the board, it is best to experiment with your model and find out which aspects are really important and worth an investment, and which aspects can be neglected (e.g., in this case, it turned out that you can leave the objects gray and skip the textures).

The second paper, in chronological order, is one of the first approaches that used synthetic *videos* for object detection, by Bochinski et al. [77]. This is one of the first attempts I could find at building a complete virtual world with the intent of making synthetic data for computer vision systems, and specifically for object detection. Bochinski et al. were also among the pioneers in using game engines for synthetic data generation. As the engine, they used *Garry's Mod*, a sandbox game on the *Source* engine designed by Valve for *Half Life* and *Counter Strike* games. For the time, *Source* had a very capable physics engine, and more than that, it supported scripting for bots, both humans and vehicles. Thus, it was relatively easy to create a simulated world for urban driving applications, complete with humans, cars, and surveillance cameras placed in realistic positions. Bochinski et al. extended the engine to be able to export bounding boxes, segmentation, and other kinds of labeling; see Fig. 6.4b for a sample human object with 3D bounding boxes and segmentation masks with occlusion. The engine also allows to vary lighting conditions (Fig. 6.4d shows a sample) and, naturally, place cameras at arbitrary positions, e.g., in realistic surveillance camera locations.

For object detection, since Bochinski et al. worked with video data, they used a simple classical technique to construct bounding boxes: background subtraction. Basically, this means that they train a Gaussian mixture model to describe the history of every pixel, and if the pixel becomes different enough, it is considered to be part of the foreground (an object) rather than background. CNNs are only used (and trained) to do classification inside the resulting bounding boxes. As a result, they achieved quite good results even on a real test set.

This work exemplifies how even relatively crude synthetic data can be helpful even for outdated pipelines: here, the bounding boxes were detected with a classical algorithm, so synthetic data was only used to train the classifier, and it still helped and resulted in a reasonable surveillance application.

For the third paper, we note an interesting method of using synthetic data for object detection proposed by Hinterstoisser et al. [341]. They note that training on purely synthetic data may give sub-par results due to the low-level differences between synthetic (rendered) images and real photographs. To avoid this, they propose to simply freeze the lower layers of, say, a pretrained object detection architecture and only train the top layers on synthetic data; in this way, basic features will remain suited for the domain of real photos while the classification part (top layers) can be fine-tuned for new classes. Otherwise, this is a straightforward test of synthetic data: Hinterstoisser et al. superimpose synthetic renderings on randomly selected backgrounds and fine-tune pretrained Faster-R-CNN [719], R-FCN [178], and Mask R-CNN [327] object detection architectures with freezed feature extraction layers. They report that freezing the layers helps significantly, and different steps in the synthetic data generation pipeline (different domain randomization steps, see also Section 9.1) help as well, obtaining results close to training on a large real dataset.

The fourth and fifth works are both devoted to multiple object detection for food and small vendor items. Before proceeding to the papers, let me briefly explain why this specific application—recognizing multiple objects on supermarket shelves or in a fridge—sounds like such a perfect fit for synthetic data. There are several reasons, and each of them is quite general and might apply to other applications as well.

First, the backgrounds and scene compositions are quite standardized (the insides of a fridge, a supermarket shelf) so it does not take too much effort to simulate them realistically. Datasets for such applications often get by with really simplistic backgrounds. Figure 6.5a shows some samples from the dataset from our first paper today, a work by Rajpura et al. [701]; the samples are available from the corresponding

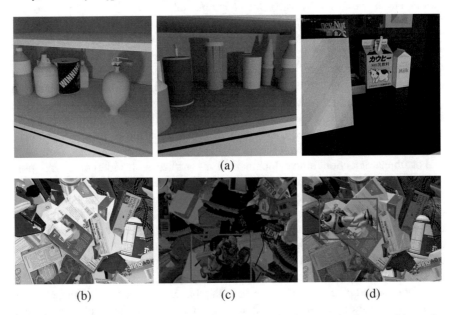

Fig. 6.5 Sample synthetic images for multiple object detection in constrained spaces: (a) samples from [701]; (b) sample background from [342] made of distractor objects; (c–d) sample synthetic images from [342].

(a)

(b)

Fig. 6.6 Sample real images for multiple object detection in constrained spaces: (a) a real sample image from SKU-110K [285]; (b) sample real images from a vending machine in [904].

GitHub repository[3]; the entire dataset is made with a few surface textures and a couple of glossy surfaces for glass shelves.

Second, while simple, the scenes and backgrounds for these problems are definitely not a common sight for *ImageNet* and other standard datasets. Standard datasets contain plenty of pictures of people enjoying outdoor picnics and dozens of different breeds of dogs (*ImageNet* has 120!) but not so many photos of the insides of a refrigerator or supermarket shelves with labeled objects. Thus, researchers cannot reuse pretrained models as easily.

Third, such scenes are perhaps unpopular in standard object detection datasets precisely because they are extremely hard to label by hand. A supermarket shelf may contain hundreds of densely packed objects; for an illustration, Figure 6.6a shows a

---

[3]https://github.com/paramrajpura/Syn2Real

sample image from a real dataset of such images called SKU-110K; it was collected only in 2019 [285].

Fourth, even now that we have a large-scale real dataset, we are not really done because new objects arrive very often. A system for a supermarket (or a fridge, since it in many ways contains the same kinds of objects) has to easily support the introduction of new object classes because new products or, even more often, new packaging for old products are introduced continuously. Thousands of new objects appear in every supermarket over a year, sometimes hundreds of new objects at once (think Christmas packaging). When you have a real dataset, adding new images takes a lot of work: it is not enough to just have a few photos of the new object, you also need to have it on the shelves, surrounded by old and new objects, in different combinations... this gets really hard really quick. In a synthetic dataset, adding a new 3D model is sufficient to create any number of scenes in any combinations you like.

The fifth and final point is that while you need a lot of objects in this application and a lot of 3D models for the synthetic dataset, most objects are relatively easy to model. They are Tetra Pak cartons, standardized bottles, paper boxes, and the like. The thousands of items in a supermarket often share relatively few different packaging options, and most of them are standard items with different labels. So once you have a 3D model for, say, a pint bottle, most beers will be covered by swapping a couple of textures, and the bottle itself is far from a hard object to model (compared to, say, a human face or a car). With all that said, object detection for small retail items does sound like a perfect fit for synthetic data.

Rajpura et al. [701] made one of the first attempts at using deep learning with synthetic data for this application. They concentrate on recognizing objects inside a refrigerator, and some samples of their synthetic data are shown in Fig. 6.5a. Their dataset is based on standard bottles and packaging from the *ShapeNet* repository that we have discussed above, and they used *Blender* (often the tool of choice for synthetic data since it is quite standard and free to use) to create simple scenes of the inside of a refrigerator and placed objects with different textures there. For object detection, Rajpura et al. used a fully convolutional version of *GoogLeNet* that generates a coverage map and a separate bounding box predictor trained on its results.

Their experimental results illustrate several important general points on the use of synthetic data in computer vision. First of all, Rajpura et al. saw significantly *improved performance for hybrid datasets*. In their comparison, using even 10% of real data and adding 90% of synthetic data to it far outperformed purely synthetic and purely real datasets. Note, however, that they only had 400 real images (since it was really hard to label such images manually), and second, the scale of synthetic data was also not so large (3600 synthetic images). Another interesting conclusion, however, is that adding more synthetic images can actually hurt. In [701], performance began to decline after 4000 synthetic images, that is, soon after the 10/90 split in favor of synthetic data.

This is probably due to overfitting to synthetic data, and it remains an important problem to this day. If the dataset uses a lot of synthetic images, the networks may begin to overfit to specific peculiarities of these synthetic images. More generally, synthetic data is different from real, and hence there is always an inherent domain transfer problem involved when you try to apply networks trained on synthetic data to real test sets (which you always ultimately want to do). This is a huge field of research, of course, and we will consider it in much more detail in Chapter 10.

The next work brings us already to 2019. Wang et al. [904] consider synthetic data generation and domain adaptation for object detection in smart vending machines. The premise looks very similar: vending machines have small food items placed there, and the system needs to find out which items are still there judging by a camera located inside the vending machine. But in terms of computer vision, there are several interesting points that differentiate it from [701] and highlight how synthetic data had progressed over these two years. Let us consider them in order.

First, data generation. In 2017, researchers took ready-made simple *ShapeNet* objects. In 2019, 3D shapes of the vending machine objects were being scanned from real objects by high-quality commercial 3D scanners, in this case one from *Shining 3D*[4]. What's more, 3D scanners still have a really hard time with specular or transparent materials, so for specular materials, Wang et al. use a whole other neural architecture (an adversarial one, actually) to transform the specular material into a diffuse one based on multiple RGB images and then restore the material during rendering (they used *Unity3D* for that). The specular-to-diffuse translation was based on a previous work devoted to this specific topic [940]. As for transparent materials, Wang et al. gave up even in 2019, saying that "although this could be alleviated by introducing some manual works, it is beyond the scope of this paper" and simply avoiding transparent objects. This has been rectified since this work, and now synthetic data can successfully help solve computer vision problems for transparent objects [751].

Second, Wang et al. introduce and apply a separate model for the deformation of resulting object meshes. Cans and packs may warp or bulge in a vending machine, so their synthetic data generation pipeline adds random deformations, complete with a (more or less) realistic energy-based model with rigidity parameters, also based on a previous work devoted specifically to this problem [906].

Third, the camera. Due to physical constraints, vending machines use fisheye cameras to be able to cover the entire area where objects are located; we show sample images from the paper in Figure 6.6b. Here is the vending machine from Wang et al. and sample images from the cameras on every shelf. 3D rendering engines usually support only the pinhole camera model, so, again, Wang et al. use a separate state-of the art camera model by Kannala and Brandt [431], calibrating it on a real fisheye camera and then introducing some random variation and noise.

Fourth, the synthetic-to-real image transfer, i.e., improving the resulting synthetic images so that they look more realistic. Wang et al. use a variation of style transfer based on CycleGAN that we will discuss in more detail in Section 10.3.

Finally, the object detection pipeline. Wang et al. compare several state of the art object detection methods, including PVANET [453], SSD [539], and YOLOv3 [711].

---

[4]https://www.shining3d.com/3d-digitizing-solutions/

Unlike all other works we have considered above, these are architectures that remain quite relevant up to this day (with some new versions released recently).

The results are consistent with previous works: while the absolute numbers and quality of the results have increased substantially since 2017, the general takeaway points remain the same. It still helps to have a hybrid dataset with both real and synthetic data. Note, by the way, that the dataset is again rather small; this time it is because the models are good enough to achieve saturation in this constrained setting with this kind of data, and more synthetic data probably does not help.

Interestingly, PVANET yields the best results, which is contrary to many other object detection applications (YOLOv3 was usually best overall in standard comparisons). This leads to another important takeaway point that remains relevant: in a specific application, it is best to redo the comparisons at least among the current state-of-the-art architectures. Usually, it does not add much to the cost of the project: in this case, Wang et al. definitely spent much more time and resources preparing and adapting synthetic data than testing two additional object detection architectures. But it can yield somewhat unexpected results (one can explain why PVANET has won in this case, but it would probably be a post-hoc explanation, and in a new application you really just do not know *a priori* which network will win) and let you choose what is best for your own project.

With that, we come to the sixth work in this case study. It comes from the same group as the third, authored by Hinterstoisser et al. [342] and aptly titled *An Annotation Saved is an Annotation Earned: Using Fully Synthetic Training for Object Instance Detections*. Similar to previous works, they consider multiple detection of small common objects, most of which are packs of food items and medicine. But the interesting thing about this paper is that they claim to achieve excellent results *without any real data at all*, by training on a purely synthetic dataset.

Their first contribution is an interesting take on domain randomization for background images (again, see Section 9.1 for more details on domain randomization). Hinterstoisser et al. try to get maximally cluttered synthetic images with the following procedure:

- take a separate dataset of distractor 3D models that represent objects that do not need to be recognized in the current application (in the paper, they had ≈ 15000 such distractor models);
- render these objects on the background in random poses and with scales roughly corresponding to the scale of the foreground objects (so they are comparable in size) while randomly varying the hues of the background object colors (this is standard domain randomization with distractor objects);
- choose and place new background objects until you have covered every pixel of the background (this is the interesting part);
- then place the foreground objects on top (we will discuss it in more detail below).

As a result of this approach, the synthetic dataset does not have to have any background images or scenes at all: the background is fully composed of distractor objects. Figure 6.5b shows a sample resulting background, and Figure 6.5c-d contains examples of the resulting synthetic images.

But that is only one part of it. Another part is how to generate the foreground layer, with objects that we actually want to recognize. Here, the contribution of Hinterstoisser et al. is that instead of placing 3D models in random poses or in poses corresponding to the background surfaces, as researchers had done before, they introduce a deterministic *curriculum* (schedule) for placing foreground objects:

- iterate over scales from largest to smallest, so that the network starts off with the easier job of recognizing large objects and then proceeds to learn to find their smaller versions; for every scale, iterate over all possible rotations;
- and then for every scale and rotation iterate through all available objects, placing them with possible overlaps and cropping at the boundaries; there is also a separate procedure to allow background distractor objects to partially occlude the foreground.

As a result, this purely synthetic approach outperforms a 2000-image real training set. Hinterstoisser et al. even estimate the costs: they report that it had taken them about 200 hours to acquire and label the real training set. This should be compared to a mere 5 hours needed for 3D scanning of the objects: as usual with synthetic data, once the pipeline is ready all you need to do to add new objects or retrain in a different setting is to scan the 3D models. The work [342] also presents a very detailed ablation study. The authors analyze which of their ideas contributed the most to their results. Interestingly (and a bit surprisingly), the largest effect is achieved by their curriculum strategy. Another interesting conclusion is that the purely synthetic cluttered background actually performs much better than a seemingly more realistic alternative strategy: take real-world background images and augment them with synthetic distractor objects (there is no doubt that distractor objects are useful anyway).

With these results, Hinterstoisser et al. have the potential to redefine how we see and use synthetic data for object detection; these conclusions most probably also extend to segmentation and possibly other computer vision problems. In essence, they show that synthetic data can be much better than real for object detection *if done right*. And we have to admit that "done right" includes virtually every single element of the synthetic data generation pipeline, even something seemingly inconsequential such as the order of poses.

Finally, the last paper in this case study is by Nowruzi et al. [635]. Titled *How much real data do we actually need: Analyzing object detection performance using synthetic and real data*, this work concentrates on a different domain of images, recognizing objects in urban outdoor environments with an obvious intent towards autonomous driving. However, the conclusions it draws appear to be applicable well beyond this specific case, and this paper has become the go-to source among experts in synthetic data.

The difference of this work from other sources is that instead of investigating different approaches to dataset generation within a single general framework, it considers various existing synthetic and real datasets, puts them in comparable conditions, and draws conclusions regarding how best to use synthetic data for object detection. Nowruzi et al. consider three real datasets:

- Berkeley Deep Drive (BDD) [981], a large-scale real dataset (100K images) with segmentation and object detection labeling;
- Kitti-CityScapes (KC), a combination of visually similar classical urban driving datasets KITTI [269] and CityScapes [167];
- NuScenes (NS) [107], a dataset with 1000 labeled video scenes, each 20 seconds long;

and three synthetic (we show sample images from them in Chapter 7):

- Synscapes (7D) [935], a synthetic dataset designed to mimic the properties of Cityscapes (see Fig. 7.3e-f);
- Playing for Benchmark (P4B) [725], a synthetic dataset with video sequences obtained from the Grand Theft Auto V game engine (see Fig. 7.3a-b);
- CARLA [203], a full-scale driving simulator that can also be used to generate labeled computer vision datasets (Fig. 7.4c-e).

To put all datasets on equal footing, the authors use only 15000 images from each (since the smallest dataset has 15K images), resize all images to 640x370 pixels, and remove annotations for objects that become too small under these conditions (less than 4% of the image height). The object detection model is also very standard: it is an SSD detector with *MobileNet* backbone, probably chosen for computational efficiency of both training and evaluation. The interesting part, of course, is the results.

First, as one would expect, adding more data helps. Training on smaller portions of each dataset significantly impedes the results. The second important question is about transfer learning: how well can object detection models perform on one dataset when trained on another? Naturally, the best results are achieved when a model is trained and tested on the same dataset; this is true for both synthetic and real datasets, but synthetic data significantly outshines real data in this comparison. This is a general theme throughout all synthetic data in computer vision: results on synthetic datasets are always better, sometimes too much so, signifying overfitting (but hopefully not in this case). But other than CARLA (which seems to be an outlier that fails all attempts to transfer), synthetic datasets fair pretty well, with transfer results clustering together with transfer from real datasets. Real datasets are still a little better, but note that Nowruzi et al. have removed one of the key advantages of synthetic data by equalizing the size of real and synthetic datasets.

But the real positive results come later. Nowruzi et al. compare two different approaches to using hybrid datasets, where synthetic data is combined with real:

- *synthetic-real data mixing*, where a small(er) amount of real data is added to a full-scale synthetic dataset, and the network is trained on the joint hybrid dataset;
- *fine-tuning on real data*, where we fully train the network on a synthetic dataset and then fine-tune on (small portions of) real datasets, so training on synthetic and real data is done separately.

The second approach is actually more convenient in practice: you can have a huge synthetic dataset and train on it once, and then adapt the resulting model to various real-life conditions by fine-tuning which is computationally much easier. And the main conclusion of [635] is that fine-tuning on real data actually performs significantly better than just mixing in real with synthetic.

Let us summarize. In this section, we have considered in some detail several studies on synthetic data for object detection. This is, of course, far from a full survey. Among other works, Bayraktar et al. [46] show improvements in object detection on a hybrid dataset in the context of robotics, extending a real dataset with images generated by the *Gazebo* simulation environment (see Section 7.4). In a recent work, Bayraktar et al. [47] test modern object recognition architectures such as *VGGNet* [477], *Inception v3* [838], *ResNet* [328], and *Xception* [152] by fine-tuning them on the *ADORESet* dataset that contains 2500 real and 750 synthetic images for each of 30 object categories in the context of robotic manipulation; they find that a hybrid dataset achieves much better recognition quality compared to purely synthetic or purely real datasets. Recent applications of synthetic data for object detection include the detection of objects in piles for training robotic arms [100], computer game objects [819], smoke detection [962], deformable part models [985], face detection in biomedical literature [185], detection of micro aerial vehicles (drones) [626, 739], and more.

However, the papers we have seen, especially the latter two influential recent papers [342, 635], have a common theme: they show how important are the exact curricula for training and the minute details of how synthetic data is generated and presented to the network. Before these studies, it had been hard to believe that simply changing the strategy of how to randomize the poses and scales of synthetic objects can improve the results by 0.2-0.3 in mean average precision (a huge difference!).

All this suggests that there is still much left to learn in the field of synthetic data, even for a relatively straightforward problem such as object detection. Using synthetic data is not quite as simple as throwing as much randomly generated stuff at the network as possible. This is a good thing, of course: harder problems with uncertain results also mean greater opportunities for research and for a deeper understanding of how neural networks work and how computer vision can be ultimately solved. And the results we have discussed in this section suggest that while synthetic data is already working well for us, there is still a fascinating and fruitful road ahead.

6.5 Other High-Level Computer Vision Problems

In the previous section, we have considered in a lot of detail the lessons learned in the use of synthetic data for object detection, a classical and often used high-level computer vision problem.

Segmentation is another classical computer vision problem with obvious benefits to be had from pixel-perfect synthetic annotations. The above-mentioned *SceneNet RGB-D* dataset by McCormac et al. [588] comes with a study showing that an RGB-only CNN for semantic segmentation pretrained from scratch on purely synthetic data can improve over CNNs pretrained on ImageNet; as far as we know, this was the first time synthetic data managed to achieve such an improvement. The dataset is actually an extension of *SceneNet* [319, 320], an annotated model generator for indoor scene understanding that can use existing datasets of 3D object models and place them in 3D environments with synthetic annotation. By now, segmentation models are commonly trained with synthetic data: semantic segmentation is the main problem for most automotive driving models (Section 7.2) and indoor navigation models (Section 7.3), Grard et al. [295] do it for object segmentation in depth maps of piles of bulk objects, and so on.

Saleh et al. [754] note that not all classes in a semantic segmentation problem are equally suited for synthetic data. Foreground classes that correspond to objects (people, cars, bikes, etc., i.e., *things* in the terminology of [334]) are well suited for object detectors (that use shape a lot) but suffer from the synthetic-to-real transfer for segmentation networks because their textures (which segmentation models usually rely upon) are hard to make photorealistic. On the other hand, background classes (grass, road surface, sky, etc., i.e., *stuff* in the terminology of [334]) look very realistic on synthetic images due to their high degree of "texture realism", and a semantic segmentation network can be successfully trained on synthetic data for background classes. Therefore, Saleh et al. propose a pipeline that combines detection-based masks by Mask R-CNN [327] for foreground classes and semantic segmentation masks by *DeepLab* [133] for background classes.

Although most works on synthetic data use large and well-known synthetic datasets, there are many efforts to bring synthetic data to novel applications by developing synthetic datasets from scratch. For instance, O'Byrne et al. [636] develop a synthetic dataset for biofouling detection on marine structures, i.e., segmenting various types of marine growth on underwater images. Ward et al. [920] improve leaf segmentation for *Arabidopsis* plants for the CVPPP Leaf Segmentation Challenge by augmenting real data with a synthetic dataset produced with *Blender*. For a parallel challenge of leaf counting, Ubbens et al. [883] produce synthetic data based on an L-system plant model; they report improved counting results. Moiseev et al. [609] propose a method to generate synthetic street signs, showing improvements in their recognition. Neff et al. [621] use GANs to produce synthetically augmented data for small segmentation datasets (see Section 10.7). We also note that researchers are also beginning to use synthetic data in other problems such as video stream summarization [12].

Another important class of applications for CGI-based synthetic data relates to problems such as 3D pose, viewpoint, and depth estimation, where manual labeling of real data is very difficult and sometimes close to impossible. One of the basic problems here is 2D-3D alignment, the problem of finding correspondences between regions in a 2D image and a 3D model (this also implies pose estimation for objects). In an early work, Aubry et al. [31] solved the 2D-3D alignment problem for chairs with a dataset of synthetic CAD models. Gupta et al. [311] train a CNN to detect and segment object instances for 3D model alignment with synthetic data with renderings of synthetic objects. Su et al. [821] learn to recognize 3D shapes from several 2D images, training their multi-view CNNs on synthetic 2D views. Triyonoputro et al. [873] train a deep neural network on multi-view synthetic images to help visual servoing for an industrial robot. Liu et al. [535] perform indoor scene modeling from a single RGB image by training on a dataset of 3D models, and in other works [533] do 2D-3D alignment from a single image for indoor basic objects. Shoman et al. [791] use synthetic data for camera localization (a crucial part of tracking and augmented reality systems), using synthetic data to cover a wide variety of lighting and weather conditions. They use an autoencoder-like architecture to bring together the features extracted from real and synthetic data and report significantly improved results.

3D position and orientation estimation for objects, known as the 6-DoF (degrees of freedom) pose estimation, is another important computer vision problem related to robotic grasping and manipulation. NVIDIA researchers Tremblay et al. [869] approach it with synthetic data: using the synthetic data generation techniques we will consider in Section 9.1, they train a deep neural network and report the first state of the art network for 6-DoF pose estimation trained purely on synthetic data. The novelty was that Tremblay et al. train on a mixture of domain randomized images, where distractor objects are placed randomly in front of a random background, and photorealistic images, where the foreground objects are placed in 3D background scenes obeying physical constraints; domain randomized images provide the diversity needed to cover real data (see Section 9.1) while realistic images provide proper context for the objects and are easier to transfer to real data. Latest results [577, 612, 695] show that synthetic data, especially with proper domain randomization for the data and domain adaptation for the features, can indeed successfully transfer 3D pose estimation from synthetic to real objects.

This also relates to depth estimation; synthetic renderings are easy to augment with pixel-perfect depth maps, and many synthetic datasets include RGB-D data. Carlucci et al. [115] created VANDAL, one of the first synthetic depth image databases, collecting 3D models from public CAD repositories for about 480 *ImageNet* categories of common objects; the authors showed that features extracted from these depth images by common CNN architectures improve object classification and are complementary to features extracted by the same architectures trained on *ImageNet*. Liebelt et al. [519] used 3D models to extract a set of 3D feature maps, then used a nearest neighbors approach to do multi-view object class detection and 3D pose estimation. Lee and Moloney [504] present a synthetic dataset with high-quality stereo pairs and show that deep neural networks for stereo vision can perform competitively with networks trained on real data. *Siemens* researchers Planche et al. [677] consider the problem of more realistic simulation of depth data from real sensors and present *DepthSynth*, an end-to-end framework able to generate realistic depth data rather than purely synthetic perfect depth maps; they show that this added realism leads to improvements with modern 2.5D recognition methods.

Easy variations and transformations provided by synthetic data can not only directly improve the results by training, but also represent a valuable tool for studying the properties of neural networks and other feature extractors. In particular, Pinto et al. [675] used synthetic data to study the invariance of different existing visual feature sets to variation in position, scale, pose, and illumination, while Kaneva et al. [429] used a photorealistic virtual environment to evaluate image feature descriptors. Peng et al. [663], Pepik et al. [667], and Aubry and Russell [32] used synthetic data to study the properties of deep convolutional networks, in particular robustness to various transformations, since synthetic data is easy to manipulate in a predefined way.

Earlier works recognized that the domain gap between synthetic and real images does not allow to expect state of the art results when training on synthetic data only, so many of them concentrated on bridging this gap by constructing hybrid datasets. In particular, Vázquez et al. [892] considered pedestrian detection and proposed a scheme based on active learning: they initially train a detector on virtual data and then use selective sampling [165] to choose a small subset of real images for manual labeling, achieving results on par with training on large real datasets while using 10x less real data.

Purely synthetic approaches were also used in early works, although mostly for problems where manual labeling would be even harder and noisier than for object detection or segmentation. The *Render for CNN* approach by Su et al. [822] outperformed real data with a hybrid synthetic+real dataset on the viewpoint estimation problem. Synthetic data helped improve 3D object pose estimation in Gupta et al. [310] and multi-view object class detection in Liebelt and Schmid [518] and Stark et al. [817]; as an intermediate step, the latter work used synthetic data to learn shape models. Hattori et al. [325] trained scene-specific pedestrian detectors on a purely synthetic dataset, superimposing rendered pedestrians onto a fixed real scene background; synthetic data has also been used for pedestrian detection by Marin et al. [582].

We finish this section by returning to an important question for direct applications: how realistic must synthetic data be in order to help with the underlying computer vision problem? Early works often argued that photorealism is not necessary for good domain transfer results; see, e.g., [826]. This question was studied in detail by Movshovitz-Attias et al. [614]. With the example of the viewpoint estimation problem for cars, they showed that photorealistic rendering does indeed help, showed that the gap between models trained on synthetic and real data can often be explained by domain adaptation (i.e., adapting from a different real dataset would be just as hard as adapting from a synthetic one), and hybrid synthetic+real datasets can significantly outperform training on real data only.

Another data point is provided by Tsirikoglou et al. [876] who present a very realistic effort for the rendering of synthetic data, including Monte Carlo-based light transport simulation and simulation of optics and sensors, within the domain of rendering outdoor scenes (see also Section 7.2, where we discuss a continuation of this work by Wrenninge and Unger [935]). They show improved results in object detection over other synthetic datasets and conclude that "a focus on maximizing variation and realism is well worth the effort".

6.6   Synthetic People

Synthetic models and images of people (both faces and full bodies) are an especially interesting subject for synthetic data. On the one hand, real datasets here are even harder to collect due to several reasons:

- there are privacy issues involved in the collection of real human faces;
- labeling for some basic computer vision problems is especially complex: while pose estimation is doable, facial keypoint detection (a key element for facial recognition and image manipulation for faces) may require to specify several dozen landmarks on a human face, which becomes very hard for human labeling [189, 843];
- finally, even if a large dataset is available, it often contains biases in its composition of genders, races, or other parameters of human subjects, sometimes famously so [470, 864].

On the other hand, there are complications as well:

- synthetic 3D models people and especially synthetic faces are much harder to create than models of basic objects, especially if sufficient fidelity is required;
- basic human-related tasks are very important in practice, so there already exist large real datasets for face recognition [112, 306, 512], pose estimation [24, 388, 543, 874], and other problems, which often limits synthetic data to covering corner cases, augmenting real datasets, or serving more exotic use cases.

This creates a tension between the quality of available synthetic faces and improvements in face recognition and other related tasks that they can provide. In this section, we review how synthetic people have been used to improve computer vision models in this domain; the datasets covered in this section are illustrated in Fig. 6.7.

In an early effort, Queiroz et al. [691] presented a pipeline for generating synthetic videos with automatic ground truth for human faces and the resulting *Virtual Human Faces Database* (VHuF) with realistic face skin textures that can be extracted from real photos. Bak et al. [34] present the *Synthetic Data for person Re-Identification* (SyRI) dataset with virtual 3D humans designed with *Adobe Fuse CC* to make the models and *Unreal Engine 4* for high-speed rendering. Interestingly, they model realistic lighting conditions by using real HDR environment maps collected with light probes and panoramic photos (Fig. 6.7).

While face recognition for full-face frontal high-quality photos has been mostly solved in 2D, achieving human and superhuman performance both for classification [843] and retrieval via embeddings [771], pose-invariant face recognition *in the wild* [197, 366, 499, 500, 902], i.e., under arbitrary angles and imperfect conditions, remains challenging. Here, synthetic data is often used to augment a real dataset, where frontal photos usually prevail, with more diverse data points; we refer to Section 10.4 for a detailed overview of the works by Huang et al. [369] and Zhao et al. [1016, 1017] on GAN-based refinement.

An interesting approach to creating synthetic data for face recognition is provided by Hu et al. [360]. In their "Frankenstein" pipeline, they combine automatically detected body parts (eyes, mouth, nose, etc.) from different subjects; interestingly, they report that the inevitable artifacts in the resulting images, both boundary effects and variations between facial patches, do not hinder training on synthetic data and may even improve the robustness of the resulting model.

There is also a related field of *3D-aided face recognition*. This approach uses a morphable synthetic 3D model of an abstract human face that has a number of free

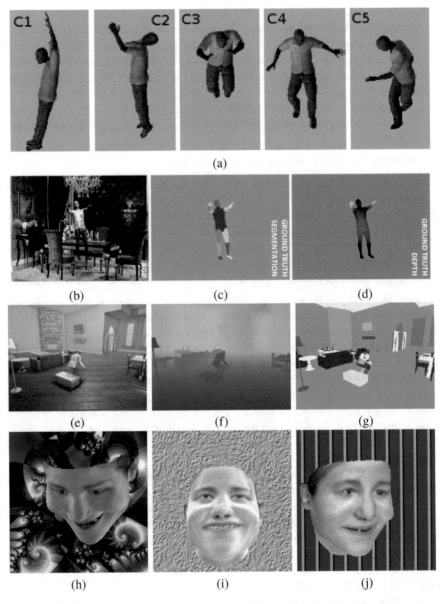

Fig. 6.7 Sample images from human-related synthetic datasets: (a) video frames from *ViHASi* [699]; (b–d) a frame with ground truth from SURREAL [889]: (b) RGB image, (c) segmentation map, (d) depth map; (e–g) a frame from PHAV [815]: (e) RGB image, (f) segmentation map, (g) depth map; (h–j) sample synthetic faces with randomized backgrounds from [471, 472] (based on the Basel Face Model).

parameters; the model learns to tune these parameters so that the 3D model fits a given photo and then uses the model and texture from the photo either to produce a frontal image or to directly recognize photos taken from other angles. This is a classic approach, dating back to late 1990s [75, 76] and developed in many subsequent works with new morphable models [361, 378], deep learning used to perform the regression for morphing parameters [865], extended to 3D face scans [74], and so on; see, e.g., the survey [197] for more details. Xu et al. [965] used synthetic data to train their 3D-aided model for pose-invariant face recognition as well. Recent works used GANs to produce synthetic data for 3D-aided face recognition [1017]; we will discuss this approach in detail in Section 10.4.

In a large-scale effort to combat dataset bias in face recognition and related problems with synthetic data, Kortylewski et al. [471, 472] have developed a pipeline to directly create synthetic faces. They use the *Basel Face Model 2017* [271], a 3D morphable model of face shape [74, 75], and take special care to randomize the pose, camera location, illumination conditions, and background. They report significantly improved results for face recognition and facial landmark detection with the *OpenFace* framework [17, 760] and state of the art models for face detection and alignment [1000] and landmark detection [703].

Human pose estimation is a very well known and widely studied problem [181, 543, 995] with many direct applications, so it is no wonder that the field does not suffer from lack of real data, with large-scale datasets available [25, 411, 525] and state of the art models achieving impressive results [823, 828, 970]. However, synthetic data still can help. Ludl et al. [562] show that in corner cases, corresponding to rare activities not covered by available datasets, existing pose estimation models produce errors, but augmenting the training set with synthetic data that covers these corner cases helps improve pose estimation.

Another specialized use-case has been considered by Rematas et al. in a very interesting application of pose estimation called *Soccer on Your Tabletop* [715]. They trained specialized pose and depth estimation models for soccer players and produced a unified model that maps 2D footage of a soccer match into a 3D model suitable for rendering on a real tabletop through augmented reality devices. For training, Rematas et al. used synthetic data captured from the *FIFA* video game series. These are model examples of how synthetic data can improve the results even when comprehensive real datasets are available. We also note the recently presented CLOTH3D dataset by Bertiche et al. [67] that contains full-body human 3D models clothed in thousands of different outfits (see Figure 6.8 for an illustration).

Moving from still images to videos, we begin with human action recognition [682, 996]. ViHASi by Ragheb et al. [699] is a virtual environment and dataset for silhouette-based human action recognition. De Souza et al. [815] present Procedural Human Action Videos (PHAV), a synthetic dataset that contains 39,982 videos with more than 1,000 examples for each action of 35 categories. Inria and MPI researchers Varol et al. [889] present the Synthetic hUmans foR REAL tasks (SURREAL) dataset. They generate photorealistic synthetic images with labeling for human part segmentation and depth estimation, producing 6.5M frames in 67.5K short clips (about 100 frames each) of 2.6K action sequences with 145 different synthetic subjects.

Fig. 6.8 Sample images from the CLOTH3D dataset [67].

Microsoft researchers Khodabandeh et al. [451] present the *DIY Human Action* generator for human actions. Their framework consists of a generative model, called the *Skeleton Trajectory GAN*, that learns to generate a sequence of frames with human skeletons conditioned on the label for the desired action, and a *Frame GAN* that generates photorealistic frames conditioned on a skeleton and a reference image of the person. As a result, they can generate realistic videos of people defined with a reference image that perform the necessary actions, and, moreover, the *Frame GAN* is trained on an unlabeled set of human action videos.

We also note here some privacy-related applications of synthetic data that are not about differential privacy (which we discuss in Section 11). For example, Ren et al. [721] present an adversarial architecture for video face anonymization; their model learns to modify the original real video to remove private information while at the same time still maximizing the performance of action recognition models.

As the problems become dynamic rather than static, e.g., as we move to recognizing human movements on surveillance cameras, synthetic data takes the form of full-scale simulated environments. This direction started a long time ago: already in 2007, The *ObjectVideo Virtual Video* (OVVV) system by Taylor et al. [853] used the *Half-Life 2* game engine with additional camera parameters designed to simulate real-world surveillance cameras to detect a variety of different events. Fernandez et al. [235] place virtual agents onto real video surveillance footage in a kind of augmented reality to simulate rare events. Qureshi and Terzopoulos [693] present a multi-camera virtual reality surveillance system.

An interesting human-related video analysis problem, important for autonomous vehicles, is to predict pedestrian trajectories in an urban environment. Anderson et al. [20] develop a method for stochastic sampling-based simulation of pedestrian trajectories. They then train the *SocialGAN* model by Gupta et al. [307] that generates pedestrian trajectories with a recurrent architecture and uses a recurrent discriminator to distinguish fake trajectories from real ones; Anderson et al. show that synthetic trajectories significantly improve the results for a predictive model such as *SocialGAN* (Fig. 6.9).

Fig. 6.9 Sample images from synthetic crowd counting datasets: (a) *GTA5 Crowd Counting* [907]; (b) *Agoraset* [171]; (c) *LCrowdV* [145].

Another important application for datasets of synthetic people is *crowd counting*. In this case, collecting ground truth labels, especially if the model is supposed to do segmentation in addition to simple counting, is especially labor-intensive since crowd counting scenes are often highly congested and contain hundreds, if not thousands of people. Existing real datasets are either relatively small or insufficiently diverse; e.g., the UCSD dataset [118] and the Mall dataset [131] have both about 50,000 pedestrians but in each case collected from a single surveillance camera, the *ShanghaiTech* dataset [1012] has about 330,000 heads but only about 1200 images, again collected on the same event, and the UCF-QNRF dataset [383], while more diverse than previous ones, is limited to extremely congested scenes, with up to 12,000 people on the same image, and has only about 1500 images.

In one of the first attempts to use synthetic datasets for crowd counting, the *LCrowdV* system [145] generated labeled crowd videos and showed that augmenting real data with the resulting synthetic dataset improved the accuracy of pedestrian detection; LCrowdV and other crowd counting datasets are illustrated in Fig. 6.9.

To provide sufficient diversity and scale, Wang et al. [907] presented a synthetic *GTA5 Crowd Counting* dataset collected with the help of the *Grand Theft Auto V* engine; the released dataset contains about 15,000 synthetic images with more than 7,5 million annotated people in a wide variety of scenes. They compare various approaches to crowd counting as a supervised problem, in particular their new spatial fully convolutional network (SFCN) model that directly predicts the density map of people on a crowded image. They report improved results when pretraining on GCC and then fine-tuning on a real dataset; they also consider GAN-based approaches that we discuss in Section 10.5. A more direct approach to generating synthetic data has been developed by Ekbatani et al. [224], who extract real pedestrians from images and add them at various locations on other backgrounds, with special improvement procedures for added realism; they also report improved counting results. Khadka et al. [448] also present a synthetic crowd dataset, showing improvements in crowd counting.

This ties into *crowd analysis*, where synthetic data is used to model crowds and train visual crowd analysis tools on rendered images [798]. Huang et al. [367] present virtual crowd models that could be used for such simulations. Courty et al. [171] present the *Agoraset* dataset for crowd analysis research that aims to provide realistic agent trajectories (simulated via the social force model by Heibling and Molnár [335]) and high-quality rendering with the Mental Ray renderer [208]; the dataset has 26 different characters and provides a variety of different scenes: corridor, flow around obstacles, escape through a bottleneck, and so on.

In general, we summarize that while the most popular problems such as frontal face recognition or human pose estimation are already being successfully solved with models trained on real datasets (because there has been sufficient interest for these problems to collect and manually label large-scale datasets), synthetic data remains very important for alleviating the effect of dataset bias in real collections, covering corner cases, and tackling other problems or basic problems with different kinds of data, e.g., in different modalities (such as face recognition with an IR sensor). I believe that there are important opportunities for synthetic data in human-related computer vision problems and expect this field to grow in the near future.

6.7 Other Vision-Related Tasks: OCR and Visual Reasoning

In this section, we discuss two more vision-related problems where the corresponding models are often trained on synthetic data. We begin with *optical character recognition* (OCR). Various tasks related to text recognition, including OCR itself, text detection, layout analysis and text line segmentation for document digitization, and others have often been attacked with the help of synthetic data, usually with synthetic text superimposed on real images.

This is a standard technique in the field because text pasted in a randomized way often looks quite reasonable even with minimal additional postprocessing. Synthetic data was used for character detection and recognition in, e.g., [10, 111, 910, 991] and for text block detection in, e.g., [392, 393]. Krishnan and Jawahar [475] use synthetic data to pretrain deep neural networks for learning efficient representations of hand-written word images. Jo et al. [404] train an end-to-end convolutional architecture that can digitize documents with a mixture of handwritten and printed text; to train the network, they produce a synthetic dataset with real handwritten text superimposed on machine-printed forms, with Otsu binarization applied before pasting.

There exist published datasets of synthetic text and software to produce them, in particular *MJSynth* [392] (see Fig. 6.10a for a sample) and *SynthText in the Wild* [308]. In the latter work, Gupta et al. use available depth estimation and segmentation solutions to find regions (planes) of a natural image suitable for placing synthetic text and find the correct rotation of text for a given plane. This process is illustrated in the top row of Fig. 6.10b, and the bottom row shows sample text inserted onto suitable regions.

Moreover, recent works have used GAN-based refinement (see Section 10.1) to make synthetic text more realistic [220]. There also exist synthetic handwriting generation models based on GANs that are conditioned on character sequences and produce excellent results [14, 398]. We note a recent work that deals with constrained OCR, specifically OCR for domain-specific languages, and also studies synthetic data generation for this task [670].

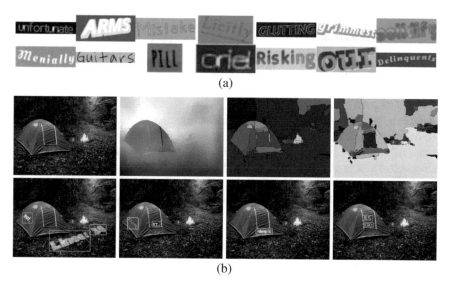

(a)

(b)

Fig. 6.10 Synthetic datasets for text recognition: (a) *MJSynth* [392]; (b) *SynthText in the Wild* [308]; left to right in the top row: RGB image, depth map, semantic segmentation, filtered regions suitable for text placement.

The second topic for this section is *visual reasoning*, the field of artificial intelligence where models are trained to reason and answer questions about visual data. It is usually studied in the form of *visual question answering* (VQA), when models are trained to answer questions about a picture such as "What is the color of the small metal sphere?" or "Is there an equal number of balls and boxes?".

There exist datasets for visual question answering based on real photographs, collected and validated by human labelers; they include the first large dataset called VQA [8] and its recent extension, VQA v2.0 [294]. However, the problem yields itself naturally to automated generation, so it is no wonder that synthetic datasets are important in the field.

The most important synthetic VQA dataset is Compositional Language and Elementary Visual Reasoning (CLEVR), created by Johnson et al. in a collaboration between *Stanford University* and *Facebook Research* [408]. It contains 100K rendered images with scenes composed of simple geometric shapes and about 1M (853K unique) automatically generated questions about these images. The intention behind this dataset was to enable detailed analysis of VQA models, simplifying visual recognition and concentrating on reasoning about the objects.

In CLEVR, scenes are represented as scene graphs [410, 474], where the nodes are objects annotated with attributes (shape, size, material, and color) and edges correspond to spatial relations between objects ("left", "right", "behind", and "in front"). A scene can be rendered based on its scene graph with randomized positions of the objects. The questions are represented as functional programs that can be executed on scene graphs, e.g., "What color is the cube to the right of the white sphere?". Different question types include querying attributes ("what color"), comparing attributes ("are they the same size"), existence ("are there any"), counting ("how many"), and integer comparison ("are there fewer"). When generating questions, special care is taken to ensure that the answer exists and is unique, and then the natural language question is generated with a relatively simple grammar. Figure 6.11 shows two sample questions and their functional programs: in Fig. 6.11a the program is a simple chain of filters, and Fig. 6.11b adds a logical connective, which makes the graph a tree.

We also note a recently published COG dataset produced by *Google Brain* researchers [968] that extends CLEVR's ideas to video processing. It also contains synthetic visual inputs and questions generated from functional programs, but now questions can refer to time (e.g., "what is the color of the latest triangle?"). The authors also released a generator that can produce synthetic video-question pairs that are progressively more challenging and that have minimal response bias, an important problem for synthetic datasets (in this case, the generator begins with a balanced set of target responses and then generates videos and questions for them rather than the other way around).

Finally, the field of visual question answering can be generalized to just question answering (QA). To create datasets for training QA models, researchers almost inevitably use synthetic data generation, extracting training sets of questions from knowledge graphs or other structured knowledge representations. But this field has only very recently started to move from simple template-based generation of ques-

(a)

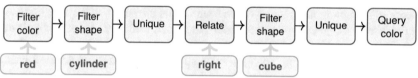

(b) Chain-structured question: *What color is the cube to the right of the red cylinder?*

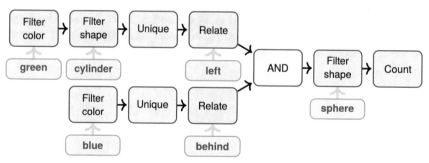

(c) Tree-structured question: *How many spheres are behind the blue thing and on the left side of the left cylinder?*

Fig. 6.11 The CLEVR dataset [408]: (a) sample image; (b–c) sample visual reasoning questions.

tions to a wider study of how best to generate synthetic data and whether existing generation approaches may lead to information leakage or other undesirable effects. We note a recent work by Lindjordet and Balog [527] that raises these questions and hope that the leakage effects and insufficiently diverse question patterns that they identify will be fixed in the nearest future.

6.8   Conclusion

This has been the first chapter in the book actually devoted to synthetic data and ways in which it can improve the training of deep learning models. We have considered basic synthetic datasets that are often used to train computer vision models and shown examples where synthetic data helps to solve fundamental computer vision problems better.

For the most basic computer vision problems such as image classification or object detection, the common theme we have seen in this chapter is that synthetic data is not that useful for general-purpose models. Indeed, it well may be that synthetic data does not help much when *ImageNet* and *OpenImages* are already available. But as soon as the problem setting changes, even slightly—be it a more detailed task in a narrow domain, a change in camera settings, or unusual objects—synthetic data immediately raises its head.

We have also discussed several applications where synthetic data is relatively easy to produce, while labeling real data is much harder, so it is almost inevitable that synthetic data will be used at least alongside real datasets. In the next chapter, we proceed to situations where synthetic data is entirely indispensable: interactive simulation environments.

Chapter 7
Synthetic Simulated Environments

In this chapter, we proceed from datasets of static synthetic images, either preren-dered or procedurally generated, to entire simulated environments that can be used either to generate synthetic datasets on the fly or provide learning environments for reinforcement learning agents. We discuss datasets and simulations for outdoor envi-ronments (mostly for autonomous driving), indoor environments, and physics-based simulations for robotics. We also make a special case study of datasets for unmanned aerial vehicles and the use of computer games as simulated environments.

7.1 Introduction

While collecting synthetic datasets is a challenging task by itself, it is insufficient to train, e.g., an autonomous vehicle such as a self-driving car or a drone, or an industrial robot. Learning to control a vehicle or robot often requires *reinforcement learning* [831], where an agent has to learn from interacting with the environment, and real-world experiments to train a self-driving car or a robotic arm by reinforcement learning are completely impractical. Fortunately, this is another field where synthetic data shines: once one has a fully developed 3D environment that can produce datasets for computer vision or other sensory readings, it is only one more step to active interaction with this environment. Therefore, in most domains considered below we can see the shift from static synthetic datasets to interactive simulation environments.

Reinforcement learning (RL) agents are commonly trained on simulations because the interactive nature of reinforcement learning makes training in the real world extremely expensive. We discuss synthetic-to-real domain adaptation in this context in Section 10.6. However, in many works, there is no explicit domain adaptation: robots are trained on simulators and later fine-tuned on real data or simply transferred to the real world.

© The Author(s), under exclusive license to Springer Nature Switzerland AG 2021
S. I. Nikolenko, *Synthetic Data for Deep Learning*, Springer Optimization
and Its Applications 174, https://doi.org/10.1007/978-3-030-75178-4_7

Table 7.1 An overview of synthetic datasets and virtual environments discussed in Section 7

Name	Year	Ref	Engine	Size / comments
Outdoor urban environments, driving				
TORCS	2014	[946]	Custom	Game-based simulation engine
Virtual KITTI	2016	[254]	Unity	5 environments, 50 videos
GTAVision	2016	[412]	GTA V	GTA plugin, 200K images
SYNTHIA	2016	[731]	Unity	213K images
GTAV	2016	[725]	GTA V	25K images
VIPER	2017	[724]	GTA V	254K images
CARLA	2017	[203]	UE	Simulator
VIES	2018	[754]	Unity3D	61K images, 5 environments
ParallelEye	2018	[857]	Esri	Procedural gen, import from OSM
VIVID	2018	[492]	UE	Urban sim with emphasis on people
DeepDrive	2018	[692]	UE	Driving sim + 8.2h of videos
PreSIL	2019	[380]	GTA V	50K images with LIDAR point clouds
AADS	2019	[513]	Custom	3D models of cars on real backgrounds
WoodScape	2019	[975]	Custom	360° panoramas with fisheye cameras
ProcSy	2019	[449]	Esri	Procedural generation with varying conditions
Robotic simulators and aerial navigation				
Gazebo	2004	[464]	Custom	Industry standard robotic sim
MuJoCo	2012	[862]	Custom	Common physics engine for robotics
AirSim	2017	[785]	UE	Sensor readings, hardware-in-the-loop
CAD^{2}RL	2017	[749]	Custom	Indoor flying sim
X-Plane	2019	[812]	X-Plane	8K landings, 114 runways
Air Learning	2019	[476]	—	Platform for flying sims
VRGym	2019	[957]	UE	VR for human-in-the-loop training
ORRB	2019	[147]	Unity	Accurate sim used to train real robots
Indoor environments				
ICL-NUIM	2014	[321]	Custom	RGB-D with noise models, 2 scenes
SUNCG	2016	[814]	Custom	45K floors, 3D models
MINOS	2017	[761]	SUNCG	Indoor sim based on SUNCG
AI2-THOR	2017	[466]	Unity3D	Indoor sim with actionable objects
House3D	2018	[943]	SUNCG	Indoor sim based on SUNCG
Habitat	2019	[579]	Custom	Indoor sim platform and library
Hypersim	2020	[726]	Custom	Photorealistic indoor sim

Table 7.1 shows a brief summary of datasets and simulators that we review in this section. To make the exposition more clear, we group together both environments and "static" synthetic datasets for outdoor (Section 7.2) and indoor (Section 7.3) scenes, including some works that use them to improve RL agents and other models.

Next, we consider synthetic robotic simulators (Section 7.4) and vision-based simulators for autonomous flying (Section 7.5), finishing with an idea of using computer games as simulation environments in Section 7.6. Reinforcement learning in virtual environments remains a common thread throughout this section.

7.2 Urban and Outdoor Environments: Learning to Drive

An important direction of applications for synthetic data is related to navigation, localization and mapping (SLAM), or similar problems intended to improve the motion of autonomous robots. Possible applications include SLAM, motion planning, and motion for control for self-driving cars (urban navigation) [239, 498, 601, 649], unmanned aerial vehicles [11, 170, 428], and more; see also general surveys of computer vision for mobile robot navigation [83, 190] and perception and control for autonomous driving [662].

Before proceeding to current state of the art, let me remind an interesting historical fact that we discussed in detail in Section 5.3: one of the first autonomous driving attempts based on neural networks, ALVINN [680], which used as input 30×32 videos supplemented with 8×32 range finder data, was already training on synthetic data. One of the first widely adopted full-scale visual simulation environments for robotics, *Gazebo* [464] (see Section 7.4), also provided both indoor and outdoor environments for robotic control training.

In a much more recent effort, *Xerox* researchers Gaidon et al. [254] presented a photorealistic synthetic video dataset *Virtual KITTI*[1] intended for object detection and multi-object tracking, scene-level and instance-level semantic segmentation, optical flow, and depth estimation. The dataset contains five different virtual outdoor environments created with the *Unity* game engine and 50 photorealistic synthetic videos. Gaidon et al. studied existing multi-object trackers, e.g., based on an improved min-cost flow algorithm [676] and on Markov decision processes [950]; they found minimal real-to-virtual gap. Note, however, that experiments in [254] were done on trackers trained on real data and evaluated on synthetic videos (and that's where they worked well), not the other way around. In general, the *Virtual KITTI* dataset is much too small to train a model on it, it is intended for evaluation, which also explains the experimental setup.

Johnson-Robertson et al. [412], on the other hand, presented a method to *train* on synthetic data. They collected a large dataset by capturing scene information from the *Grand Theft Auto V* video game that provides sufficiently realistic graphics and at the same time stores scene information such as depth maps and rough bounding boxes in the GPU stencil buffer, which can also be captured; the authors developed an automated pipeline to obtain tight bounding boxes. Three datasets were generated, with 10K, 50K, and 200K images, respectively. The main positive result of [412] is

---

[1]The name comes from the KITTI dataset [269, 598] created in a joint project of the Karlsuhe Institute of Technology and Toyota Technological Institute at Chicago.

that a standard Faster R-CNN architecture [719] trained on 50K and 200K images outperformed on a real validation set (KITTI) the same architecture trained on a real dataset. The real training set was *Cityscapes* [167] that contains 2,975 images, so while the authors used more synthetic data than real, the difference is only 1-2 orders of magnitude. The VIPER and GTAV datasets by Richter et al. [724, 725] were also captured from *Grand Theft Auto V*; the latter provides more than 250K 1920 × 1080 images fully annotated with optical flow, instance segmentation masks, 3D scene layout, and visual odometry.

The SYNTHIA dataset presented by Ros et al. [731] provides synthetic images of urban scenes labeled for semantic segmentation. It consists of renderings of a virtual New York City constructed by the authors with the *Unity* platform and includes segmentation annotations for 13 classes such as pedestrians, cyclists, buildings, roads, and so on. The dataset contains more than 213,000 synthetic images covering a wide variety of scenes and environmental conditions; experiments in [731] show that augmenting real datasets with SYNTHIA leads to improved segmentation. Later, Hernandez-Juarez et al. [338] presented SYNTHIA-SF, the San Francisco version of SYNTHIA. We illustrate SYNTHIA with a sample frame (that is, two frames since the dataset contains two cameras) from the SYNTHIA-SF dataset in Figure 7.1.

Saleh et al. [754] presented a *Unity3D* framework called virtual environment for instance segmentation (VEIS); while not very realistic, it worked well with their detection-based pipeline (see Section 6.5). Li et al. [511] present a synthetic dataset with foggy images to simulate difficult driving conditions. We note the work of Lopez et al. [548] whose experiments suggest that the level of realism achieved in SYNTHIA and GTAV is already sufficient for successful transfer of object detection methods (Fig. 7.2).

Tian et al. [857] present the *ParallelEye* synthetic dataset for urban outdoor scenes. Their approach is rather flexible and relies on previously developed *Esri CityEngine* framework [954] that provides capabilities for batch generation of 3D city scenes based on terrain data. In [857], this data was automatically extracted from the *OpenStreetMap* platform[2]. The 3D scene is then imported into the *Unity3D* game engine, which helped add urban vehicles on the roads, set up traffic rules, and add support for different weather and lighting conditions. Tian et al. showed improvements in object detection quality for state-of-the-art architectures trained on *ParallelEye* and tested on the real KITTI test set as compared to training on the real KITTI training set.

Li et al. [513] develop the *Augmented Autonomous Driving Simulation* (AADS) environment that is able to insert synthetic traffic on real-life RGB images. Starting from the real-life *ApolloScape* dataset for autonomous driving [372] that contains LIDAR point clouds, the authors remove moving objects, restore backgrounds by inpainting, estimate illumination conditions, simulate traffic conditions and trajectories of synthetic cars, preprocess the textures of the models according to lighting and other conditions, and add synthetic cars in realistic places on the road. In this way, a single real image can be reused many times in different synthetic traffic situations.

---

[2]https://www.openstreetmap.org/.

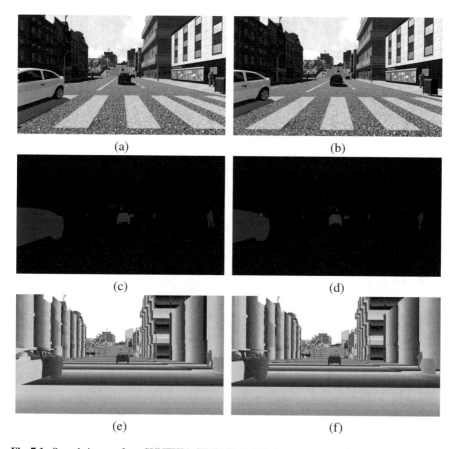

Fig. 7.1 Sample images from SYNTHIA-SF [338]: (a–b) RGB ground truth (left and right camera); (c–d) ground truth segmentation maps; (e–f) depth maps (depth is encoded in the color as $R + 256 \cdot G + 256^2 \cdot B$).

This is similar to the approach of Abu Alhaija et al. [5] (recall Section 6.3) but due to available 3D information AADS can also change the observation viewpoint and even be used in a closed-loop simulator such as CARLA or *AirSim* (see below). We do not go into details on the already large and diverse field of virtual traffic simulation and refer to a recent survey [123].

Wrenninge and Unger [935] present the *Synscapes* dataset that continues the work of Tsirikoglou et al. [876] (see Section 6.5) and contains accurate photorealistic renderings of urban scenes (Fig. 7.3e-f), with unbiased path tracing for rendering, special models for light scattering effects in camera optics, motion blur, and more. They find that their additional efforts for photorealism do indeed result in significant improvements in object detection over GTA-based datasets, even though the latter has a wider variety of scenes and pedestrian and car models.

Fig. 7.2 Sample images from synthetic outdoor datasets: (a) VEIS [754]; (b) *Esri CityEngine* Venice sample scene [954]; (c) AADS [513] (part of a frame from a showcase video).

Khan et al. [449] introduce *ProcSy*, a procedurally generated synthetic dataset aimed at semantic segmentation (we showed a sample frame on Fig. 1.7c-d). It is modeling a real-world urban environment, and its main emphasis is on simulating various weather and lighting conditions for the same scenes. The authors show that, e.g., adding a mere 3% of rainy images in the training set improves the mIoU of a state-of-the-art segmentation network (in this case, Deeplab v3+ [135]) by as much

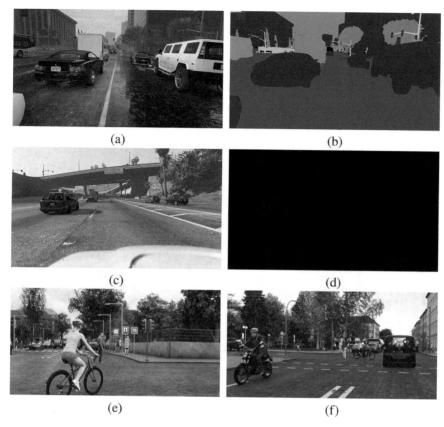

Fig. 7.3 Sample images from synthetic outdoor datasets: (a–b) GTAV [725]: (a) RGB image, (b) ground truth segmentation; (c–d) VIPER [724]: (c) RGB image, (d) ground truth segmentation; (e–f) *Synscapes* [935].

as 10% on rainy test images. This again supports the benefits from using synthetic data to augment real datasets and cover rare cases; for a discussion of the procedural side of this work see Section 12.1.

Synthetic datasets with explicit 3D data (with simulated sensors) for outdoor environments are less common, although such sensors seem to be straightforward to include into self-driving car hardware. In their development of the *SqueezeSeg* architecture, Wu et al. [936, 937] added a LiDAR simulator to *Grand Theft Auto V* and collected a synthetic dataset from the game. *SynthCity* by Griffiths and Boehm [298] is a large-scale open synthetic dataset which is basically a huge point cloud of an urban/suburban environment. It simulates *Mobile Laser Scanner* (MLS) readings with a *Blender* plugin [300] and is specifically intended for pretraining deep neural networks.

Yogamani et al. [975] present *WoodScape*, a multi-camera fisheye dataset for autonomous driving that concentrates on getting 360° sensing around a vehicle

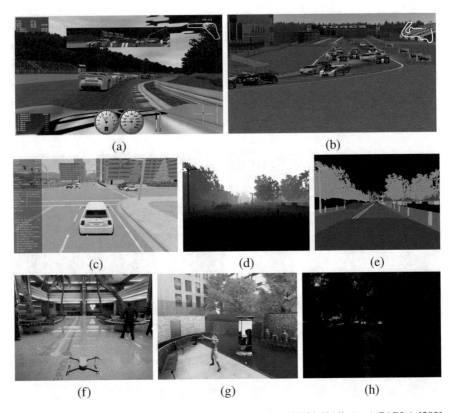

Fig. 7.4 Sample images from outdoor environments: (a–b) TORCS [946]; (c–e) CARLA [203]; (f–h) VIVID [492].

through panoramic fisheye images with a large field of view. They record 4 fisheye cameras with 190° horizontal field of view, a rotating LiDAR, GNSS and IMU sensors, and odometry signals with 400K frames with depth labeling and 10K frames with semantic segmentation labeling. Importantly for us, together with their real dataset they also released a synthetic part (10K frames) that matches their fisheye cameras, with the explicit purpose of helping synthetic-to-real transfer learning. This further validates the importance of synthetic data in autonomous driving.

Simulated environments rather than datasets are, naturally, also an important part of the outdoor navigation scene; below we describe the main players in the field and also refer to surveys [430, 824] for a more in-depth analysis of some of them. There is also a separate line of work related to developing more accurate modeling in such simulators, e.g., sensor noise models [605], that falls outside the scope of this book.

TORCS[3] (The Open Racing Car Simulator) [946] is an open-source 3D car racing simulator that started as a game for *Linux* in the late 1990s but became increasingly popular as a virtual simulation platform for driving agents and intelligent control systems for various car components. TORCS provides a sufficiently involved simulation of racing physics, including accurate basic properties (mass, rotational inertia), mechanical details (suspension types, etc.), friction profiles of tyres, and a realistic aerodynamic model, so it is widely accepted as a useful source of synthetic data. TORCS has become the basis for the annual Simulated Car Racing Championship [544] and has been used in hundreds of works on autonomous driving and control systems (see Section 7.4); TORCS and other outdoor driving simulators are illustrated in Fig. 7.4.

CARLA (CAR Learning to Act) [203] is an open simulator for urban driving, developed as an open-source layer over *Unreal Engine 4* [434]. Technically, it operates similarly to [412], extending *Unreal Engine 4* by providing sensors in the form of RGB cameras (with customizable positions), ground truth depth maps, ground truth semantic segmentation maps with 12 semantic classes designed for driving (road, lane marking, traffic sign, sidewalk, and so on), bounding boxes for dynamic objects in the environment, and measurements of the agent itself (vehicle location and orientation). *DeepDrive* [692] is a simulator designed for training self-driving AI models, also developed as an *Unreal Engine* plugin; it provides 8 RGB cameras with 512×512 resolution at close to real-time rates (20Hz), as well as a generated 8.2-hour video dataset.

VIVID (VIrtual environment for VIsual Deep learning), developed by Lai et al. [492], tackles a more ambitious problem: adding people interacting in various ways and a much wider variety of synthetic environments, they present a universal dataset and simulator of outdoor scenes such as outdoor shooting, forest fires, drones patrolling a warehouse, pedestrian detection on the roads, and more. VIVID is also based on the *Unreal Engine* and uses the wide variety of assets available for it; for example, NPCs acting in the scenes are programmed using *Blueprint*, an *Unreal* scripting engine, and the human models are animated by the *Unreal* animation editor. VIVID provides the ability to record video simulations and can communicate with deep learning libraries via the TCP/IP protocol, through the *Microsoft Remote Procedure Call* (RPC) library originally developed for *AirSim* (see Section 7.4).

As for reinforcement learning (RL) in autonomous driving, the original paper on the CARLA simulator [203] also provides a comparison on synthetic data between conditional imitation learning, deep reinforcement learning, and a modular pipeline with separated perception, local planning, and continuous control, with limited success but generally best results obtained by the modular pipeline.

Many works on autonomous driving use TORCS [946] as a testbed, both in virtual autonomous driving competitions and simply as a well-established research platform. I do not aim to provide a full in-depth survey of the entire field and only note that despite its long history TORCS is being actively used for research purposes up to this day. In particular, Sallab et al. [756, 757] use it in their deep reinforcement learning

---

[3]http://torcs.sourceforge.net/.

frameworks for lane keeping assist and autonomous driving, Xiong et al. [960] add safety-based control on top of deep RL, Wang et al. [908] train a deep RL agent for autonomous driving in TORCS, Barati et al. [41] use it to add multi-view inputs for deep RL agents, Li et al. [509] develop *Visual TORCS*, a deep RL environment based on TORCS, Ando, Lubashevsky et al. [21, 560] use TORCS to study the statistical properties of human driving, Glassner et al. [279] shift the emphasis to trajectory learning, Luo et al. [564] use TORCS as the main test environment for a new variation of the policy gradient algorithm, Liu et al. [534] make use of the multimodal sensors available in TORCS for end-to-end learning, Xu et al. [844] train a segmentation network and feed segmentation results to the RL agent in order to unify synthetic imagery from TORCS and real data, and so on.

In an interesting recent work, Choi et al. [149] consider the driving experience transfer problem but consider a transfer not from a synthetic simulator to the real domain but from one simulator (TORCS) to another (GTA V). Tai et al. [842] learn continuous control for mapless navigation with asynchronous deep RL in virtual environments.

Synthetic data in autonomous driving extends to other sensor modalities as well. Thieling et al. [855] discuss the issues of physically realistic simulation for various robot sensors. Yue et al. [983] present a LIDAR point cloud generator based on the *Grand Theft Auto V* (GTA V) engine, showing significant improvements in point cloud segmentation when augmenting the KITTI dataset with their synthetic data. Sanchez et al. [840] generate synthetic 3D point clouds with the robotic simulator *Gazebo* (see Section 7.4). Wang et al. [901] develop a separate open-source plugin for LIDAR point cloud generation. Fang et al. [231] present an augmented LIDAR point cloud simulator that can generate simulated point clouds from real 3D scanner data, extending it with synthetic objects (additional cars).

The work based on GTA V has recently been continued by Hurl et al. [380] who have developed a precise LIDAR simulator within the GTA V engine and published the *PreSIL* (Precise Synthetic Image and LIDAR) dataset with over 50000 frames with depth information, point clouds, semantic segmentation, and detailed annotations; we use *PreSIL* to showcase the modalities available in modern synthetic datasets in Fig. 7.5. There also exist works on synthesizing specific elements of the environment, thus augmenting real data with synthetic elements; for example, Bruls et al. [95] generate synthetic road marking layouts which improves road marking segmentation, especially in corner cases.

Outdoor simulated environments go beyond driving, however, with simulators and synthetic datasets successfully used for autonomous aerial vehicles. Most of them are intended for unmanned aerial vehicles (UAVs) and have an emphasis on plugging in robotic controllers, possibly even with hardware-in-the-loop approaches; we discuss these simulators in Sections 7.4 and 7.5.

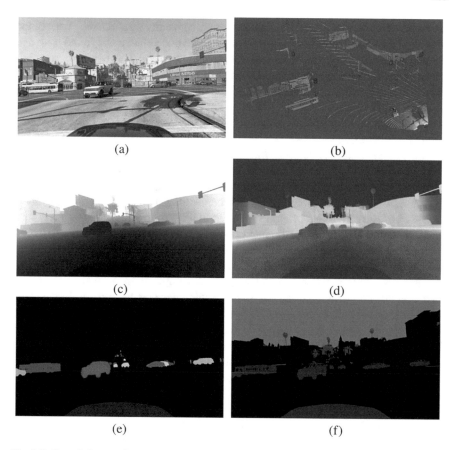

Fig. 7.5 Sample images from the *PreSIL* dataset [380]: (a) RGB image, (b) point cloud, (c) black-and-white depth map, (d) color depth map, (e) segmentation map, (f) stencil buffer.

7.3 Datasets and Simulators of Indoor Scenes

Although, as we have seen in the previous section, the main emphasis of many influential applications remains in the outdoors, *indoor navigation* is also an important field where synthetic datasets are required. The main problems remain the same SLAM and navigation—but the potential applications are now more in the field of home robotics, industrial robots, and embodied AI [584, 1023]. There exist large-scale efforts to create real annotated datasets of indoor scenes [121, 176, 796, 813, 947, 951], but synthetic data is increasingly being used in the field [1001].

Historically, the main synthetic dataset for indoor navigation has been SUNCG[4] presented by Song et al. [814]. It contains over 45,000 different scenes (floors of private houses) with manually created realistic room layouts, 3D models of the fur-

---

[4]http://suncg.cs.princeton.edu/

niture, realistic textures, and so on. All scenes are semantically annotated at the object level, and the dataset provides synthetic depth maps and volumetric ground truth data for the scenes. The original paper [814] presented state of the art results in semantic scene completion, but, naturally, SUNCG has been used for many different tasks related to depth estimation, indoor navigation, SLAM, and others [2, 143, 318, 568, 688], and it often serves as the basis for scene understanding competitions [825].

Interestingly, at the time of writing the SUNCG website was down and the dataset itself unavailable due to a legal controversy over the data[5]; that is why Fig. 7.6c shows a standard showcase picture from [814] instead of data samples. While synthetic data can solve a lot of legal issues with real data (as the main example of such issues, see Chapter 11 for a discussion of privacy concerns), the data, and especially handmade or manually collected 3D models, are still intellectual property and can bring about problems of its own unless properly released to the public domain.

SUNCG has given rise to a number of simulation environments. Before SUNCG, we note the *Gazebo* platform mentioned above [464] (see also Section 7.4) and the V-REP robot simulation framework [728] that not only provided visual information but also simulated a number of actual robot types, further simplifying control deployment and development. Handa et al. [321] provide their own simulated living room environment and dataset ICL-NUIM (Imperial College London and National University of Ireland Maynooth) with special emphasis on visual odometry and SLAM; they render high-quality RGB-D images with ray tracing and take special care to model the noise in both depth and RGB channels.

MINOS by Savva et al. [761] is a multimodal simulator for indoor navigation (later superceded by *Habitat*, see below). Wu et al. [943] made SUNCG into a full-scale simulation environment named *House3D*[6], with high-speed rendering suitable for large-scale reinforcement learning. In *House3D*, a virtual agent can freely explore 3D environments taken from SUNCG while providing all the modalities of SUNCG.

House3D has been famously used for navigation control with natural language: Wu et al [943] presented *RoomNav*, a task of navigation from natural language instructions and some models able to do it, while Das et al. [183] presented an embodied question answering model, where a robot is supposed to answer natural language questions by navigating an indoor environment. The AI2-THOR (The House Of inteRactions) framework [466] provides near photorealistic interactive environments with actionable objects (doors that can be opened, furniture that can be moved, etc.) based on the *Unity3D* game engine (see an example of the same scene after some actions have been applied to objects on Fig. 7.6f-g).

Zhang et al. [1009] studied the importance of synthetic data realism for various indoor vision tasks. They fixed some problems with 3D models from SUNCG, improved their geometry and materials, sampled a diverse set of cameras for each scene, and compared OpenGL rendering against physically-based rendering (with Metropolis light transport models [893] and the Mitsuba renderer[7]) across a variety

---

[5]https://futurism.com/tech-suing-facebook-princeton-data

[6]http://github.com/facebookresearch/House3D

[7]http://www.mitsuba-renderer.org/

Fig. 7.6 Sample images from indoor datasets and simulation environments: (a-b) ICL-NUIM [321]; (c) SUNCG [814]; (d) House3D [943]; (e) Habitat [579]; (f–g) AI2THOR [466].

of lighting conditions. Their main conclusion is that, again, added realism is worth the effort: the quality gains are quite significant.

The two most recent (and also very impressive) advances in the field are *Habitat* released by *Facebook* and *Hypersim* released by *Apple*. *Habitat* is a simulation platform for embodied AI developed by Facebook researchers Savva et al. [579]. Its simulator, called *Habitat-Sim*, presents a number of important improvements over previous work that we have surveyed in this section:

- dataset support: *Habitat-Sim* supports synthetic datasets such as SUNCG [814] and real-world datasets such as Matterport3D [121] and Gibson [947];
- rendering performance: *Habitat-Sim* can render thousands of frames per second, 10-100x faster than previous simulators; the authors claim that "it is often faster to generate images using *Habitat-Sim* than to load images from disk"; this is important because simulation stops being a bottleneck in large-scale model training;
- humans-as-agents: humans can function as agents in the simulated environment, which allows to use real human behaviour in agent training and evaluation;
- accompanying library: the *Habitat-API* library defines embodied AI tasks and implements metrics for easy agent development.

Savva et al. also provide a large-scale experimental study of various state of the art agents, arriving at the conclusion (counter to previous research) that agents based on reinforcement learning outperform SLAM-based ones, and RL agents generalize best across datasets, including the synthetic-to-real generalization from SUNCG to Matterport3D and Gibson. This is an important finding for synthetic data in indoor navigation, and we expect it to be validated in later studies.

Hypersim is a dataset of photorealistic indoor images generated by *Apple* researchers Roberts and Paczan and released in late 2020 [726]. Moving further than just straightforward rendering of 3D scenes, they develop a novel computational CGI pipeline that uses triangle meshes for annotation (in order to, in particular, avoid re-rendering if a different annotation is required) and special algorithms for sampling the camera trajectory. This has allowed Roberts and Paczan to create a nearly 2TB dataset that includes 3D scenes, camera trajectories, lossless high-dynamic-range images, and plenty of different types of annotations, as illustrated with a sample image from *Hypersim* in Fig. 7.7. In addition to the types of annotation shown in Fig. 7.7, the dataset also includes 3D bounding boxes and mesh annotations.

I definitely expect *Habitat*, *Hypersim*, and their successors to become the new standard for indoor navigation and embodied AI research.

7.4 Robotic Simulators

We have seen autonomous driving sims that mostly concentrate on accurately reflecting the outside world, modeling additional sensors such as LIDAR, and physics of the driving process. Simulators for indoor robots and unmanned aerial vehicles (UAV) add another complication: embedded hardware for such robots may be relatively

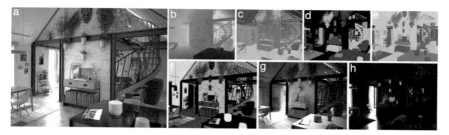

Fig. 7.7 Sample image from the *Hypersim* dataset [726]: (a) photorealistic rendering; (b) depth; (c) surface normals; (d-e) instance-level semantic segmentation; (f) diffuse reflectance; (g) diffuse illumination; (h) non-diffuse residual image that captures lighting effects.

weak and needs to be taken into account. Hence, robotic simulators usually support the *Robot Operating System*[8] (ROS), a common framework for writing robot software. In some cases, simulators go as far as provide *hardware-in-the-loop* capabilities, where a real hardware controller can be plugged into the simulator; for example, hardware-in-the-loop approaches to testing UAVs have been known for a long time and represent an important methodology in flight controller development [7, 683, 899]. We also refer to the surveys [337, 575].

For a brief review, we highlight four works, starting with two standard references. *Gazebo*, originally presented by Koenig and Howard [464] and now being developed by OSRF (Open Source Robotics Foundation), is probably the best-known robotic simulation platform. It supports ROS integration out of the box, has been used in the DARPA Robotics Challenge [9], NASA Space Robotics Challenge, and others, and has been instrumental for thousands of research and industrial projects in robotics. *Gazebo* uses a realistic physical engine (actually, several different engines) that supports illumination and lighting effects, gravity, inertia, and so on; it can be integrated with robotic hardware via ROS and provides realistic simulation that often leads to successful transfer to the real world.

MuJoCo (Multi-Joint Dynamics with Contact) developed by Todorov [862] is a physics engine specializing in contact-rich behaviours, which abound in robotics. Both *Gazebo* and *MuJoCo* have become industry standards for robotics research, and surveying the full range of their applications goes far beyond the scope of this work. There are, of course, other platforms as well. For example, Gupta and Jarvis [309] present a simulation platform for training mobile robots based on the *Half-Life 2* game engine.

The other two works are, on the contrary, very recent and may well define a new industry standard in the near future. First, Xie et al. present *VRGym* [957], a virtual reality testbed for physical and interactive AI agents. Their main difference from previous work is the support of human input via VR hardware integration. The rendering and physics engine are based on *Unreal Engine 4*, additional multisensor hardware is capable of full-body sensing and integration of human subjects to

---

[8]https://www.ros.org/.

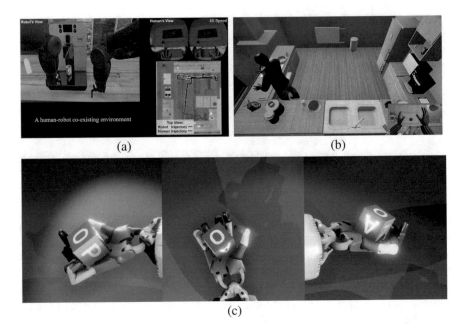

Fig. 7.8 Sample images from robotic simulation environments: (a) *VRGym* [957]; (b) *VRKitchen* [261]; (c) ORRB [147].

virtual environments, while a ROS bridge allows to easily communicate with robotic hardware. Xie et al. benchmark RL algorithms and show possibilities for socially aware models and intention prediction; *VRGym* is already being further extended and specialized into, e.g., *VRKitchen* by Gao et al. [261] designed to learn cooking tasks (Fig. 7.8).

The second work is the *OpenAI Remote Rendering Backend* (ORRB) developed by OpenAI researchers Chociej et al. [147], able to render robotic environments and provide depth and segmentation maps in its renderings. ORRB emphasizes diversity, aiming to provide domain randomization effects (see Section 9.1); it is based on *Unity3D* for rendering and *MuJoCo* for physics, and it supports distributed cloud environments. ORRB has been used to train *Dactyl* [646], a robotic hand for multi-finger small object manipulation developed by OpenAI; *Dactyl*'s RL-based policies have been trained entirely in ORRB simulation and then have been successfully transferred to the physical robot. We expect more exciting developments in such synthetic-to-real transfer for robotic applications in the near future.

Fig. 7.9 Sample images from flight simulators: (a) *AirSim* [785] drone demo with depth, segmentation, and RGB drone view on the bottom; (b) CAD^{2}RL [749]; (c) *Air Learning* [476] (depth image and drone camera view on the bottom); (d–e) XPlane dataset [812].

7.5 Vision-Based Applications in Unmanned Aerial Vehicles

A long line of research deals with vision-based approaches to operating unmanned aerial vehicles (UAV) [11]. Since in this case it is almost inevitable to use simulated environments for training, and often testing the model is also restricted to synthetic simulators (real-world experiments are expensive outdoors and simply prohibited in urban environments), almost the entire field uses some kind of synthetic datasets or simulators; some of the datasets and simulators we cover in this section are illustrated in Fig. 7.9. Classical vision-related problems for UAVs include three major tasks that UAVs often solve with computer vision:

- localization and pose estimation, i.e., estimating the UAV position and orientation in both 2D (on the map) and in the 3D space; for this problem, real datasets collected by real-world UAVs are available [82, 727, 779, 820], and the field is advanced enough to use actual field tests, so synthetic-only results are viewed with suspicion, but existing research still often employs synthetic simulators, either handmade for a specific problem [921] or based on professional flight simulators [26];
- obstacle detection and avoidance, where real-world experiments are often too expensive even for testing [106];
- visual servoing, i.e., using feedback from visual sensors in order to maintain position, stability, and perform maneuvers; here synthetic data has usually been used in the form of hardware-in-the-loop simulators for the developed controllers [503], sometimes augmented with full-scale flight simulators [485] or specially designed "virtual reality" environments [763].

During the latest years, these classical applications of synthetic data for UAVs have been extended and taken to new heights with modern approaches to synthetic data generation. For example, Marcu et al. [581] considered the problem of locating safe landing areas for UAVs, which requires depth estimation and segmentation into "horizontal", "vertical", and "uncertain" regions to distinguish horizontal areas that would be safe for landing. To train a convolutional architecture for this segmentation and depth estimation task, the authors propose an interesting approach to generating synthetic data: they begin with *Google Earth* data[9] and extract 3D scenes from it. However, since 3D meshes in *Google Earth* are far from perfect, the authors then map textures to the 3D scenes to obtain less realistic-looking images but ones for which the depth maps are known perfectly. Marcu et al. showed that from a bird's eye view the resulting images look quite realistic, and compare different segmentation architectures on the resulting synthetic dataset.

Oyuki Rojas-Perez et al. [729] also compared the results obtained by training on synthetic and real datasets, with the results in favor of synthetic data due to the availability of depth maps for synthetic images. In a recent work, Castagno et al. [116] solve the landing site selection problem with a high-fidelity visual synthetic model of Manhattan rooftops, rendered with the *Unreal Engine* and provided to the simulated robots via *AirSim*.

Gazebo has been used for UAV simulations [251], but there are more popular specialized simulators in the field. *AirSim* [785] by *Microsoft* is a flight simulator that operates more like a robotic simulator such as *Gazebo* than an accurate flight sim such as *Microsoft Flight Simulator* or *X-Plane*: it contains a detailed physics engine

---

[9]https://www.google.com/earth/

designed to interact with specific flight controllers and providing a realistic physics-based vehicle model but also supports ROS to interact with the drones' software and/or hardware. Apart from visualizations rendered with *Unreal Engine 4*, *AirSim* provides sensor readings, including a barometer, gyroscope, accelerometer, magnetometer, and GPS. *AirSim* is also available as a plugin for *Unreal Engine 4* which enables extensions and new projects based on *AirSim* simulations of autonomous vehicles.

There are plenty of extensions and projects that use *AirSim* and provide interesting synthetic simulated environments (see also the survey [575]), in particular:

- Chen et al. [134] add realistic forests to use UAVs for forest visual perception;
- Bondi et al. [81] concentrate on wildlife preservation, extending the engine with realistic renderings of, e.g., the African savanna;
- in two different projects, Smyth et al. [807, 808] and Ullah et al. [885] simulate critical incidents (such as chemical, biological, or nuclear incidents or attacks) with an explicit goal of training autonomous drones to collect data in these virtual environments;
- Huang et al. [364] extend *AirSim* with a natural language understanding model to simulate natural language commands for drones, and so on.

We also note that *Microsoft* itself recently extended *AirSim* to include autonomous car simulations, making the engine applicable to all the tasks we discussed in Section 7.2.

While *AirSim* and *Gazebo* have been the most successful simulation frameworks, we also note other frameworks for multiagent simulation that have been used for autonomous vehicles, usually without a realistic 3D rendered environment (see also a comparison of the frameworks in [615] and of their perception systems in [737]): JaSIM [257], a multiagent 3D environment model based on the *Janus* platform [266], *Repast Simphony* [634], an agent-based simulation library for complex adaptive systems, FLAME [459] that concentrates on parallel simulations, and JADE [56], a popular library for multiagent simulations.

Sadeghi and Levine [749] present the CAD^2RL framework for avoiding collisions while flying in indoor environments. They motivate the use of synthetic data by the domain randomization idea (see Section 9.1) and produce a wide variety of indoor simulated environments constructed in *Blender*. They do not use any real images during training, learning the Q-function for the reinforcement learning agent entirely on simulated data, with a fully convolutional neural network. The authors report improvements in a number of complex settings such as flying around corners, navigating through narrow corridors, flying up a staircase, avoiding dynamic obstacles, and so on.

The *Air Learning* platform recently presented by Krishnan et al. [476] is an end-to-end simulation environment for autonomous aerial robots that combines pluggable environment and physics engine (such as *AirSim* or *Gazebo*), learning algorithms, policies for robot control (implemented in, e.g., *TensorFlow* or *PyTorch*), and "hardware-in-the-loop" controllers, where a real flight controller can be plugged in and evaluated on the *Air Learning* platform. As an example, Krishnan et al. benchmark several reinforcement learning approaches to point-to-point obstacle avoidance tasks and arrive at important conclusions: for instance, it turns out that having more onboard computational power (a desktop CPU vs. a *Rapsberry Pi*) can significantly improve the result, producing almost 2x shorter trajectories.

Among vision-based synthetic datasets, we note the work by Solovev et al. [812] who present a synthetic dataset for airplane landing based on the XPlane flight simulator [493], which has been used for other UAV simulations as well [73, 262]. The main intention of Solovev et al. is to present a benchmark for representation learning by combining different modalities (RGB images and various sensors), but the resulting synthetic dataset is sufficiently large and diverse to be used for other applications: 93GB of images and sensor readings from 8K landings on 114 different runways.

Another interesting application where real and synthetic data come together is provided by Madaan et al. [571], who solve a truly life-or-death problem for UAVs: wire detection. They use real background images and superimpose them with renderings of realistic 3D models of wires. The authors vary different properties of the wires (material, wire sag, camera angle, etc.) but do not make any attempts to adapt the wires to the semantics of the background image itself, simply pasting wires onto the images. Nevertheless, Madaan et al. report good results with training first on synthetic data and then fine-tuning on a small real dataset; synthetic pretraining proves to be helpful.

7.6   Computer Games as Virtual Environments

Computer games and game engines have been a very important source of problems and virtual environments for deep RL and AI in general [113, 417, 766]. Much of the foundational work in modern deep RL has been done in the environments of 2D arcade games, usually classical *Atari* games [55, 607]. First, many new ideas for RL architectures or training procedures for robotic navigation were introduced as direct applications of deep RL to navigation in complex synthetic environments, in particular to navigating mazes in video games such as *Doom* [69] and *Minecraft* [638] or specially constructed 3D mazes, e.g., from the *DeepMind Lab* environment [49, 780]. *Doom* became something of an industry standard, with the *VizDoom* framework developed by Kempka et al. [443] and later used in many important works [15, 495, 705, 942]; it is illustrated in Fig. 7.10.

Games of other genres, in particular real-time strategy games [19, 851], also represent a rich source of synthetic environments for deep RL; we note the *TorchCraft* library for machine learning research in *StarCraft* [834], synthetic datasets extracted from *StarCraft* [526, 599, 939], the *ELF* (Extensive, Lightweight, and Flexible) library platform with three simplified real-time strategy games [856], and the SC2LE (StarCraft II Learning Environment) library for *Starcraft II* [896]. Racing games have been used as environments for training end-to-end RL driving agents [672].

While data from computer games could technically be considered synthetic data, we do not go into further details on game-related research and concentrate on virtual

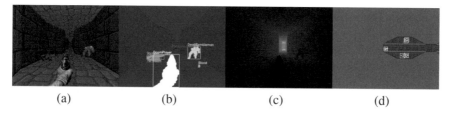

Fig. 7.10 A sample snapshot of the VizDOOM environment [443]: (a) RGB image; (b) object detection labeling; (c) depth map; (d) map of the environment.

environments specially designed for machine learning and/or transfer to real-world tasks. Note, however, that synthetic data is already making inroads even into learning to play computer games: Justesen et al. [418] show that using procedurally generated levels improves generalization and final results for *Atari* games and can even produce models that work well when evaluated on a completely new level every time they play.

All of the above does not look too much in line with the general topic of synthetic data: while game environments are certainly "synthetic", there is usually no goal to transfer, say, an RL agent playing *StarCraft* to a real armed conflict (thankfully). However, recent works suggest that there is potential in this direction. For example, while navigation in a first-person shooter is very different from real robotic navigation, successful attempts at transfer learning from computer games to the real world are already starting to appear. Karttunen et al. [442] present an RL agent for navigation trained on the *Doom* environment and transferred to a real-life Turtlebot by freezing most of the weights in the DQN network and only fine-tuning a small subset of them. As computer games get more realistic, we expect such transfer to become easier.

7.7 Conclusion

This has been the second chapter on primary applications of synthetic data. We have discussed simulation environments, ranging from classical robotic simulators to computer games. In such environments, the produced data is necessarily synthetic, but in many domains and applications, such as reinforcement learning, they represent the only practical way to allow models to train.

The next two chapters will be shorter and will sweep up most other direct applications of synthetic data. In Chapter 8, we will consider applications of synthetic data outside computer vision and simulated environments; in other machine learning domains such as natural language processing synthetic data is much less prominent, but we will find several important cases. Chapter 9 will present several directions in synthetic data development that may allow for better synthetic datasets in the future. And then we will proceed to the second main topic of this book: synthetic-to-real domain adaptation with generative models.

Chapter 8
Synthetic Data Outside Computer Vision

While computer vision remains the main focus of synthetic data applications, other fields also begin to use synthetic datasets, with some directions entirely dependent on synthetic data. In this chapter, we survey some of these fields. Specifically, Section 8.1 discusses how structured synthetic data is used for fraud and intrusion detection and other applications in the form of network and/or system logs; in Section 8.2, we consider neural programming; Section 8.3 discusses synthetic data generation and use in bioinformatics, and Section 8.4 reviews the (admittedly limited) applications of synthetic data in natural language processing.

8.1 Synthetic System Logs for Fraud and Intrusion Detection

In this chapter, we discuss domains other than computer vision where synthetic data can also be helpful. First, let us consider a very wide domain of datasets that span many different fields of application: tabular data, that is, user records, logs of various technical or computer systems, network traffic traces, and so on. There are important cases where the need for synthetic tabular data is motivated by privacy concerns; for instance, it is impossible to publish real electronic medical records, and open datasets have to be generated synthetically with privacy guarantees. We will discuss such applications in Section 11.5.

Apart from privacy concerns, another source of motivation for synthetic tabular data is the fact that a system needs to recognize rare events that occur in real datasets insufficiently often. For a specific case study, let us consider in this section rare events in security applications such as fraud detection or intrusion detection. Fraud and intrusion detection systems run based on network logs (traces) and system call logs, so this is the kind of structured data that needs to be synthesized in this case.

© The Author(s), under exclusive license to Springer Nature Switzerland AG 2021
S. I. Nikolenko, *Synthetic Data for Deep Learning*, Springer Optimization
and Its Applications 174, https://doi.org/10.1007/978-3-030-75178-4_8

Since real data on proven intrusions is very scarce, this field has used synthetic data for a long time, but usually the scenario was more akin to the early days of computer vision that we discussed in Section 5.2: synthetic data was generated to serve as a testbed to compare approaches that were not even necessarily based on machine learning. One notable example here is the 1998 and 1999 Off-Line Intrusion Detection Evaluation organized by the Lincoln Laboratory of MIT and sponsored by DARPA [528, 574]. The test data in this competition contained network traffic and system log files from a large computer network which was completely simulated, and both attacks and background "normal operation" data were entirely synthetic. The competition organizers used real network traces to model the background data after, and attacks were generated from a database of known attacks by running the corresponding simulation scripts [445].

However, the 1998 and 1999 competitions were criticized precisely for the quality of their synthetic dataset. McHugh [590] pointed out that the test dataset in these competitions was not sufficiently validated, and the whole process of data generation was not even explained in sufficient detail. Indeed, the papers cited above only give vague descriptions that say that they have generated data "similar" to real traffic (in fact, to traffic seen on operational Air Force bases, no less). However, real data on the Internet is highly variable, contains a lot of seemingly anomalous but legitimate traffic (including random garbage), often comes in bursts that are similar to malicious flooding, and so on. The synthetic attacks were also questionable: while the attack simulation scripts seemed to be realistic enough, it was again unclear what their distribution was and how well it reflected attacks in the real world. All of these points indeed represent design choices that need to be made in the generation of synthetic data, and they have to be explicitly and adequately addressed.

Therefore, since the early 2000s researchers in this field started paying closer attention to the process of generating synthetic data. Let us highlight the work by Lundin et al. [563] who developed a methodology for synthetic fraud data generation. Noting the deficiencies of the ad-hoc approach, they design a general framework that highlights the necessary design choices; we show it in Figure 8.1. The framework identifies the following steps:

- *data collection* that should yield real data after which synthetic data is supposed to be modeled;
- *data analysis* that identifies important properties of collected data, including user statistics, a classification of users and attackers, and system profiles;
- *profile generation* that identifies various user profiles to be used in synthetic data generation;
- *user and attack modeling* that produces (usually simplified) models of user and attacker behaviour, e.g., by a finite state machine as in [186], with user profiles serving as parameters for these simulators;
- *system modeling* that produces a simulator for the system's reactions to user actions; one can use the real system if it's available in software, but in general it is sufficient to limit the modeling to the aspects relevant for fraud or intrusion detection.

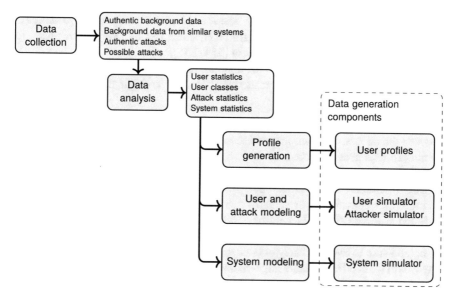

Fig. 8.1 Synthetic data generation methodology by Lundin et al. [563].

As a result of going through these steps, we obtain a set of user profiles that can be fed to user and attacker simulators, and the behaviour produced by these simulators is fed into the system simulator. Note that while this methodology may look trivial at first glance, the point is actually to highlight which design choices one has to make in constructing a synthetic data generation system; from this point of view, this methodology remains relevant to this day.

Apart from network fraud/intrusion detection, this methodology was put to work in, for instance, fraud detection in mobile transfer services [252] and generation of synthetic electronic health records [592].

Another similar field where synthetic data in the form of structured system logs is needed for fraud detection is financial security [552]. Here, real-life financial records usually represent sensitive information, and cases of fraud in real data are often hard to label, which also leads to the need for a synthetic generation. Here, we can highlight a sequence of works by two researchers from Blekinge Institute of Technology, E.A. Lopez-Rojas and S. Axelsson. In [550], they constructed a multiagent-based simulator (MABS) for synthetic data designed to train fraud detection systems. In subsequent works, they have applied this agent-based approach to several different domains where financial fraud needs to be detected:

- the original work [550] applied to money laundering detection, and later the same simulator was applied by the authors to detecting suspicious financial transactions in a mobile money system [551];

- in [553], Lopez-Rojas and Axelsson developed *RetSim*, an agent-based simulator of a shoe store based on real data from a Swedish retailer; *RetSim* was used for fraud detection in retail transactions in [554, 555];
- in [549], they extended the original simulator to *PaySim*, an agent-based simulator that simulates mobile money transactions based on an original dataset.

The basic idea of agent-based simulators is to generate entities corresponding to clients (e.g., users of a mobile payment system), set up the parameters of transactions between the clients, and then proceed to generate synthetic data with the purpose of training fraud detection models on it.

In general, structured synthetic data in the form of simulated system logs or synthetic electronic records has found plenty of applications in domains where real data is either unavailable or sensitive. In my personal opinion, however, there is an inherent trade-off in this kind of approach. With simplistic simulators that are programmed by hand or with tuning only a few distribution parameters from real data, we can be sure to avoid privacy concerns. But these simulators have limited value: why would we need to generate large-scale synthetic datasets to detect behaviour that we ourselves already know how to program? So to provide additional value on top of what we already know, synthetic data generation systems need to be able to learn patterns from real data that we cannot simply program into an agent's behaviour. And as soon as we rely on learning to uncover implicit patterns, privacy issues raise their heads.

So while some of these relatively early attempts we have discussed in this section can learn to reflect the original data distribution quite well, these methods actually have no inherent security needed to protect sensitive original data on which they are trained. Therefore, we will return to this form of data in more detail in Chapter 11, where we will discuss privacy guarantees in synthetic data. And now it is time to move forward to a completely different domain where the data is definitely not sensitive, and highly realistic (one could even say real) data can be quite easily generated.

8.2   Synthetic Data for Neural Programming

One interesting domain where synthetic data is paramount is neural program synthesis and neural program induction. The basic idea of teaching a machine learning model to program can be broadly divided into two subfields:

- *program induction* aims to train an end-to-end differentiable model to capture an algorithm [191];
- *program synthesis* tries to teach a model the semantics of a domain-specific language (DSL), so that the model is able to generate programs according to given specifications [432].

Basically, in program induction the network *is* the program, while in program synthesis the network *writes* the program.

Naturally, both tasks require large datasets of programs together with their input–output pairs. Since no such large datasets exist, and generating synthetic programs and running them in this case is relatively easy (arguably even easier than generating synthetic data for computer vision), all modern works use synthetic data to train the "neural computers".

In program induction, the tasks are so far relatively simple, and synthetic data generation does not present too many difficulties. For example, Joulin and Mikolov [415] present a new architecture (stack-augmented recurrent networks) to learn regularities in algorithmically generated sequences of symbols; the training data, as in previous such works [273, 350, 354, 926], is synthetically generated by handcrafted simple algorithms, including generating sequences from a pattern such as $a^n b^{2n}$ or $a^n b^m c^{n+m}$, binary addition (supervised problem asking to continue a string such as $110 + 10 =$), and similar formalizations of simple algorithmic tasks.

Zaremba and Sutskever [987] train a model to execute programs, i.e., map their textual representations to outputs; they generate training data as Python-style programs with addition, subtraction, multiplication, variable assignments, if-statements, and for-loops (but without nested loops), each ending with a `print` statement; see Fig. 8.2 for an illustration. Neural RAM machines [483] are trained on a number of simple tasks (access, increment, copy, etc.) whose specific instances are randomly synthesized. The same goes for neural Turing machines [296] and neural GPUs [241, 421]: they are trained and evaluated on synthetic examples generated for simple basic problems such as copying, sorting, and arithmetic operations.

In neural program synthesis, again, the programs are usually simple and also generated automatically; attempts to collect natural datasets for program synthesis have begun only very recently [988]. Learning-based program synthesis (earlier attempts had been based on logical inference [578]) began with learning string manipulation programs [304] and soon branched into deep learning, using recurrent architectures to guide search-based methods [894] or to generate programs in encoder–decoder architectures [138, 194]. In [101, 794], reinforcement learning is added on top of a recurrent architecture in order to alleviate the *program aliasing* problem, i.e., the fact that many different equivalent programs provide equally correct answers while the training set contains only one. All of the above-mentioned models were trained on synthetic datasets of randomly generated programs (usually in the form of abstract syntax trees) run on randomized sets of inputs.

Input:

```
j=8584
for x in range(8):
    j+=920
b=(1500+j)
print((b+7567))
```

Target: 25011.

Input:

```
i=8827
c=(i-5347)
print((c+8704) if 2641<8500 else 5308)
```

Target: 12184.

Fig. 8.2 Sample synthetic programs from [987].

As a separate thread, we mention works on program synthesis for visual reasoning and, generally speaking, question answering [362, 409, 759]. To do visual question answering [8], models are trained to compose a short program based on the natural language question that will lead to the answer, usually in the form of an execution graph or a network architecture. This line of work is based on neural module networks [22, 120] and similar constructions [23, 363], where the network learns to create a composition of modules (in QA, based on parsing the question) that are also neural networks, all learned jointly (see, however, the critique in [776]). Latest works use architectures based on self-attention mechanisms that have proven their worth across a wide variety of NLP tasks [507]. Naturally, most of these works use the CLEVR synthetic dataset for evaluation (see Section 6.7).

Generation of synthetic data itself has not been given much attention in this field until very recently, but the interest is rising. The work by Shin et al. [790] presents a study of contemporary synthetic data generation techniques for neural program synthesis and concludes that the resulting models do not capture the full semantics of a programming language, even if they do well on the test set. Shin et al. show that synthetic data generation algorithms, including the one from the popular *tensor2tensor* library [890], have biases that fail to cover important parts of the program space and deteriorate the final result. To fix this problem, they propose a novel methodology for generating distributions over the space of datasets for program induction and synthesis, showing significant improvements for two important domains: *Calculator* (which computes the results of arithmetic expressions) and *Karel* (which achieves a given objective with a virtual robot moving on a two-dimensional grid with walls and markers). We expect more research into synthetic data generation for neural program induction and synthesis to follow in the near future.

8.3 Synthetic Data in Bioinformatics

We use examples from the heathcare and biomedical domains throughout this book; see, e.g., Sections 10.7 and 11.5. In this section, we concentrate on applications of synthetic data in bioinformatics that fall outside either producing synthetic medical images (usually done with the help of generative models; see Section 10.7) or providing privacy guarantees for sensitive data through synthetic datasets (Section 11.5). It turns out that there are still plenty, and synthetic data is routinely used and generated throughout bioinformatics; see also a survey in [146]. Note that in this section, we will present some sample applications of generative models, and a full treatment of what deep generative models are will follow in Chapter 4.

For many of such methods in bioinformatics and medicine, generated synthetic data is the end goal rather than a tool to improve machine learning models. In particular, *de novo* drug design [324, 769] is a field that searches for molecules with desirable properties in a search space of about 10^{60} synthesizable molecules [267, 722]. The goal is to find (which in a space of this size rather means to *generate*) candidate molecules that would later have to be explored further in lab studies and then clinical trials.

First modern attempts at *de novo* drug design used rule-based methods that simulate chemical reactions [769], but the field soon turned to generative models, in particular based on deep learning [130, 268]. In this context, molecules are usually represented in the SMILES format [923] that encodes molecular graphs as strings in a certain formal grammar, which makes it possible to use sequence learning models to generate new SMILES strings. Segler et al. [773] used LSTM-based RNNs to learn a chemical language model, while Gómez-Bombarelli et al. [286] trained a variational autoencoder (VAE) which is already a generative model capable of generating new candidate molecules. To further improve generation, Kusner et al. [487] developed a novel extension of VAEs called Grammar Variational Autoencoders that can take into account the rules of the SMILES formal grammar and make VAE outputs conform to the grammar. To then use the models to obtain molecules with desired properties, researchers used either a small set of labeled positive examples [773] or augment the RNN training procedure with reinforcement learning [397]. In particular, Olivecrona et al. [640] use a recurrent neural network trained to generate SMILES representations: first, a prior network (3 layers of 1024 GRU) is trained in a supervised way on the RDKit subset [496] of ChEMBL database [267], then an RL agent (with the same structure, initialized from the prior network) is fine-tuned with the REINFORCE algorithm to improve the resulting SMILES encoding. A similar architecture, but with a stack-augmented RNN [416] as the basis, which enables more long-term dependencies, was presented by Popova et al. [681].

We especially note a series of works by *Insilico* researchers Kadurin, Polykovskiy, and others who applied different generative models to this problem:

- Kadurin et al. [419] train a supervised adversarial autoencoder [576] with the condition (in this case, growth inhibition percentage for tumor cells after treatment) added as a separate neuron to the latent layer;
- in [420], Kadurin et al. compared adversarial autoencoders (AAE) with variational autoencoders (VAE) for the same problem, with new modifications to the architecture that resulted in improved generation;
- in [679], Polykovskiy et al. introduced a new AAE modification, the entangled conditional adversarial autoencoder, to ensure the disentanglement of latent features; in this case, which is still quite rare for deep learning in drug discovery, a newly discovered molecule (a new inhibitor of Janus kinase 3) was actually tested in the lab and showed good activity and selectivity *in vitro*;
- Kuzmynikh et al. [488] presented a novel 3D molecular representation based on the wave transform that led to improved performance for CNN-based autoencoders and improved MACCS fingerprint prediction;
- Polykovskiy et al. [678] presented Molecular Sets (MOSES), a benchmarking platform for molecular generation models, which implemented and compared various generative models for molecular generation including CharRNN [774], VAE, AAE, and Junction Tree VAE [403], together with a variety of evaluation metrics for generation results.

The above works can be thought of as generation of synthetic data (in this case, molecular structures) that could be of direct use for practical applications.

Johnson et al. [406] undertake an ambitious project: they learn a generative model of the variation in cell and nuclear morphology based on fluorescence microscopy images. Their model is based on two adversarial autoencoders [576], one learning a probabilistic model of cell and nuclear shape and the other learning the interrelations between subcellular structures conditional on an encoding of the cell and nuclear shape from the first autoencoder. The resulting model produces plausible synthetic images of the cell with known localizations of subcellular structures.

One interesting variation of "synthetic data" in bioinformatics concerns learning from live experiments on synthetically generated biological material. For example, Rosenberg et al. [734] study alternative RNA splicing, in particular the functional effects of genetic variation on the molecular phenotypes through alternative splicing. To do that, they created a large-scale gene library with more than two million randomly generated synthetic DNA sequences, then used massively parallel reporter assays (MPRA) to measure the isoform ratio for all mini-genes in the experiment, and then used it to learn a (simple linear) machine learning model for alternative splicing. It turned out that this approach significantly improved prediction quality, outperforming state-of-the-art deep learning models for alternative splicing trained on the actual human genome [959] in predicting the results of *in vivo* experiments.

We also note a related field of imitational modeling for bioinformatics data that often results in realistic synthetic generators. For example, van den Bulcke et al. [99] provide a generator for synthetic gene expression data, able to produce synthetic transcriptional regulatory networks and simulated gene expression data while closely matching real statistics of biological networks.

8.4   Synthetic Data in Natural Language Processing

Synthetic data has not been widely used in natural language processing (NLP), at least by far not as widely as in computer vision. In our opinion, there is a conceptual reason for this. Compare with computer vision: there, the process of synthetic data generation can be done separately from learning the models, and the essence of what the models are learning is, in a way, "orthogonal" to the difference between real and synthetic data. If I show you a cartoon ad featuring a new bottle of soda, you will be able to find it in a supermarket even though you would never confuse the cartoon with a real photo.

In natural language processing, on the other hand, text generation is the hard problem itself. The problem of generating meaningful synthetic text with predefined target variables such as topic or sentiment is the subject of many studies in NLP; we are still a long way to go before it is solved entirely, and it is quite probable that when text generation finally reaches near-human levels, discriminative models for the target variables will easily follow from it, rendering synthetic data useless. In fact, recent developments in natural language processing, especially the GPT family of language models and BERT family of models for learning language representations based on self-attention, are already proving this: recent works show that these models can learn to perform complex NLP-related tasks with very little or no supervision [94, 698].

Nevertheless, there have been works that use data augmentation for NLP in a fashion that borders on using synthetic data. There have been simple augmentation approaches such as to simply drop out certain words [778]. A development of this idea shown in [913] switches out certain words, replacing them with random words from the vocabulary. The work [958] develops methods of data noising for language models, adding noise to word counts in a way reminiscent of smoothing in language models based on n-grams.

A more developed approach is to do data augmentation with synonyms: to expand a dataset, one can replace words with their synonyms, getting "synthetic sentences" that can still preserve target variables such as the topic of the text, its sentiment, and so on. Zhang et al. [1004] used this method directly to train a character-level network for text classification, while Galinsky et al. [256] tested augmentation with synonyms for morphology-rich languages such as Russian. In [230], augmentation with synonyms was used for low-resource machine translation, with an auxiliary LSTM-based language model used to recognize whether the synonym substitution is correct. Wang et al. [912], who concentrated on studying tweets, proposed to use embedding-based data augmentation, using neighboring words in the word vector space as synonyms. Kobayashi [461] extends augmentation with synonyms by replacing words in sentences with other words in paradigmatic relations with the original words, as predicted by a bidirectional language model at the word positions.

Techniques for generating synthetic text are constantly evolving. First, modern language models based on multi-head self-attention from the Transformer family, starting from the Transformer itself [891] and then further developed by BERT [192], OpenAI GPT [697], Transformer-XL [179], OpenAI GPT-2 [698], GROVER [990], and GPT-3 [94] generate increasingly coherent text. Actually, Zellers et al. [990] showed that their GROVER model for conditional generation (e.g., generating the text of a news article given its title, domain, and author), based on GPT-2, outperforms human-generated text in the "fake news"/"propaganda" category in terms of style and content (evaluated by humans), and GPT-3 is even better.

Moreover, recently developed models allow to generate text with GANs. This is a challenging problem because unlike, say, images, text is discrete and hence the generator output is not differentiable. There are several approaches to solving this problem:

- training with the REINFORCE algorithm and other techniques from reinforcement learning that are able to handle discrete outputs; this path has been taken in the pioneering model named SeqGAN [982], LeakGAN for generating long text fragments [305], and MaskGAN that learns to fill in missing text with an actor-critic conditional GAN [233], among others;
- approximating discrete sampling with a continuous function; these approaches include the Gumbel Softmax trick [486], TextGAN that approximates the arg max

function [1007], TextKD-GAN that uses an autoencoder to smooth the one-hot representation into a softmax output [313], and more;

- generating elements of the latent space for an autoencoder instead of directly generating text; this field started with adversarially regularized autoencoders by Zhao et al. [1018] and has been extended into text style transfer by disentangling style and content in the latent space [405], disentangling syntax and semantics [40], DialogWAE for dialog modeling [301], Bilingual-GAN able to generate parallel sentences in two languages [704], and other works.

This abundance of generative models for text has not, however, led to any significant use of synthetic data for training NLP models; this has been done only in very restricted domains such as electronic medical records [302] (we discuss this field in detail in Section 11.5). We have suggested the reasons for this in the beginning of this section, and so far the development of natural language processing supports our view.

8.5 Conclusion

Overall, in this chapter we have seen a bird's eye overview of using synthetic data in several different domains that do not usually spring to mind in discussions of synthetic data: tabular data such as system logs, data for neural programming, and synthetic data for natural language processing and for bioinformatics other than medical imaging. Sometimes this kind of synthetic data generation is still restricted (as in NLP), and sometimes it has different purposes (as in, e.g., *de novo* drug design), but all of these examples show how synthetic data is starting to gain traction even in less-than-obvious applications.

In the next chapter, we will consider several directions intended to improve synthetic data generation itself rather than find new applications for the result. We will see methods that aim to improve the CGI-based process of making synthetic data or find completely different ways to produce synthetic data: compose it from real data points or produce directly by generative models.

Chapter 9
Directions in Synthetic Data Development

In this chapter, we outline the main directions that we believe to represent promising ways to further improve synthetic data, making it more useful for a wide variety of applications in computer vision and other fields. In particular, we discuss the idea of domain randomization (Section 9.1) intended to improve the applications of synthetic datasets, methods to improve CGI-based synthetic data generation itself (Section 9.2), ways to create synthetic data from real images by cutting and pasting (Section 9.3), and finally possibilities to produce synthetic data by generative models (Section 9.4). The latter means generating useful synthetic data from scratch rather than domain adaptation and refinement, which we consider in a separate Chapter 10.

9.1 Domain Randomization

Domain randomization is one of the most promising approaches to make straightforward transfer learning from synthetic-to-real data actually work. The basic idea of domain randomization had been known since the 1990s [394], but was probably first explicitly presented and named in [861]. Consider a model that is supposed to train on a synthetic dataset $D_{syn} \sim p_{syn}$, where p_{syn} denotes the distribution of synthetic data, and later be applied to a real dataset $D_{real} \sim p_{real}$, where p_{real} is the distribution of real data. The idea is simple: let us try to make the synthetic data distribution p_{syn} sufficiently wide and varied so that the model trained on p_{syn} will be robust enough to work well on p_{real}.

Ideally, we would like to cover p_{real} with p_{syn}, but in reality this is never achieved directly. Instead, synthetic data in computer vision can be randomized and made more diverse in a number of different ways at the level of either constructing a 3D scene or rendering 2D images from it

S. I. Nikolenko, *Synthetic Data for Deep Learning*, Springer Optimization and Its Applications 174, https://doi.org/10.1007/978-3-030-75178-4_9

227

Fig. 9.1 Sample images generated by the domain randomization approach by Tremblay et al. [867] for an outdoor driving dataset.

- at the scene construction level, a synthetic data generator (SDG) can randomize the number of objects, its relative and absolute positions, number and shape of distractor objects, contents of the scene background, textures of all objects participating in the scene, and so on;
- at the rendering level, SDG can randomize lighting conditions, in particular, the position, orientation, and intensity of light sources, change the rendering quality by modifying image resolution, rendering type such as ray tracing or other options, add random noise to the resulting images, and so on.

Tobin et al. [861] made the first steps to show that domain randomization works well; they used simple geometric shapes (polyhedra) as both target and distractor objects, random textures such as gradient fills or checkered patterns. The authors found that synthetic pretraining is indeed very helpful when only a small real training set is available, but helpful only if sufficiently randomized, in particular, when using a large number of random textures.

This approach was subsequently applied to a more ambitious domain by NVIDIA researchers Tremblay et al. [867], who trained object detection models on synthetic data with the following procedure:

- create randomized 3D scenes, adding objects of interest on top of random surfaces in the scenes;
- add so-called "flying distractors", diverse geometric shapes that are supposed to serve as negative examples for object detection;
- add random textures to every object, randomize the camera settings, lighting, and other parameters.

The resulting images are completely unrealistic (see Fig. 9.1 for a few samples that are supposed to represent outdoor scenes to train the detection of cars), yet diverse enough that the networks have to concentrate on the shape of the objects in question. Tremblay et al. report improved car detection results for R-FCN [178] and SSD [539] architectures (but failing to improve Faster R-CNN [719]) on their dataset compared to Virtual KITTI (see Section 7.2), as well as improved results on hybrid datasets (adding a domain-randomized training set to COCO [525]), a detailed ablation study, and extensive experiments showing the effect of various hyperparameters.

Since then, domain randomization has been used and further developed in many works. Borrego et al. [85] aim to improve object detection for common objects, showing that domain randomization in the synthetic part of the dataset significantly improves the results. Tobin et al. [860] consider robotic grasping, a problem where the lack of real data is especially dire (see also Sections 7.4 and 10.6). They use domain randomization to generate a wide variety of unrealistic procedurally generated object meshes and textured objects for grasping, so that a model trained on them would generalize to real objects as well. They show that a grasping model trained entirely on non-realistic procedurally generated objects can be successfully transferred to realistic objects.

Up until recently, domain randomization had operated under the assumption that realism is not necessary in synthetic data. Prakash et al. [684] take the next logical step, continuing this effort to *structured* domain randomization. They still randomize all of the settings mentioned above, but only within realistic ranges, taking into account the structure and context of a specific scene.

Finally, another important direction is learning *how* to randomize. Van Vuong et al. [897] provide one of the first works in this direction, concentrating on picking the best possible domain randomization parameters for sim-to-real transfer of reinforcement learning policies. They show that the parameters that control sampling over Markov decision processes are important for the quality of transferring the learned policy to a real environment and that these parameters can be optimized. We mark this as a first attempt and expect more works devoted to structuring and honing the parameters of domain randomization.

9.2 Improving CGI-Based Generation

The basic workflow of synthetic data in computer vision is relatively straightforward: prepare the 3D models, place them in a controlled scene, set up the environment (camera type, lighting etc.), and render synthetic images to be used for training. However, some works on synthetic data present additional ways to enhance the data not by domain adaptation/refinement to real images (we will discuss these approaches in Section 10), but directly on the stage of CGI generation.

There are two different directions for this kind of added realism in CGI generation. The first direction is to make more realistic objects. For example, Wang et al. [904] recognize retail items in a smart vending machine; to simulate natural deformations in the objects, they use a surface-based mesh deformation algorithm proposed in [906], introducing and minimizing a global energy function for the object's mesh that accounts for random deformations and rigidity properties of the material (Wang et al. also use GAN-based refinement, see Section 10.3). Another approach, initiated by Rozantsev et al. [738], is to estimate the rendering parameters required to synthetize similar images from data; this approach ties into the synthetic data generation feedback loop that we discuss in Section 12.2.

The second direction is to make more realistic "sensors", introducing synthetic data postprocessing that mimics the noise characteristics of real cameras/sensors. For example, we discussed *DepthSynth* by Planche et al. [677] (see Section 6.5), a system that makes simulated depth data more realistic, more similar to real depth sensors, while the OVVV system by Taylor et al. [853] (Section 6.6), and the ICL-NUIM dataset by Handa et al. [321] (Section 7.2) take special care to simulate the noise of real cameras. There is even a separate area of research completely devoted to better modeling of the noise and distortions in real-world cameras [72]

Apart from added realism on the level of images, there is also the question of high-level coherence and realism of the scenes. While there is no problem with coherence when the scenes are done by hand, the scale of modern datasets requires to automate scene composition as well. We note a recent joint effort in this direction by NVIDIA, University of Toronto, and MIT: Kar et al. [433] present *Meta-Sim*, a general framework that learns to generate synthetic urban environments (see also Section 7.2). *Meta-Sim* represents the composition of a 3D scene with a *scene graph* and a *probabilistic scene grammar*, a common representation in computer graphics [1026]. The goal is to learn how to transform samples coming from the probabilistic grammar so that the distribution of synthetic scenes becomes similar to the distribution of scenes in a real dataset; this is known as bridging the *distribution gap*. What's more, *Meta-Sim* can also learn these transformations with the objective of improving the performance of networks trained on the resulting synthetic data for a specific task such as object detection (see also Section 12.2).

There are also a number of domain-specific developments that improve synthetic data generation for specific fields. For example, Cheung et al. [145] present *LCrowdV*, a generation framework for crowd videos that combines a procedural simulation framework that concentrates of movements and human behaviour and a rendering framework for image/video generation, while Anderson et al. [20] develop a method for stochastic sampling-based simulation of pedestrian trajectories (see Section 6.6).

In general, while computer graphics is increasingly using machine learning to speed up rendering (by, e.g., learning approximations to complex computationally intensive transformations [424, 651]) and improve the resulting 3D graphics, works on synthetic data seldom make use of these advances; a need to improve CGI-based synthetic data is usually considered in the direction of making it more realistic with refinement models (see Section 10.1). However, we do expect further interesting developments in specific domains, especially in situations where the characteristics of specific sensors are important (such as, e.g., LIDARs in autonomous vehicles).

9.3 Compositing Real Data to Produce Synthetic Datasets

Another notable line of work that, in our opinion, lies at the boundary between synthetic data and data augmentation is to use combinations and fusions of different real images to produce a larger and more diverse set of images for training. This does not require the use of CGI for rendering the synthetic images, but does require a dataset of real images.

Early works in this direction were limited by the quality of segmentation needed to cut out real objects. For some problems, however, it was easy enough to work. For example, Eggert et al. [221] concentrate on company logo detection. To generate synthetic images, they use a small number of real base images where the logos are clearly visible and supplied with segmentation masks, apply random warping, color transformations, and blurring, and then paste the modified (segmented) logo onto a new background image. Training on this extended dataset yielded improvements in logo detection results. In Section 6.6, we have discussed the "Frankenstein" pipeline for compositing human faces [360].

The field started in earnest with the *Cut, Paste, and Learn* approach by Dwibedi et al. [213], which is based on the assumption that only *patch-level realism* is needed to train, e.g., an object detector. They take a collection of object instance images, cut them out with a segmentation model (assuming that the instance images are simple enough that segmentation will work almost perfectly), and paste them onto randomized background scenes, with no regard to preserving scale or scene composition. Dwibedi et al. compare different classical computer vision blending approaches (e.g., Gaussian and Poisson blending [669]) to alleviate the influence of boundary artifacts after the paste; they report improved instance detection results. The work on cut-and-paste was later extended with GAN-based models (used for more realistic pasting and inpainting) and continued in the direction of unsupervised segmentation by Remez et al. [716] and Ostyakov et al. [648].

Subsequent works extend this approach for generating more realistic synthetic datasets. Dvornik et al. [212] argue that an important problem for this type of data augmentation is to preserve visual context, i.e., make the environment around the objects more or less realistic. They describe a preliminary experiment where they placed segmented objects at completely random positions in new scenes and not only did not see significant improvements for object detection on the VOC'12 dataset, but actually saw the performance deteriorate, regardless of the distractors or strategies used for blending and boundary artifact removal. Therefore, they added a separate model (also a CNN) that predicts what kind of objects can be placed in a given bounding box of an image from the rest of the image with this bounding box masked out; then the trained model is used to evaluate potential bounding boxes for data augmentation, choose the ones with the best object category score, and then paste a segmented object of this category in the bounding box. The authors report improved object detection results on VOC'12.

Wang et al. [903] develop this into an even simpler idea of *instance switching*: let us switch only instances of the same class between different images in the training set; in this way, the context is automatically right, and shape and scale can also be taken into account. Wang et al. also propose to use instance switching to adjust the distribution of instances across classes in the training set and account for class importance by adding more switching for classes with lower scores. The resulting PSIS (Progressive and Selective Instance Switching) system provides improved results on the MS COCO dataset for various object detectors including Faster-RCNN [719], FPN [523], Mask R-CNN [327], and SNIPER [803].

For a detailed consideration, let us consider a recent work by Jin and Rinard [402] who take this basic cut-and-paste approach to the next level. In essence, they still use the same basic pipeline:

- take an object space O consisting of synthetic objects placed in random poses and subjected to a number of different augmentations;
- take a context space C consisting of background images;
- superimpose objects from O against backgrounds from C at random;
- train a neural network on the resulting composite images.

However, Jin and Rinard consider this approach in detail and introduce several important tricks that allow this simple approach to provide some of the very best results available in domain adaptation and few-shot learning.

First, the sampling. One common pitfall of computer vision is that when you have relatively few examples of a class, they cannot come in a wide variety of backgrounds. Hence, in a process akin to overfitting the networks might start learning the characteristic features of the backgrounds rather than the objects in this class.

What is the easiest way out of this? How can we tell the classifier that it's the object that's important and not the background? With synthetic images, it's easy: let us place several different objects on the same background! Then, since the labels are different, the classifier will be forced to learn that backgrounds are not important and it is the objects that differentiate between classes. Therefore, Jin and Rinard take care to introduce *balanced sampling* of objects and backgrounds. The basic procedure samples a random biregular graph so that every object is placed on an equal number of backgrounds and vice versa, every background is used with the same number of objects.

The other idea used by Jin and Rinard stems from the obvious fact that the classifier must learn to distinguish between different objects. Therefore, it would be beneficial for training to concentrate on the hard cases where the classifier might confuse two objects. In [402], this idea comes in two flavors. First, specifically for images the authors suggest to superimpose one object on top of another, so that the previous object provides a maximally confusing context for the next one. Second, they use *robustness training*, a method basically equivalent to self-adversarial training that we discussed in Section 3.4 but applied to synthetic images here. The idea is that if we are training on synthetic image that might look a little unrealistic and might not be hard enough to confuse even an imperfect classifier, we can try to make it harder for the classifier by turning it into an adversarial example.

With all these ideas combined, Jin and Rinard obtain a relatively simple pipeline that is able to achieve state-of-the-art results by training with only *a single synthetic image* of each object class. Note that there is no complex domain adaptation here: all ideas can be thought of as smart augmentations similar to the ones we considered in Section 3.4.

With the development of conditional generative models, this field has blossomed into more complex conditional generation, usually called *image fusion*, that goes beyond cut-and-paste; we discuss these extensions in Section 10.4.

9.4   Synthetic Data Produced by Generative Models

Generative models, especially generative adversarial networks (GAN) [290] that we will discuss in detail in Chapter 4, are increasingly being used for domain adaptation, either in the form of refining synthetic images to make them more realistic or in the form of "smart augmentation", making nontrivial transformations on real data. We discuss these techniques in Chapter 10. Producing synthetic data directly from random noise for classical computer vision applications generally does not sound promising: GANs can only try to approximate what is already in the data, so why can't the model itself do it? However, in a number of applications synthetic data produced by GANs directly from random noise, usually with an abstract condition such as a segmentation mask, can help; in this section, we consider several examples of these approaches.

Counting (objects on an image) is a computer vision problem that, formally speaking, reduces to object detection or segmentation but in practice is significantly harder: to count correctly the model needs to detect all objects on the image, missing not a single one. Large datasets are helpful for counting, and synthetic data generated with a GAN conditioned on the number of objects or a segmentation mask with known number of objects, either produced at random or taken from a labeled real dataset, proves to be helpful. In particular, there is a line of work that deals with leaf counting on images of plants: ARIGAN by Giuffrida et al. [278] generates images of arabidopsis plants conditioned on the number of leaves, Zhu et al. generate the same conditioned on segmentation masks [1028], and Kuznichov et al. [490] generate synthetically augmented data that preserves the geometric structure of the leaves; all works report improved counting.

Santana and Hotz [758] present a generative model that can learn to generate realistic looking images and even videos of the road for potential training of self-driving cars. Their model is a VAE+GAN autoencoder based on the architecture from [497] that is combined with a recurrent transition model that learns realistic transitions in the embedded space. The resulting model produces synthetic videos that preserve road texture, lane markings, and car edges, keeping the road structure for at least 100 frames of the video. This interesting approach, however, has not yet led to any improvements in the training of actual driving agents.

It is hard to find impressive applications where synthetic data is generated purely from scratch by generative models; as we have discussed, this may be a principled

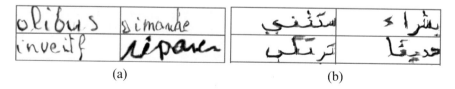

Fig. 9.2   Sample handwritten text generated by Alonso et al. [14]: (a) French; (b) Arabic.

limitation. Still, even a small amount of additional supervision may do. For example, Alonso et al. [14] consider adversarial generation of handwritten text (see also Section 6.7). They condition the generator on the text itself (sequence of characters), generate handwritten instances for various vocabulary words, and augment the real RIMES dataset [299] with the resulting synthetic dataset (Fig. 9.2). Alonso et al. report improved character recognition performance in terms of both edit distance and word error rate. This example shows that synthetic data does not need to involve complicated 3D modeling to work and improve results; in this case, all information Alonso et al. provided for the generative model was a vocabulary of words.

A related but different field considers unsupervised approaches to segmentation and other computer vision problems based on adversarial architectures, including learning to segment via cut-and-paste [716], unsupervised segmentation by moving objects between pairs of images with inpainting [648], segmentation learned from unannotated medical images [1011], and more [70]. While this is not synthetic data *per se*, in general we expect unsupervised approaches to computer vision to be an important trend in the use of synthetic data.

At this point, we have seen many examples and applications of synthetic data. Most synthetic data generation that we have encountered has involved manual components: for instance, in computer vision, the 3D scene is usually set up by hand, with manually crafted 3D objects. However, we have already seen a few cases where synthetic data can be produced automatically with generative models. What's even more important in the context of synthetic data applications, generative models can help adapt synthetic data to make it more realistic, or adapt models for downstream tasks to work well on real data after training on synthetic. We have already introduced generative models and specifically GAN-based architectures in Chapter 4, and in the next chapter, it is time to put them to work for synthetic-to-real domain adaptation.

Chapter 10
Synthetic-to-Real Domain Adaptation and Refinement

Domain adaptation is a set of techniques aimed to make a model trained on one domain of data to work well on a different target domain. In this chapter, we give a survey of domain adaptation approaches that have been used for synthetic-to-real adaptation, that is, methods for making models trained on synthetic data work well on real data, which is almost always the end goal. We distinguish two main approaches. In *synthetic-to-real refinement* input synthetic data is modified, usually to be made more realistic, and we can actually see the modified data. In *model-based domain adaptation*, it is the training process or the model structure that changes to ensure domain adaptation, while the data remains as synthetic as it has been. We will discuss neural architectures for both approaches, including many models based on generative adversarial networks.

10.1 Synthetic-to-Real Domain Adaptation and Refinement

So far, we have discussed direct applications where synthetic data has been used to augment real datasets of insufficient size or to create virtual environments for training. In this chapter, we proceed to methods that can make the use of synthetic data much more efficient. *Domain adaptation* is a set of techniques designed to make a model trained on one domain of data, the *source* domain, work well on a different, *target* domain. This is a natural fit for synthetic data: in almost all applications, we would like to train the model in the source domain of synthetic data but then apply the results in the target domain of real data.

In this chapter, we give a survey of domain adaptation approaches that have been used for such synthetic-to-real adaptation. We broadly divide the methods outlined in this chapter into two groups. Approaches from the first group operate on the data level, which makes it possible to extract synthetic data "refined" in order to work better on real data, while approaches from the second group operate directly on the model, its feature space, or training procedure, leaving the data itself unchanged. We concentrate mostly on recent work related to deep neural networks and refer to, e.g., the survey [660] for an overview of earlier work.

In Section 10.1, we discuss synthetic-to-real refinement, where a model learns to make synthetic "fake" data more realistic with an adversarial framework; we begin with a case study on gaze estimation (Section 10.2), which was one of the first applications for this field, and then proceed to other applications of such refiners (Section 10.3) and GAN-based models that work in the opposite direction, making real data more "synthetic-like" (Section 10.4). In Section 10.5, we discuss domain adaptation at the feature and model level, i.e., methods that perform synthetic-to-real domain adaptation but do not necessarily yield more realistic synthetic data as a by-product. Section 10.6 is devoted to domain adaptation in control and robotics, and in Section 10.7 we present a case study of adversarial architectures for medical imaging, one of the fields where synthetic data produced with GANs can significantly improve results.

We concentrate mostly on recent work related to deep neural networks and refer to, e.g., the survey [660] for an overview of earlier work.

The first group of approaches for synthetic-to-real domain adaptation work with the data itself. The models below can take a synthetic image and "refine" it, making it better for subsequent model training. Note that while in most works we discuss here the objective is basically to make synthetic data more realistic (and it is supported by discriminators that aim to distinguish refined synthetic data from real samples), this does not necessarily have to be the case; some early works on synthetic data concluded that, e.g., synthetic imagery may work better if it is less realistic, resulting in better generalization of the models; we have discussed this, e.g., in Section 9.1.

We begin with a case study on a specific problem that kickstarted synthetic-to-real refinement and then proceed to other approaches, both refining already existing synthetic data and generating new synthetic data from real by generative manipulation.

10.2 Case Study: GAN-Based Refinement for Gaze Estimation

One of the first successful examples of straightforward synthetic-to-real refinement was given by Apple researchers Shrivastava et al. in [793], so we begin by considering this case study in more detail and show how the research progressed afterwards. The underlying problem here is *gaze estimation*: recognizing the direction where a human eye is looking. Gaze estimation methods are usually divided into *model-based*, which model the geometric structure of the eye and adjacent regions, and *appearance-based*, which use the eye image directly as input; naturally, synthetic data is made and refined for the latter class of approaches.

Before [793], this problem had already been tackled with synthetic data. Wood et al. [933, 934] presented a large dataset of realistic renderings of human eyes and showed improvements on real test sets over previous work done with the *MPIIgaze*

Fig. 10.1 Synthetic images used to train gaze estimation models: (a) sample images from *UnityEyes* [934]; (b) sample images from UnityEyes (top) refined by SimGAN (bottom) [793].

dataset of real labeled images [1003]. Note that the usual increase in scale here is manifested as an increase in variability: *MPIIgaze* contains about 214K images, and the synthetic training set was only about 1M images, but all images in *MPIIgaze* come from the same 15 participants of the experiment, while the *UnityEyes* system developed in [934] can render every image in a different randomized environment, which makes the model significantly more robust. Sample images from the *UnityEyes* dataset are shown in Fig. 10.1a.

Shrivastava et al. further improve upon this result by presenting a GAN-based system trained to improve synthesized images of eyes, making them more realistic. They call this idea *Simulated+Unsupervised learning*, learning a transformation implemented with a *Refiner* network with the *SimGAN* adversarial architecture. Sim-GAN consists of a generator (refiner) G_θ^{REF} with parameters θ and a discriminator D_ϕ^{REF} with parameters ϕ; see Fig. 10.2 for an illustration. The discriminator learns to distinguish between real and refined images with standard binary classification loss function

$$\mathcal{L}_D^{\text{REF}}(\phi) = -\mathbb{E}_S \left[\log D_\phi^{\text{REF}}(\hat{\mathbf{x}}_S) \right] - \mathbb{E}_T \left[\log \left(1 - D_\phi^{\text{REF}}(\mathbf{x}_T) \right) \right],$$

where $\hat{\mathbf{x}}_S = G_\theta^{\text{REF}}(\mathbf{x}_S)$ is the refined version of $\mathbf{x}_S$ produced by G_θ^{REF}. The generator, in turn, is trained with a combination of the realism loss $\mathcal{L}_{\text{real}}^{\text{REF}}$ that makes G_θ^{REF} learn to fool D_ϕ^{REF} and regularization loss $\mathcal{L}_{\text{reg}}^{\text{REF}}$ that captures the similarity between the refined image and the original one in order to preserve the target variable (gaze direction in [793]):

$$\mathcal{L}_G^{\text{REF}}(\theta) = \mathbb{E}_S \left[\mathcal{L}_{\text{real}}^{\text{REF}}(\theta; \mathbf{x}_S) + \lambda \mathcal{L}_{\text{reg}}^{\text{REF}}(\theta; \mathbf{x}_S) \right],$$

where

$$\mathcal{L}_{\text{real}}^{\text{REF}}(\theta; \mathbf{x}_S) = -\log \left(1 - D_\phi^{\text{REF}}(G_\theta^{\text{REF}}(\mathbf{x}_S)) \right),$$

$$\mathcal{L}_{\text{reg}}^{\text{REF}}(\theta; \mathbf{x}_S) = \left\| \psi(G_\theta^{\text{REF}}(\mathbf{x}_S)) - \psi(\mathbf{x}_S) \right\|_1,$$

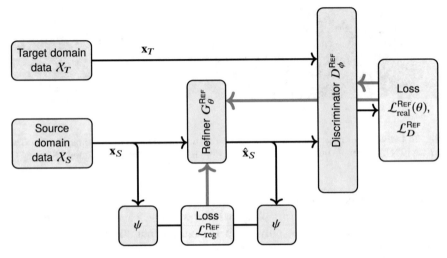

Fig. 10.2 The architecture of SimGAN, a GAN-based refiner for synthetic data [793].

where $\psi(\mathbf{x})$ is a mapping to a feature space (that can contain the image itself, image derivatives, statistics of color channels, or features produced by a fixed extractor such as a pretrained CNN), and $\|\cdot\|_1$ denotes the L_1 distance. On Fig. 10.2, black arrows denote the data flow and green arrows show the gradient flow (on subsequent pictures, we omit the gradient flow to avoid clutter); $\mathcal{L}_{\mathrm{real}}^{\mathrm{REF}}(\boldsymbol{\theta})$ and $\mathcal{L}_D^{\mathrm{REF}}(\boldsymbol{\phi})$ are shown in the same block since it is the same loss function differentiated with respect to different weights for G and D, respectively.

In SimGAN, the generator is a fully convolutional neural network that consists of several ResNet blocks [328] and does not contain any striding or pooling, which makes it possible to operate on pixel level while preserving the global structure. The training proceeds by alternating between minimizing $\mathcal{L}_G^{\mathrm{REF}}(\boldsymbol{\theta})$ and $\mathcal{L}_D^{\mathrm{REF}}(\boldsymbol{\phi})$, with an additional trick of drawing training samples for the discriminator from a stored history of refined images in order to keep it effective against all versions of the generator. Another important feature is the locality of adversarial loss: D_{ϕ}^{REF} outputs a probability map on local patches of the original image, and $\mathcal{L}_D^{\mathrm{REF}}(\boldsymbol{\phi})$ is summed over the patches. Sample results are shown in Fig. 10.1b; it is clear that SimGAN results (shown on the bottom in Fig. 10.1b) look more realistic than original synthetic images (on top in Fig. 10.1b).

SimGAN's ideas were later picked up and extended in many works. A direct successor of SimGAN, GazeGAN developed by Sela et al. [777], applied to synthetic data refinement the idea of CycleGAN for unpaired image-to-image translation [1025] (recall Section 4.7). The structure of GazeGAN contains four networks: G^{GZ} is the generator that learns to map images from the synthetic domain S to the real domain R, F^{GZ} learns the opposite mapping, from R to S, and two discriminators D_S^{GZ} and D_R^{GZ} learn to distinguish between real and fake images in the synthetic and real domains, respectively.

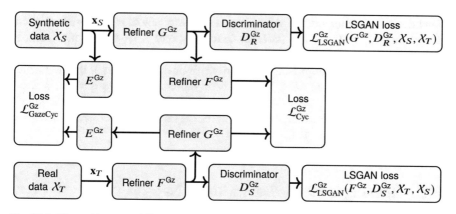

Fig. 10.3 The architecture of GazeGAN [777]. Blocks with identical labels have shared weights.

An overview of the GazeGAN architecture is shown in Fig. 10.3. It uses the following loss functions:

- the LSGAN [580] loss for the generator with label smoothing to 0.9 [668] to stabilize training:

$$\mathcal{L}_{\text{LSGAN}}^{\text{Gz}}(G, D, S, R) = \mathbb{E}_{\mathbf{x}_S \sim p_{\text{syn}}}\left[(D(G(\mathbf{x}_S)) - 0.9)^2\right] + \mathbb{E}_{\mathbf{x}_T \sim p_{\text{real}}}\left[D(\mathbf{x}_T)^2\right];$$

this loss is applied to both directions, as $\mathcal{L}_{\text{LSGAN}}^{\text{Gz}}(G^{\text{Gz}}, D_R^{\text{Gz}}, \mathcal{X}_S, \mathcal{X}_T)$ and also as $\mathcal{L}_{\text{LSGAN}}^{\text{Gz}}(F^{\text{Gz}}, D_S^{\text{Gz}}, \mathcal{X}_T, \mathcal{X}_S)$;
- the *cycle consistency loss* [1025] designed to make sure both $F \circ G$ and $G \circ F$ are close to identity:

$$\mathcal{L}_{\text{Cyc}}^{\text{Gz}}(G^{\text{Gz}}, F^{\text{Gz}}) = \mathbb{E}_{\mathbf{x}_S \sim p_{\text{syn}}}\left[\|F^{\text{Gz}}(G^{\text{Gz}}(\mathbf{x}_S)) - \mathbf{x}_S\|_1\right] + \\ \mathbb{E}_{\mathbf{x}_T \sim p_{\text{real}}}\left[\|G^{\text{Gz}}(F^{\text{Gz}}(\mathbf{x}_T)) - \mathbf{x}_T\|_1\right];$$

- finally, a special *gaze cycle consistency loss* to preserve the gaze direction (so that the target variable can be transferred with no change); for this, the authors train a separate gaze estimation network E^{Gz} designed to overfit and predict the gaze very accurately on synthetic data; the loss makes sure E^{Gz} still works after applying $F \circ G$:

$$\mathcal{L}_{\text{GazeCyc}}^{\text{Gz}}(G^{\text{Gz}}, F^{\text{Gz}}) = \mathbb{E}_{\mathbf{x}_S \sim p_{\text{syn}}}\left[\|E^{\text{Gz}}(F^{\text{Gz}}(G^{\text{Gz}}(\mathbf{x}_S))) - E^{\text{Gz}}(\mathbf{x}_S)\|_2^2\right].$$

Sela et al. report improved gaze estimation results. Importantly for us, they operate not on the 30×60 grayscale images as in [793], but on 128×128 color images, and GazeGAN actually refines not only the eye itself but parts of the image (e.g., nose and hair) that were not part of the 3D model of the eye.

Finally, a note of caution: GAN-based refinement is not the only way to go. Kan et al. [427] compared three approaches to data augmentation for pupil center point detection, an important subproblem in gaze estimation: affine transformations of real images, synthetic images from *UnityEyes*, and GAN-based refinement. In their experiments, real data augmentation with affine transformations was a clear winner, with the GAN improving over *UnityEyes* but falling short of the augmented real dataset. This is one example of a general common wisdom: in cases where a real dataset is available, one should squeeze out all the information available in it and apply as much augmentation as possible, regardless of whether the dataset is augmented with synthetic data or not.

10.3 Refining Synthetic Data with GANs

Gaze estimation is a convenient problem for GAN-based refining because the images of eyes used for gaze estimation have relatively low resolution, and scaling GANs up to high-resolution images has proven to be a difficult task in many applications. Nevertheless, in this section, we consider a wider picture of other GAN-based refiners applied for synthetic-to-real domain adaptation.

We begin with an early work in refinement, parallel to [793], which was done by *Google* researchers Bousmalis et al. [87]. They train a GAN-based architecture for pixel-level domain adaptation (PixelDA), using a basic style transfer GAN (basically *pix2pix* that we discussed in Section 4.7), i.e., by alternating optimization steps they solve

$$\min_{\theta_G, \theta_T} \max_{\phi} \lambda_1 \mathcal{L}_{\text{dom}}^{\text{PIX}}(D^{\text{PIX}}, G^{\text{PIX}}) + \lambda_2 \mathcal{L}_{\text{task}}^{\text{PIX}}(G^{\text{PIX}}, T^{\text{PIX}}) + \lambda_3 \mathcal{L}_{\text{cont}}^{\text{PIX}}(G^{\text{PIX}}),$$

where

- $\mathcal{L}_{\text{dom}}^{\text{PIX}}(D^{\text{PIX}}, G^{\text{PIX}})$ is the domain loss,

$$\mathcal{L}_{\text{dom}}^{\text{PIX}}(D^{\text{PIX}}, G^{\text{PIX}}) = \mathbb{E}_{\mathbf{x}_S \sim p_{\text{syn}}} \left[\log \left(1 - D^{\text{PIX}}(G^{\text{PIX}}(\mathbf{x}_S; \theta_G); \phi) \right) \right] + \mathbb{E}_{\mathbf{x}_T \sim p_{\text{real}}} \left[\log D^{\text{PIX}}(\mathbf{x}_T; \phi) \right];$$

- $\mathcal{L}_{\text{task}}^{\text{PIX}}(G^{\text{PIX}}, T^{\text{PIX}})$ is the task-specific loss, which in [87] was the image classification cross-entropy loss provided by a classifier $T^{\text{PIX}}(\mathbf{x}; \theta_T)$ which is also trained as part of the model:

$$\mathcal{L}_{\text{task}}^{\text{PIX}}(G^{\text{PIX}}, T^{\text{PIX}}) = \\ \mathbb{E}_{\mathbf{x}_S, \mathbf{y}_S \sim p_{\text{syn}}} \left[-\mathbf{y}_S^\top \log T^{\text{PIX}}(G^{\text{PIX}}(\mathbf{x}_S; \theta_G); \theta_T) - \mathbf{y}_S^\top \log T^{\text{PIX}}(\mathbf{x}_S; \theta_T) \right];$$

- $\mathcal{L}_{\text{cont}}^{\text{PIX}}(G^{\text{PIX}})$ is the content similarity loss, intended to make G^{PIX} preserve the parts of the image related to target variables; in [87], $\mathcal{L}_{\text{cont}}^{\text{PIX}}$ was used to preserve foreground objects (that would later need to be classified) with a mean squared error applied to their masks:

$$\mathcal{L}_{\text{cont}}^{\text{PIX}}(G^{\text{PIX}}) = \mathbb{E}_{\mathbf{x}_S \sim p_{\text{syn}}} \left[\frac{1}{k} \left\| (\mathbf{x}_S - G^{\text{PIX}}(\mathbf{x}_S; \boldsymbol{\theta}_G)) \odot \mathbf{m}(\mathbf{x}) \right\|_2^2 - \right.$$
$$\left. - \frac{1}{k^2} \left((\mathbf{x}_S - G^{\text{PIX}}(\mathbf{x}_S; \boldsymbol{\theta}_G))^\top \mathbf{m}(\mathbf{x}) \right)^2 \right],$$

where $\mathbf{m}(\mathbf{x}_S)$ is a segmentation mask for the foreground object extracted from the synthetic data renderer; note that this loss does not "insist" on preserving pixel values in the object but rather encourages the model to change object pixels in a consistent way, preserving their pairwise differences.

Bousmalis et al. applied this GAN to the *Synthetic Cropped LineMod* dataset, a synthetic version of a small object classification dataset [931], doing both classification and pose estimation for the objects. They report improved results in both metrics compared to training on purely synthetic data and to a number of previous approaches to domain adaptation.

Many modern approaches to synthetic data refinement include the ideas of Cycle-GAN [1025]. The most direct application is the *GeneSIS-RT* framework by Stein and Roy [818] that refines synthetic data directly with a CycleGAN trained on unpaired datasets of synthetic and real images. They show that a training set produced by image-to-image translation learned by CycleGAN improves the results of training machine learning systems for real-world tasks such as obstacle avoidance and semantic segmentation.

T^2Net by Zheng et al. [1020] uses synthetic-to-real refinement for depth estimation from a single image. This work also uses the general ideas of CycleGAN with a translation network that makes the images more realistic. The new idea here is that T^2Net asks the synthetic-to-real generator G_S^{T2} not only to translate one specific domain (synthetic data) to another (real data) but also to work across a number of different input domains, making the input image "more realistic" in every case, as shown in Figure 10.4.

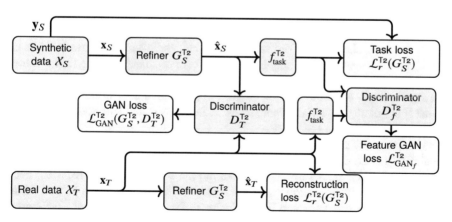

Fig. 10.4 The architecture of T^2Net [1020]. Blocks with identical labels have shared weights.

In essence, this means that G_S^{T2} aims to learn the minimal transformation necessary to make an image realistic, in particular, it should not change real images much. In total, T^2Net has the generator loss function

$$\mathcal{L}^{T2} = \mathcal{L}_{GAN}^{T2}(G_S^{T2}, D_T^{T2}) + \lambda_1 \mathcal{L}_{GAN_f}^{T2}(f_{task}^{T2}, D_f^{T2}) + \lambda_2 \mathcal{L}_r^{T2}(G_S^{T2})$$
$$+ \lambda_3 \mathcal{L}_t^{T2}(f_{task}^{T2}) + \lambda_4 \mathcal{L}_s^{T2}(f_{task}^{T2}),$$

where

- $\mathcal{L}_{GAN}^{T2}(G_S^{T2}, D_T^{T2})$ is the usual GAN loss for synthetic-to-real transfer with discriminator D_T^{T2}:

$$\mathcal{L}_{GAN}^{T2}(G_S^{T2}, D_T^{T2}) = \mathbb{E}_{\mathbf{x}_S \sim p_{syn}} \left[\log(1 - D_T^{T2}(G_S^{T2}(\mathbf{x}_S))) \right] + \mathbb{E}_{\mathbf{x}_T \sim p_{real}} \left[\log D_T^{T2}(\mathbf{x}_T) \right] ;$$

- $\mathcal{L}_{GAN_f}^{T2}(f_{task}^{T2}, D_T^{T2})$ is the feature-level GAN loss for the features extracted from translated and real images with discriminator D_f^{T2}:

$$\mathcal{L}_{GAN_f}^{T2}(f_{task}^{T2}, D_T^{T2}) = \mathbb{E}_{\mathbf{x}_S \sim p_{syn}} \left[\log D_f^{T2}(f_{task}^{T2}(G_S^{T2}(\mathbf{x}_S))) \right] + \mathbb{E}_{\mathbf{x}_T \sim p_{real}} \left[\log(1 - D_f^{T2}(f_{task}^{T2}(\mathbf{x}_T))) \right] ;$$

- $\mathcal{L}_r^{T2}(G_S^{T2}) = \left\| G_S^{T2}(\mathbf{x}_T) - \mathbf{x}_T \right\|_1$ is the reconstruction loss for real images;
- $\mathcal{L}_r^{T2}(f_{task}^{T2}) = \left\| f_{task}^{T2}(\hat{\mathbf{x}}_S) - \mathbf{y}_S \right\|_1$ is the *task loss* for depth estimation on synthetic images, namely the L_1-norm of the difference between the predicted depth map for a translated synthetic image $\hat{\mathbf{x}}_S$ and the original ground truth synthetic depth map $\mathbf{y}_S$; this loss ensures that translation does not change the depth map;
- $\mathcal{L}_s^{T2}(f_{task}^{T2}) = \left| \partial_x f_{task}^{T2}(\mathbf{x}_T) \right|^{-|\partial_x \mathbf{x}_T|} + \left| \partial_y f_{task}^{T2}(\mathbf{x}_T) \right|^{-|\partial_y \mathbf{x}_T|}$, where ∂_x and ∂_y are image gradients, is the task loss for depth estimation on real images; since ground truth depth maps are not available now, this regularizer is a locally smooth loss intended to optimize object boundaries, a common tool in depth estimation models [284].

Zheng et al. show that T^2Net can produce realistic images from synthetic ones, conclude that end-to-end training is preferable over separated training (of the translation network and depth estimation network), and note that T^2Net can achieve good results for depth estimation with no access to real paired data, even outperforming some (but not all) supervised approaches. A few sample outdoor scenes processed by T^2Net are shown in Fig. 10.5.

We note a few more interesting applications of refiner-based architectures. Wang et al. [919] use a classical refiner modeled after [793] for human motion synthesis and control. Their model first generates a motion sequence from a recurrent neural network and then refines it with a GAN; since the goal is to model and refine sequences, both generator and discriminator in the refiner also have RNN-based architectures. Dilipkumar [196] applied SimGAN to improve handwriting recognition. They generated synthetic handwriting images and applied SimGAN to refine

Fig. 10.5 Synthetic-to-real refinement by T²*Net* [1020]: (a) input synthetic images; (b) refined images; (c) real images (for comparison).

them, with significantly improved recognition of real handwriting after training on the resulting hybrid dataset.

A recent example that applies GAN-based refinement in a classical computer vision setting is provided by Wang et al. [904], in a work whose other aspects we already discussed in Section 6.4. They consider the problem of recognizing objects inside an automatic vending machine; this is a basic functionality needed for monitoring the state of supplies and is usually done based on object detection.

Wang et al. begin by scanning the objects, adding random deformations to the resulting 3D models (see Section 9.2), setting up scenes and rendering with settings matching the fisheye cameras used in smart vending machines. Then they refine rendered images with virtual-to-real style transfer done by a CycleGAN-based architecture (Fig. 10.6). The novelty here is that Wang et al. separate foreground and background losses, arguing that style transfer needed for foreground objects is very different from (much stronger than) the style transfer for backgrounds. Thus, they use the overall loss function

$$\mathcal{L}^{OD} = \mathcal{L}_{GAN}^{OD}(G^{OD}, D_T^{OD}, \mathcal{X}_S, \mathcal{X}_T) + \mathcal{L}_{GAN}^{OD}(F^{OD}, D_S^{OD}, \mathcal{X}_T, \mathcal{X}_S) + \\ + \lambda_1 \mathcal{L}_{cyc}^{OD}(G^{OD}, F^{OD}) + \lambda_2 \mathcal{L}_{bg}^{OD} + \lambda_3 \mathcal{L}_{fg}^{OD}, \quad \text{where}$$

- $\mathcal{L}_{GAN}^{OD}(G, D, X, Y)$ is the standard adversarial loss for generator G mapping from domain X to domain Y and discriminator D distinguishing real images from fake ones in the domain Y;
- $\mathcal{L}_{cyc}^{OD}(G, F)$ is the cycle consistency loss as used in CycleGAN [1025] and detailed above;
- $\mathcal{L}_{bg}^{OD}$ is the *background loss*, which is the cycle consistency loss computed only for the background part of the images as defined by the mask $\mathbf{m}_{bg}$:

$$\mathcal{L}_{bg}^{OD} = \mathbb{E}_{\mathbf{x}_T \sim p_{real}} \left[\left\| \left(G^{OD}(F^{OD}(\mathbf{x}_T)) - \mathbf{x}_T\right) \odot \mathbf{m}_{bg}(\mathbf{x}_T) \right\|_2 \right] \\ + \mathbb{E}_{\mathbf{x}_S \sim p_{syn}} \left[\left\| \left(F^{OD}(G^{OD}(\mathbf{x}_S)) - \mathbf{x}_S\right) \odot \mathbf{m}_{bg}(\mathbf{x}_S) \right\|_2 \right];$$

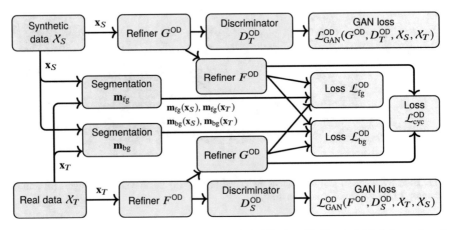

Fig. 10.6 The architecture of the refiner used in [904]. Blocks with identical labels have shared weights.

- $\mathcal{L}_{\mathrm{fg}}^{\mathrm{OD}}$ is the *foreground loss*, similar to $\mathcal{L}_{\mathrm{bg}}^{\mathrm{OD}}$ but computed only for the hue channel in the HSV color space (the authors argue that color and profile are the most critical for recognition and thus need to be preserved the most), as denoted by $\cdot^{H}$ below

$$\mathcal{L}_{\mathrm{fg}}^{\mathrm{OD}} = \mathbb{E}_{\mathbf{x}_T \sim p_{\mathrm{real}}} \left[\left\| \left(G^{\mathrm{OD}}(F^{\mathrm{OD}}(\mathbf{x}_T))^H - \mathbf{x}_T^H \right) \odot \mathbf{m}_{\mathrm{fg}}(\mathbf{x}_T) \right\|_2 \right]$$
$$+ \mathbb{E}_{\mathbf{x}_S \sim p_{\mathrm{syn}}} \left[\left\| \left(F^{\mathrm{OD}}(G^{\mathrm{OD}}(\mathbf{x}_S))^H - \mathbf{x}_S^H \right) \odot \mathbf{m}_{\mathrm{fg}}(\mathbf{x}_S) \right\|_2 \right].$$

Segmentation into foreground and background is done automatically in synthetic data and is made easy in [904] for real data since the camera position is fixed, and the authors can collect a dataset of real background templates from the vending machines they used in the experiments and then simply subtract the backgrounds to get the foreground part.

As a result, Wang et al. report significantly improved results when using hybrid datasets of real and synthetic data for all three tested object detection architectures: PVANET [453], SSD [539], and YOLOv3 [711]. Even more importantly, they report a comparison between basic and refined synthetic data with clear gains achieved by refinement across all architectures.

Wang et al. [907][1] discuss synthetic-to-real domain adaptation in the context of crowd counting, a domain where synthetic data has been successfully used for a long time. They collect synthetic data with the *Grand Theft Auto V* engine, producing the so-called *GTA5 Crowd Counting* (GCC) dataset (see also Section 6.6). They also use a CycleGAN-based refiner from the domain of synthetic images $\mathcal{X}_S$ to the domain of real images $\mathcal{X}_T$ but remark that CycleGAN can easily lose local patterns and textures which is exactly what is important for crowd counting. Therefore, they modify the

---

[1] That's a different person, Qi Wang from Northwestern Polytechnical University in Shaanxi, China, while the previous paper's first author was Kai Wang from *CloudMinds Technologies*.

CycleGAN loss function with the structural similarity index (SSIM) [1024] that computes the similarity between images in terms of local patterns. Their final loss function is

$$\mathcal{L}^{SSIM} = \mathcal{L}_{GAN}^{SSIM}(G^{SSIM}, D_T^{SSIM}, \mathcal{X}_S, \mathcal{X}_T) + \mathcal{L}_{GAN}^{SSIM}(F^{SSIM}, D_S^{SSIM}, \mathcal{X}_T, \mathcal{X}_S)$$
$$+ \lambda \mathcal{L}_{cyc}^{SSIM}(G^{SSIM}, F^{SSIM}, \mathcal{X}_S, \mathcal{X}_R) + \mu \mathcal{L}_{SE}^{SSIM}(G^{SSIM}, F^{SSIM}, \mathcal{X}_S, \mathcal{X}_R),$$

where $G^{SSIM} : \mathcal{X}_S \to \mathcal{X}_T$ is the generator from synthetic to real domains, $F^{SSIM} : \mathcal{X}_T \to \mathcal{X}_S$ works in the opposite direction, D_T^{SSIM} and D_S^{SSIM} are the corresponding discriminators, $\mathcal{L}_{GAN}^{SSIM}$ is a standard GAN loss function, and $\mathcal{L}_{cyc}^{SSIM}$ is the cycle consistency loss as defined above, while $\mathcal{L}_{SE}^{SSIM}$ is a special loss function designed to improve SSIM:

$$\mathcal{L}_{SE}^{SSIM}(G^{SSIM}, F^{SSIM}, \mathcal{X}_S, \mathcal{X}_R) = \mathbb{E}_{\mathbf{x}_S \sim p_{syn}}\left[1 - SSIM(\mathbf{x}_S, F^{SSIM}(G^{SSIM}(\mathbf{x}_S)))\right]$$
$$+ \mathbb{E}_{\mathbf{x}_T \sim p_{data}}\left[1 - SSIM(\mathbf{x}_T, G^{SSIM}(F^{SSIM}(\mathbf{x}_T)))\right].$$

Bak et al. [34] (see Section 6.6) use domain translation with synthetic people as a method for person re-identification. The domains in this model are represented by illumination conditions for the images; the model has access to M real source domains and N synthetic domains, with $N \gg M$, and the objective is to perform re-identification in an unknown target domain. To achieve this, Bak et al. first learned a generic feature representation from all domains, but the resulting model, even trained on a hybrid dataset, did not generalize well.

Therefore, Bak et al. proceeded to domain adaptation done as follows: first choose the nearest synthetic domain with a separately trained domain identification network fine-tuned for illumination classification, then use the CycleGAN architecture (as shown above) to do domain translation from this synthetic domain to the target domain, and then use it to fine-tune the re-identification network. Bak et al. report improved results from the entire pipeline as compared to any individual parts in the ablation study.

In robotics, domain adaptation of synthetic imagery is not yet common, but some applications have appeared there as well. For example, Pecka et al. [661] train a CycleGAN-based domain adaptation model to learn a transformation from data observed in a non-differentiable physics simulator to the domain of data collected from a real robotic platform, showing improved sim-to-real policy transfer results.

10.4 Making Synthetic Data from Real with GANs

So far, we have discussed the refinement of synthetic data with a purpose to make it more realistic or more suitable for training computer vision models. Now let us discuss an interesting variation of GAN-based refinement that goes in the inverse direction: the models *turn real data into synthetic*. Why would we ever want to do that if the final goal is always to make the model work on real data rather than synthetic? In this section, we will see two examples from different domains that show both why and how.

The first application of this idea is to start from real data and produce other realistic images that have been artificially changed in some respects. This approach could either simply serve as a "smart augmentation" to extend the dataset (recall Section 3.4, where we discussed some other "smart augmentations") or, even more interestingly, could "fill in the holes" in the data distribution, obtaining synthetic data for situations that are lacking in the original dataset.

As the first example, let us consider the work of Zhao et al. [1016, 1017] who concentrated on applying this idea to face recognition in the wild, with different poses rather than by a frontal image. They continued the work of Tran et al. [866] (we do not review it in detail) and Huang et al. [369], who presented a TP-GAN (two-pathway GAN) architecture for frontal view synthesis: given a picture of a face, generate a frontal view picture. TP-GAN's generator G_θ^{TP} has two pathways: a global network $G_{\theta g}^{\mathrm{TP}}$ that rotates the entire face and four local patch networks $G_{\theta_i^l}^{\mathrm{TP}}$, $i = 1, \ldots, 4$, that process local textures around four facial landmarks (eyes, nose, and mouth). Both $G_{\theta g}^{\mathrm{TP}}$ and $G_{\theta_i^l}^{\mathrm{TP}}$ have encoder-decoder architectures with skip connections for multi-scale feature fusion. The discriminator D_ϕ^{TP} learns to distinguish real frontal face images $\mathbf{x}_{\mathrm{front}} \sim \mathcal{D}_{\mathrm{front}}$ from synthesized images $G_\theta^{\mathrm{TP}}(\mathbf{x})$, $\mathbf{x} \sim p_{\mathrm{data}}$:

$$\min_\theta \max_\phi \left[\mathbb{E}_{\mathbf{x}_{\mathrm{front}} \sim \mathcal{D}_{\mathrm{front}}} \left[\log D_\phi^{\mathrm{TP}}(\mathbf{x}_{\mathrm{front}}) \right] + \mathbb{E}_{\mathbf{x} \sim p_{\mathrm{data}}} \left[\log \left(1 - D_\phi^{\mathrm{TP}}(G_\theta^{\mathrm{TP}}(\mathbf{x})) \right) \right] \right].$$

The synthesis loss function in TP-GAN is a sum of four loss functions:

- pixel-wise L_1-loss between the ground truth frontal image and rotated image:

$$\mathcal{L}_{\mathrm{pixel}}^{\mathrm{TP}}(\mathbf{x}, \mathbf{x}^{\mathrm{gt}}) = \frac{1}{W \times H} \sum_{i=1}^{W} \sum_{j=1}^{H} \left| G_\theta^{\mathrm{TP}}(\mathbf{x})_{i,j} - \mathbf{x}_{i,j}^{\mathrm{gt}} \right|,$$

where $\mathbf{x}^{\mathrm{gt}}$ is the ground truth frontal image corresponding to $\mathbf{x}$, and W and H are an image's width and height; this loss is measured at several different places of the network: output of $G_{\theta g}^{\mathrm{TP}}$, outputs of $G_{\theta_i^l}^{\mathrm{TP}}$, and the final output of G_θ^{TP};

- symmetry loss

$$\mathcal{L}_{\mathrm{sym}}^{\mathrm{TP}} = \frac{1}{W/2 \times H} \sum_{i=1}^{W/2} \sum_{j=1}^{H} \left| G_\theta^{\mathrm{TP}}(\mathbf{x})_{i,j} - G_\theta^{\mathrm{TP}}(\mathbf{x})_{W-(i-1),j} \right|,$$

intended to preserve the symmetry of human faces;

- adversarial loss

$$\mathcal{L}_{\mathrm{adv}}^{\mathrm{TP}} = \frac{1}{N} \sum_{n=1}^{N} - \log D_\phi^{\mathrm{TP}}(G_\theta^{\mathrm{TP}}(\mathbf{x}_n));$$

- identity preserving loss

$$\mathcal{L}_{\text{ip}}^{\text{TP}} = \sum_l \frac{1}{W_l \times H_l} \sum_{i=1}^{W_l} \sum_{j=1}^{H_l} \left| F^l(G_\theta^{\text{TP}}(\mathbf{x}))_{i,j} - F^l(\mathbf{x}^{\text{gt}})_{i,j} \right|,$$

where F^l denotes the output of the lth layer of a face recognition network applied to $\mathbf{x}$ and $\mathbf{x}^{\text{gt}}$ (Huang et al. used Light CNN [941], and only used its last two layers in $\mathcal{L}_{\text{ip}}^{\text{TP}}$); this idea is based on *perceptual losses* [407], a popular idea in GANs designed to preserve high-level features when doing low-level transformations; in this case, it serves to preserve the person's identity when rotating the face.

As usual, the final loss function is a linear combination of the four losses above and a regularization term.

In [1017], Zhao et al. propose the dual-agent GAN (DA-GAN) model that also works with faces but in the opposite scenario: while TP-GAN rotates every face into the frontal view, DA-GAN aims to fill in the "holes" in the real data distribution, rotating real faces so that the distribution of angles becomes more uniform. They begin with a 3D morphable model (see also Section 6.6) from [1027], extracting 68 facial landmarks with the recurrent attentive-refinement (RAR) model [952] and estimating the transformation matrix with 3D-MM [74]. However, the authors report that simulation quality dramatically decreases for large yaw angles, necessitating further improvement with the DA-GAN framework.

Again, DA-GAN's generator G_θ^{DA} maps a synthetized image to a refined one, $\hat{\mathbf{x}}_S = G_\theta^{\text{DA}}(\mathbf{x}_S)$. It is trained on a linear combination of three loss functions

$$\mathcal{L}_G^{\text{DA}} = -\mathcal{L}_{\text{adv}}^{\text{DA}} + \lambda_1 \mathcal{L}_{\text{ip}}^{\text{DA}} + \lambda_2 \mathcal{L}_{\text{pp}}^{\text{DA}},$$

and the discriminator D_ϕ^{DA} consists of two parallel branches (agents) that optimize $\mathcal{L}_{\text{adv}}^{\text{DA}}$ and $\mathcal{L}_{\text{ip}}^{\text{DA}}$, respectively. The loss functions are defined as follows:

- the adversarial loss $\mathcal{L}_{\text{adv}}^{\text{DA}}$ follows the BEGAN architecture introduced in [66]: this branch of D_ϕ^{DA} is an autoencoder that minimizes the Wasserstein distance with a boundary equilibrium regularization term:

$$\mathcal{L}_{\text{adv}}^{\text{DA}} = \sum_j \left| \mathbf{x}_{T,j} - D_\phi^{\text{DA}}(\mathbf{x}_T)_j \right| - k_t \sum_i \left| \hat{\mathbf{x}}_{S,i} - D_\phi^{\text{DA}}(\hat{\mathbf{x}}_S)_i \right|,$$

where, again, $\mathbf{x}_T$ is a real image, $\hat{\mathbf{x}}_S$ is a refined image, and k_t is a boundary equilibrium regularization term continuously trained to maintain the equilibrium

$$\mathbb{E}\left[\sum_i \left| \hat{\mathbf{x}}_{S,i} - D_\phi^{\text{DA}}(\hat{\mathbf{x}}_S)_i \right| \right] = \gamma \mathbb{E}\left[\sum_j \left| \mathbf{x}_{T,j} - D_\phi^{\text{DA}}(\mathbf{x}_T)_j \right| \right]$$

for some diversity ratio γ (see [66, 1017] for more details); in general, $\mathcal{L}_{adv}^{DA}$ is designed to keep the refined face in the manifold of real faces;

- the identity preservation loss $\mathcal{L}_{ip}^{DA}$, similar to $\mathcal{L}_{ip}^{TP}$, aims to make the refinement respect the identities, but does it in a different way; here the idea is to put both $\mathbf{x}_T$ and $\mathbf{x}_S$ through the same (relatively simple) face recognition network and bring its features together; DA-GAN uses for this purpose a classifier C_ϕ^{DA} trained on the bottleneck layer of D_ϕ^{DA}:

$$\mathcal{L}_{ip}^{DA} = \frac{1}{N} \sum_j \left[-\mathbf{y}_j \log C_\phi^{DA}(\mathbf{x}_{T,j}) + (1 - \mathbf{y}_j) \log(1 - C_\phi^{DA}(\mathbf{x}_{T,j})) \right]$$

$$+ \frac{1}{N} \sum_j \left[-\mathbf{y}_j \log C_\phi^{DA}(\hat{\mathbf{x}}_{S,j}) + (1 - \mathbf{y}_j) \log(1 - C_\phi^{DA}(\hat{\mathbf{x}}_{S,j})) \right],$$

where $\mathbf{y}_j$ is the ground truth label;

- the pixel-wise loss $\mathcal{L}_{pp}^{DA}$ is the L_1-loss intended to make sure that the pose (angle of inclination for the head) remains the same after refinement:

$$\mathcal{L}_{pp}^{DA} = \frac{1}{W \times H} \sum_{i=1}^{W} \sum_{j=1}^{H} \left| \mathbf{x}_{S,i,j} - \hat{\mathbf{x}}_{S,i,j} \right|.$$

In total, during training DA-GAN alternatively optimizes G_θ^{DA} and D_ϕ^{DA} with loss functions $\mathcal{L}_G^{DA}$ and $\mathcal{L}_D^{DA} = \mathcal{L}_{adv}^{DA} + \lambda_1 \mathcal{L}_{ip}^{DA}$. Following [66], to measure convergence DA-GAN tests the reconstruction quality together with proportion control theory, evaluating

$$\mathcal{L}_{con}^{DA} = \sum_j \left| \mathbf{x}_{T,j} - D_\phi^{DA}(\mathbf{x}_{T,j}) \right| + \left| \gamma \sum_j \left| \mathbf{x}_{T,j} - D_\phi^{DA}(\mathbf{x}_{T,j}) \right| - \sum_i \left| \hat{\mathbf{x}}_{S,j} - D_\phi^{DA}(\hat{\mathbf{x}}_{S,j}) \right| \right|.$$

Apart from experiments done by the authors, DA-GAN was verified in a large-scale NIST IJB-A competition [229] where a model based on DA-GAN won the face verification and face identification tracks. This result heavily supports the general premise of using synthetic data: augmenting the dataset and balancing out the training data distribution with synthetic images proved highly beneficial in this case.

Inoue et al. [386] try to find a middle ground between synthetic and real data. They use two variational autoencoders (VAE) to reduce both synthetic and real data to a common pseudo-synthetic image space, and then train CNNs on images from this common space. The training sequence is as follows:

- train $VAE_1 : \mathcal{X}_S \to \mathcal{X}_S$ as an autoencoder;
- train $VAE_2 : \mathcal{X}_T \to \mathcal{X}_S$ where the decoder is fixed and shares weights with the decoder from VAE_1; as a result, VAE_2 has to learn to generate pseudo-synthetic images from real images;

- train a CNN for the task in question on synthetic data, using VAE_1 to map it to the common image space;
- during inference, use a composition of VAE_2 and CNN.

Another idea for generating synthetic data from real is to compose parts of real images to produce synthetic ones. We have discussed the cut-and-paste approaches in Section 9.3; a natural continuation of these ideas would be to use more complex, semantic conditioning with a GAN-based architecture. For example, Joo et al. [413] provide a GAN-based architecture for generating a fusion image, where, say, one input $\mathbf{x}$ provides the identity of a person, another input $\mathbf{y}$ provides the shape (pose) of a person, and the result is $\hat{\mathbf{x}}$ which has the identity of $\mathbf{x}$ and the shape of $\mathbf{y}$. Their FusionGAN architecture extends CycleGAN-like ideas to losses that distinguish between identity and shape of an image, introducing the concepts of *identity loss* and *shape loss*. FusionGAN relies on a dataset that has several images with different shapes but the same identity (e.g., the same person in different poses; a dataset of videos can serve as a simple example); its overall loss function is

$$\mathcal{L}^{\text{FUS}} = \mathcal{L}_I^{\text{FUS}} + \lambda \mathcal{L}_S^{\text{FUS}}, \text{ where}$$

- $\mathcal{L}_I^{\text{FUS}}$ is the *identity loss*

$$\mathcal{L}_I^{\text{FUS}}(G^{\text{FUS}}, D^{\text{FUS}}) = \mathbb{E}_{\mathbf{x}, \mathbf{x}' \sim p_{\text{real}}(\mathbf{x})} \left[\left\| 1 - D^{\text{FUS}}(\mathbf{x}, \mathbf{x}') \right\|_2 \right]$$
$$+ \mathbb{E}_{\mathbf{x} \sim p_{\text{real}}(\mathbf{x}), \mathbf{y} \sim p_{\text{real}}(\mathbf{y})} \left[\left\| D^{\text{FUS}}(\mathbf{x}, G^{\text{FUS}}(\mathbf{x}, \mathbf{y})) \right\|_2 \right],$$

  i.e., the discriminator D^{FUS} learns to distinguish real pairs of images $(\mathbf{x}, \mathbf{x}')$ with the same identity (but different shapes) and fake pairs of images $(\mathbf{x}, G^{\text{FUS}}(\mathbf{x}, \mathbf{y}))$ where G^{FUS} is supposed to take the identity from $\mathbf{x}$;
- $\mathcal{L}_S^{\text{FUS}}$ is the *shape loss* defined as

$$\mathcal{L}_{S_1}^{\text{FUS}}(G^{\text{FUS}}) = \mathbb{E}_{\mathbf{x}, \mathbf{x}' \sim p_{\text{real}}(\mathbf{x})} \left[\left\| \mathbf{x}' - G^{\text{FUS}}(\mathbf{x}, \mathbf{x}') \right\|_1 \right]$$

when $\mathbf{x}$ and $\mathbf{x}'$ have the same identity, and

$$\mathcal{L}_{S_2a}^{\text{FUS}}(G^{\text{FUS}}) = \mathbb{E}_{\mathbf{x} \sim p_{\text{real}}(\mathbf{x}), \mathbf{y} \sim p_{\text{real}}(\mathbf{y})} \left[\left\| \mathbf{y} - G^{\text{FUS}}(\mathbf{y}, G^{\text{FUS}}(\mathbf{x}, \mathbf{y})) \right\|_1 \right],$$
$$\mathcal{L}_{S_2b}^{\text{FUS}}(G^{\text{FUS}}) = \mathbb{E}_{\mathbf{x} \sim p_{\text{real}}(\mathbf{x}), \mathbf{y} \sim p_{\text{real}}(\mathbf{y})} \left[\left\| G^{\text{FUS}}(\mathbf{x}, \mathbf{y}) - G^{\text{FUS}}(G^{\text{FUS}}(\mathbf{x}, \mathbf{y}), \mathbf{y}) \right\|_1 \right],$$

  i.e., $G^{\text{FUS}}(\mathbf{y}, G^{\text{FUS}}(\mathbf{x}, \mathbf{y}))$ should be the same as $\mathbf{y}$, with identity from $\mathbf{y}$ and shape also from $\mathbf{y}$, and $G^{\text{FUS}}(G^{\text{FUS}}(\mathbf{x}, \mathbf{y}), \mathbf{y}))$ should be the same as $G^{\text{FUS}}(\mathbf{x}, \mathbf{y})$, with identity from $\mathbf{x}$ and shape from $\mathbf{y}$.

Similar ideas have been extended to animating still images [795], motion transfer [119], and image-to-image translation [566]. In general, these works belong to an interesting field of generative semantic manipulation with GANs. Important works in this direction include Mask-Contrasting GAN [517] that can modify an object to a different suitable category inside its segmentation mask (e.g., replace a cat with a dog), Attention-GAN [139] that performs the same task with an attention-based architecture, IterGAN [255] that attempts iterative small-scale 3D manipulations such as rotation from 2D images, and others.

However, while this field produces very interesting works, so far we have not seen direct applications of such architectures to generating synthetic data. I believe that ideas similar to TP-GAN can also be fruitful in other domains, especially in situations where one- or few-shot learning is required so "smart augmentations" such as rotation can bring significant improvements.

The second approach to using real-to-synthetic refinement deals with a completely different idea of using the same transfer direction. Why do we need domain adaptation at all? Because we want models that have been trained on synthetic data to transfer to real inputs. The idea is to reverse this logic: let us transfer real data into the synthetic domain, where the model is already working great!

In the context of robotics, this kind of *real-to-sim* approach was implemented by Zhang et al. [999] in a framework called *VR-Goggles for Robots*. It is based on the CycleGAN ideas as they were continued in CyCADA [352], a popular domain adaptation model that adds semantic losses to CycleGAN. The *VR-Goggles* model has two generators, $G_S^{\text{VRG}} : \mathcal{X}_T \to \mathcal{X}_S$ with discriminator D_S^{VRG} that distinguishes fake synthetic images and $G_T^{\text{VRG}} : \mathcal{X}_S \to \mathcal{X}_T$ with discriminator D_T^{VRG} that is defined in the domain of real images. The overall loss function is

$$
\begin{aligned}
\mathcal{L}^{\text{VRG}} =& \mathcal{L}_{\text{GAN}}^{\text{VRG}}(G_T^{\text{VRG}}, D_T^{\text{VRG}}; \mathcal{X}_S, \mathcal{X}_T) + \mathcal{L}_{\text{GAN}}^{\text{VRG}}(G_S^{\text{VRG}}, D_S^{\text{VRG}}; \mathcal{X}_T, \mathcal{X}_S) \\
&+ \lambda_1 \left(\mathcal{L}_{\text{cyc}}^{\text{VRG}}(G_S^{\text{VRG}}, G_T^{\text{VRG}}; \mathcal{X}_T) + \mathcal{L}_{\text{cyc}}^{\text{VRG}}(G_T^{\text{VRG}}, G_S^{\text{VRG}}; \mathcal{X}_S) \right) \\
&+ \lambda_2 \left(\mathcal{L}_{\text{sem}}^{\text{VRG}}(G_S^{\text{VRG}}; \mathcal{X}_T, f_S^{\text{VRG}}) + \mathcal{L}_{\text{sem}}^{\text{VRG}}(G_S^{\text{VRG}}; \mathcal{X}_S, f_S^{\text{VRG}}) \right) \\
&+ \lambda_3 \left(\mathcal{L}_{\text{shift}}^{\text{VRG}}(G_T^{\text{VRG}}; \mathcal{X}_S) + \mathcal{L}_{\text{shift}}^{\text{VRG}}(G_S^{\text{VRG}}; \mathcal{X}_T) \right), \text{ where}
\end{aligned}
$$

- $\mathcal{L}_{\text{GAN}}^{\text{VRG}}$ is the standard GAN loss:

$$
\begin{aligned}
\mathcal{L}_{\text{GAN}}^{\text{VRG}}(G_T^{\text{VRG}}, D_T^{\text{VRG}}; \mathcal{X}_S, \mathcal{X}_T) =& \mathbb{E}_{\mathbf{x}_T \sim p_{\text{real}}} \left[\log D_T^{\text{VRG}}(\mathbf{x}_T) \right] + \\
& \mathbb{E}_{\mathbf{x}_S \sim p_{\text{syn}}, \mathbf{z}} \left[\log \left(1 - D_T^{\text{VRG}}(G_T^{\text{VRG}}(\mathbf{x}_S)) \right) \right]
\end{aligned}
$$

and similarly for $\mathcal{L}_{\text{GAN}}^{\text{VRG}}(G_S^{\text{VRG}}, D_S^{\text{VRG}}; \mathcal{X}_T, \mathcal{X}_S)$;
- $\mathcal{L}_{\text{sem}}^{\text{VRG}}$ is the semantic loss as introduced in CyCADA [352]; the idea is that if we have ground truth labels for the synthetic domain $\mathcal{X}_S$ (in this case, we are doing semantic segmentation), we can train a network f_S^{VRG} on $\mathcal{X}_S$ and then use it to generate pseudolabels for the domain $\mathcal{X}_T$ where ground truth is not available; the semantic loss now makes sure that the results (segmentation maps) remain the same after image translation:

$$\mathcal{L}_{\text{sem}}^{\text{VRG}}(G_S^{\text{VRG}}; \mathcal{X}_T, f_S^{\text{VRG}}) = \mathbb{E}_{\mathbf{x}_T \sim p_{\text{real}}}\left[\text{CE}\left(f_S^{\text{VRG}}(\mathbf{x}_T), f_S^{\text{VRG}}(G_S^{\text{VRG}}(\mathbf{x}_T))\right)\right],$$
$$\mathcal{L}_{\text{sem}}^{\text{VRG}}(G_T^{\text{VRG}}; \mathcal{X}_S, f_S^{\text{VRG}}) = \mathbb{E}_{\mathbf{x}_S \sim p_{\text{syn}}}\left[\text{CE}\left(f_S^{\text{VRG}}(\mathbf{x}_S), f_S^{\text{VRG}}(G_T^{\text{VRG}}(\mathbf{x}_S))\right)\right],$$

where CE denotes cross-entropy;

- $\mathcal{L}_{\text{shift}}^{\text{VRG}}$ is the *shift loss* that makes the image translation result invariant to shifts:

$$\mathcal{L}_{\text{shift}}^{\text{VRG}}(G_T^{\text{VRG}}; \mathcal{X}_S) = \mathbb{E}_{\mathbf{x}_S, i, j}\left[\left\| G_T^{\text{VRG}}(\mathbf{x}_S)_{\left[\begin{smallmatrix} x \to i \\ y \to j \end{smallmatrix}\right]} - G_T^{\text{VRG}}\left(\mathbf{x}_{S,\left[\begin{smallmatrix} x \to i \\ y \to j \end{smallmatrix}\right]}\right)\right\|_2^2\right],$$

$$\mathcal{L}_{\text{shift}}^{\text{VRG}}(G_S^{\text{VRG}}; \mathcal{X}_T) = \mathbb{E}_{\mathbf{x}_T, i, j}\left[\left\| G_S^{\text{VRG}}(\mathbf{x}_T)_{\left[\begin{smallmatrix} x \to i \\ y \to j \end{smallmatrix}\right]} - G_S^{\text{VRG}}\left(\mathbf{x}_{T,\left[\begin{smallmatrix} x \to i \\ y \to j \end{smallmatrix}\right]}\right)\right\|_2^2\right],$$

where $\mathbf{x}_{\left[\begin{smallmatrix} x \to i \\ y \to j \end{smallmatrix}\right]}$ denotes the shifting operation by i pixels along the X-axis and j pixels along the Y-axis, and i and j are chosen uniformly at random up to the total downsampling factor of the network K (since the result will always be invariant to shifts of multiples of K).

Zhang et al. test their solution on the CARLA navigation benchmark [203] and show significant improvements.

James et al. [396] consider the same kind of approach for robotic grasping. Their model, *Randomized-to-Canonical Adaptation Networks* (RCAN), learn to map heavily randomized simulation images (with random textures) to a canonical (much simpler) rendered image and also map real images to canonical rendered images; interestingly, they achieve good results with a much simpler GAN architecture where additional losses simply bring together the segmentation masks and depth maps for simulated images, and there are no cycle consistency losses.

An even simpler approach is taken by Yang et al. [969] who introduce *domain unification* for autonomous driving. Their model, called *DU-Drive*, consists of a generator that translates real images to simplified synthetic images and a discriminator that distinguishes them from actual synthetic images; the driving policy is then trained in the simulator.

Why might this inverse real-to-sim direction be a good fit for robotics specifically, and why haven't we seen this approach in other domains before? The reason is that in the real-to-sim approach, we need to use domain adaptation models during inference, as part of using the model. In most regular computer vision applications, this would be a great hindrance. In computer vision, if some kind of preprocessing is only part of the training process it is usually assumed to be free (we have discussed some rather complicated examples in Section 3.4), and inference time is precious.

Robotics is a very different setting: robots and controllers are often trained in simulation environments with reinforcement learning, which implies a lot of computational resources needed for training. The simulation environment needs to be responsive and cheap to support, and if every frame of the training needs to be translated via a GAN-based model it may add up to a huge cost that would make RL-based training infeasible. Adding an extra model during inference, on the other hand, may be admissible: yes, we reduce the number of processed frames per second, but if it stays high enough for the robot to react in real time, that is fine.

With this, we conclude the part of this chapter that deals with *refinement*, i.e., domain adaptation techniques that operate on the data level, translating data points from one domain to another. In the next section, we discuss model-based domain adaptation, that is, approaches that change the model itself and leave the data in place.

10.5 Domain Adaptation at the Feature/Model Level

In previous sections, we have considered models that perform domain adaptation (DA) at the data level, i.e., one can extract a part of the model that takes as input a data point from the source domain (in our case, a synthetic image) and map it to the target domain (domain of real images). However, the final goal of model design rarely involves the generation of more realistic synthetic images; they are merely a stepping stone to producing models that work better, e.g., in the absence of supervision in the target domain. Therefore, to make better use of synthetic data it makes sense to also consider *feature-level* or *model-level* domain adaptation, i.e., methods that work in the space of features or model weights and never go back to change the actual data.

The simplest approach to domain adaptation would be to share the weights among networks operating on different domains or learn an explicit mapping between them [155, 283]. While we mostly discuss other approaches, we note that simpler techniques based on weight sharing remain relevant for domain adaptation. In particular, Rozantsev et al. [739] recently presented a domain adaptation approach where two similar networks are trained on the source and target domain with special regularizers that bring their weights together; the authors evaluate their approach on synthetic-to-real domain adaptation for drone detection with promising results.

Another approach to model-level domain adaptation is related to mining relatively strong priors from real data that can then inform a model trained on synthetic data, helping fix problematic cases or incongruencies between the synthetic and real datasets. For example, Zhang et al. [1005, 1006] present a curriculum learning approach to domain adaptation for semantic segmentation of urban scenes. They train a segmentation network on synthetic data (specifically on the GTA dataset; see also Section 7.2) but with a special component in the loss function related to the general label distribution in real images:

$$\mathcal{L}^{\text{CURR}} = \frac{1}{|\mathcal{X}_S|} \sum_{\mathbf{x}_S \in \mathcal{X}_S} \mathcal{L}\left(\mathbf{y}_S, \hat{\mathbf{y}}_S\right) + \lambda \frac{1}{|\mathcal{X}_T|} \sum_{\mathbf{x}_T \in \mathcal{X}_T} \sum_k C\left(p^k(\mathbf{x}_T), \hat{p}^k(\mathbf{x}_T)\right),$$

where $\mathcal{L}\left(\mathbf{y}_S, \hat{\mathbf{y}}_S\right)$ is the pixel-wise cross-entropy, a standard segmentation loss, and $C\left(p^k(\mathbf{x}_T), \hat{p}^k(\mathbf{x}_T)\right)$ is the cross-entropy between the distribution of labels $\hat{p}(\mathbf{x}_T)$ in a real image $\mathbf{x}_T$ that the network produces and $p(\mathbf{x}_T)$ is the real label distribution (superscript k denotes different kinds of label distributions).

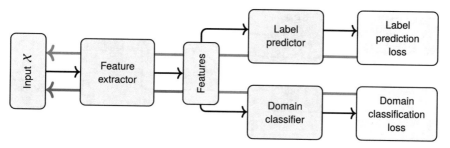

Fig. 10.7 High-level architecture of model-level domain adaptation from [259, 260]: the gradient flow (green) from the domain classification loss is reversed (becomes red) at the features.

Note that $p(\mathbf{x}_T)$ is not available in the real data, so this is where curriculum learning comes in: the authors first train a simpler model on synthetic data to estimate $p(\mathbf{x}_T)$ from image features and then use it to inform the segmentation model. Recent developments of this interesting direction shift from merely enforcing the label distribution to matching features on multiple different levels [368]. In particular, recent works [515, 984] have introduced the so-called *pyramid consistency loss* instead of $C\left(p(\mathbf{x}_T), \hat{p}(\mathbf{x}_T)\right)$ that tries to enforce consistency across domains on the activation maps of later layers of the network.

One of the main directions in model-level domain adaptation was initiated by Ganin and Lempitsky [259] who presented a generic framework for unsupervised domain adaptation. Their approach, illustrated in Figure 10.7, consists of

- a *feature extractor*,
- a *label predictor* that performs the necessary task (e.g., classification) on extracted features, and
- a *domain classifier* that takes the same features and attempts to classify which domain the original input belonged to.

The idea is to train the label predictor to perform as well as possible and at the same time make the domain classifier perform as badly as possible; this is achieved with *gradient reversal*, i.e., multiplying the gradients by a negative constant as they pass from the domain classifier to the feature extractor. In a subsequent work, Ganin et al. [260] generalized this domain adaptation approach to arbitrary architectures and experimented with DA in different domains, including image classification, person re-identification, and sentiment analysis. We also note extensions and similar approaches to domain adaptation developed in [545, 546, 881] and the domain confusion metric that helps produce domain-invariant representations [882], but proceed to highlight the works that perform specifically synthetic-to-real domain adaptation.

Many general model-level domain adaptation approaches have been validated on or subsequently extended to synthetic-to-real domain adaptation. Xu et al. [964] consider the pedestrian detection problem (this work is a continuation of [892], see Section 6.5). They adapt detectors trained on virtual datasets with a boosting-based procedure, assigning larger weights to samples that are similar to target domain

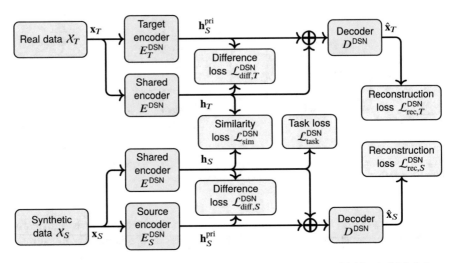

Fig. 10.8 Architecture of the domain separation network [88]. Blocks with identical labels have shared weights.

ones. Sun and Saenko [826] propose a domain adaptation approach based on decorrelating the features of a classifier, both in unsupervised and supervised settings. Later, López et al. [548] extended the SA-SVVM domain adaptation used in [892] to train deformable part-based models, using synthetic pedestrians from the SYNTHIA dataset (see Section 7.2) as the main example. In a parallel paper, the authors of SYNTHIA Ros et al. [732] used a simple domain adaptation technique called Balanced Gradient Contribution [733], where training on synthetic data is regularized by the gradient obtained on a (small) real dataset, to further improve their results on segmentation aided by synthetic data. Ren et al. [720] perform cross-domain self-supervised multitask learning with synthetic images: their model predicts several parameters of an image (surface normal, depth, and instance contour) and at the same time tries to minimize the difference between synthetic and real data in feature space.

Domain separation networks by Bousmalis et al. [88], illustrated in Fig. 10.8, explicitly separate the shared and private components of both source and target domains. Specifically, they introduce a shared encoder $E^{\text{DSN}}(\mathbf{x})$ and two private encoders, E_S^{DSN} for the source domain and E_T^{DSN} for the target domain. The total objective function for a domain separation network is

$$\mathcal{L}^{\text{DSN}} = \mathcal{L}_{\text{task}}^{\text{DSN}} + \lambda_1 \mathcal{L}_{\text{rec}}^{\text{DSN}} + \lambda_2 \mathcal{L}_{\text{diff}}^{\text{DSN}} + \lambda_1 \mathcal{L}_{\text{sim}}^{\text{DSN}},$$

where

- $\mathcal{L}_{\text{task}}^{\text{DSN}}$ is the supervised task loss in the source domain, e.g., for the image classification task it is

$$\mathcal{L}_{\text{task}}^{\text{DSN}} = -\mathbb{E}_{\mathbf{x}_S \sim p_{\text{syn}}} \left[\mathbf{y}^\top \log f^{\text{DSN}}(E^{\text{DSN}}(\mathbf{x}_S)) \right],$$

where f^{DSN} is the classifier operating on the output of the shared encoder;
- $\mathcal{L}_{\mathrm{rec}}^{\mathrm{DSN}}$ is the reconstruction loss defined as the difference between original samples $\mathbf{x}_S$ and $\mathbf{x}_T$ and the results of a shared decoder D^{DSN} that tries to reconstruct the images from a combination of shared and private representations:

$$\mathcal{L}_{\mathrm{rec}}^{\mathrm{DSN}} = - \mathbb{E}_{\mathbf{x}_S \sim p_{\mathrm{syn}}} \left[\mathcal{L}_{\mathrm{sim}}(\mathbf{x}_S, D^{\mathrm{DSN}}(E^{\mathrm{DSN}}(\mathbf{x}_S) + E_S^{\mathrm{DSN}}(\mathbf{x}_S))) \right]$$
$$- \mathbb{E}_{\mathbf{x}_T \sim p_{\mathrm{real}}} \left[\mathcal{L}_{\mathrm{sim}}(\mathbf{x}_T, D^{\mathrm{DSN}}(E^{\mathrm{DSN}}(\mathbf{x}_T) + E_T^{\mathrm{DSN}}(\mathbf{x}_T))) \right]$$

for some similarity metric $\mathcal{L}_{\mathrm{sim}}$;
- $\mathcal{L}_{\mathrm{diff}}^{\mathrm{DSN}}$ is the difference loss that encourages the hidden shared representations of instances from the source and target domains $E^{\mathrm{DSN}}(\mathbf{x}_S)$ and $E^{\mathrm{DSN}}(\mathbf{x}_T)$ to be orthogonal to their corresponding private representations $E_S^{\mathrm{DSN}}(\mathbf{x}_S)$ and $E_T^{\mathrm{DSN}}(\mathbf{x}_T)$; in [88], the difference loss is defined as

$$\mathcal{L}_{\mathrm{diff}}^{\mathrm{DSN}} = \left\| H_S^{\top} H_S^{\mathrm{pri}} \right\|_F^2 + \left\| H_T^{\top} H_T^{\mathrm{pri}} \right\|_F^2,$$

where H_S is the matrix of $E^{\mathrm{DSN}}(\mathbf{x}_S)$, H_S^{pri} is the matrix of $E_S^{\mathrm{DSN}}(\mathbf{x}_S)$, and similarly for H_T and H_T^{pri};
- $\mathcal{L}_{\mathrm{sim}}^{\mathrm{DSN}}$ is the similarity loss that encourages the hidden shared representations from the source and target domains $E^{\mathrm{DSN}}(\mathbf{x}_S)$ and $E^{\mathrm{DSN}}(\mathbf{x}_T)$ to be similar to each other, i.e., indistinguishable by a domain classifier trained through the gradient reversal layer as in [259]; in [88], this loss is composed of the cross-entropy for the domain classifier and maximal mean discrepancy (MMD) [297] for the hidden representations themselves.

Bousmalis et al. evaluate their model on several synthetic-to-real scenarios, e.g., on synthetic traffic signs from [609] and synthetic objects from the *LineMod* dataset [931].

Domain separation networks became one of the first major examples in domain adaptation with *disentanglement*, where the hidden representations are domain-invariant and some of the features can be changed to transition from one domain to another. Further developments include asymmetric training for unsupervised domain adaptation [750], DistanceGAN for one-sided domain mapping [60], co-regularized alignment [482], cross-domain autoencoders [288], multisource domain adversarial networks [1015], continuous cross-domain translation [531], face recognition adaptation from images to videos with the help of synthetic augmentations [811], and more [998]; all of these advances may be relevant for synthetic-to-real domain adaptation but we will highlight some works that are already doing adaptation between these two domains.

The popular and important domains for feature-based domain adaptation are more or less the same as in domain adaptation on the data level, but feature-based DA may be able to handle higher dimensional inputs and more complex scenes because the adaptation itself is done in an intermediate lower dimensional space. As an illustrative example, let us consider feature-based DA for computer vision problems for outdoor scenes (see also Section 7.2). In their *FCNs in the Wild* model, Hoffman et al. [353] consider feature-based DA for semantic segmentation with fully convolutional networks (FCNs) where ground truth is available for the source domain (synthetic data) but unavailable for the target domain (real data). Their unsupervised domain adaptation framework contains a feature extractor f^{FCNW} and the joint objective function

$$\mathcal{L}^{\text{FCNW}} = \mathcal{L}_{\text{seg}}^{\text{FCNW}} + \mathcal{L}_{\text{DA}}^{\text{FCNW}} + \mathcal{L}_{\text{MI}}^{\text{FCNW}},$$

where

- $\mathcal{L}_{\text{seg}}^{\text{FCNW}}$ is the standard supervised segmentation objective on the source domain, where supervision is available;
- $\mathcal{L}_{\text{DA}}^{\text{FCNW}}$ is the domain alignment objective that minimizes the observed source and target distance in the representation space by training a discriminator (domain classifier) to distinguish instances from source and target domains; an interesting new idea here is to take as an instance for this objective not the entire image but a cell from a coarse grid that corresponds to the scale of high-level features that domain adaptation is supposed to bring together;
- $\mathcal{L}_{\text{MI}}^{\text{FCNW}}$ is the multiple instance loss that encourages pixels to be assigned to class c in such a way that the percentage of an image labeled with c remains within the expected range derived from the source domain.

Another direction of increasing the input dimension is to move from images to videos. Xu et al. [962] use adversarial domain adaptation to transfer object detection models—single-shot multi-box detector (SSD) [539] and multi-scale deep CNN (MSCNN) [108]—from synthetic samples to real videos in the smoke detection problem.

Chen et al. [141] construct the *Cross City Adaptation* model that brings together features from different domains, again with semantic segmentation of outdoor scenes in mind. Their framework optimizes the joint objective function

$$\mathcal{L}^{\text{CCA}} = \mathcal{L}_{\text{task}}^{\text{CCA}} + \mathcal{L}_{\text{global}}^{\text{CCA}} + \mathcal{L}_{\text{class}}^{\text{CCA}},$$

where

- $\mathcal{L}_{\text{task}}^{\text{CCA}}$ is the task loss, in this case cross-entropy between predicted and ground truth segmentation masks in the source domain;
- $\mathcal{L}_{\text{global}}^{\text{CCA}}$ is the global domain alignment loss, again defined as fooling the domain discriminator similar to *FCNs in the Wild*;
- $\mathcal{L}_{\text{class}}^{\text{CCA}}$ is the class-wise domain alignment loss, where grid cells are assigned soft class labels (extracted from the truth in the source domain and predicted in the target domain), and the domain classifiers and discriminators are trained and applied *class-wise*, separately.

As the title suggests, *Cross City Adaptation* is intended to adapt outdoor segmentation models trained on one city to other cities, but Chen et al. also apply it to synthetic-to-real domain adaptation from SYNTHIA to Cityscapes (see Section 7.2), achieving noticeable gains in segmentation quality.

Hong et al. [355] provide one of the most direct and most promising applications of feature-level synthetic-to-real domain adaptation. In their *Structural Adaptation Network*, the conditional generator $G_\theta^{\text{SDA}}(\mathbf{x}_S, \mathbf{z})$ takes as input the features $f_l^{\text{SDA}}(\mathbf{x}_S)$ from a low-level layer of the feature extractor (i.e., features with fine-grained details) and random noise $\mathbf{z}$ and produces transformed feature maps that should be similar to feature maps extracted from real images. To achieve that, G^{SDA} produces a noise map $\hat{G}_\theta^{\text{SDA}}(f_l^{\text{SDA}}(\mathbf{x}_S), \mathbf{z})$ and then adds it to high-level features: $G_\theta^{\text{SDA}}(\mathbf{x}_S, \mathbf{z}) = f_h^{\text{SDA}}(\mathbf{x}_S) + \hat{G}_\theta^{\text{SDA}}(f_l^{\text{SDA}}(\mathbf{x}_S), \mathbf{z})$.

The optimization problem is

$$\min_{\theta, \theta'} \max_{\phi} \left(\mathcal{L}_{\text{GAN}}^{\text{SDA}}(G_\theta^{\text{SDA}}, D_\phi^{\text{SDA}}) + \lambda \mathcal{L}_{\text{task}}^{\text{SDA}}(G_\theta^{\text{SDA}}, T_{\theta'}^{\text{SDA}}) \right),$$

where

- $\mathcal{L}_{\text{GAN}}^{\text{SDA}}$ is the GAN loss in the feature space:

$$\mathcal{L}_{\text{GAN}}^{\text{SDA}}(G_\theta^{\text{SDA}}, D_\phi^{\text{SDA}}) = \mathbb{E}_{\mathbf{x}_T \sim p_{\text{real}}} \left[\log D_\phi^{\text{SDA}}(\mathbf{x}_T) \right] + \\ \mathbb{E}_{\mathbf{x}_S \sim p_{\text{syn}}, \mathbf{z}} \left[\log \left(1 - D_\phi^{\text{SDA}}(G_\theta^{\text{SDA}}(\mathbf{x}_S, \mathbf{z})) \right) \right];$$

- $\mathcal{L}_{\text{task}}^{\text{SDA}}$ is the task loss for the pixel-wise classifier $T_{\theta'}^{\text{SDA}}$ which is trained end-to-end, together with the rest of the architecture; the task loss is defined as pixel-wise cross-entropy between the segmentation mask $T_{\theta'}^{\text{SDA}}(G_\theta^{\text{SDA}}(\mathbf{x}_S, \mathbf{z}))$ produced by $T_{\theta'}^{\text{SDA}}$ on adapted features and the ground truth synthetic segmentation mask $\mathbf{y}_S$.

Hong et al. compare the *Structural Adaptation Network* with other state of the art approaches, including *FCNs in the Wild* [353] and *Cross City Adaptation* [141], with source domain datasets SYNTHIA and GTA and target domain dataset *Cityscapes*; they conclude that this adaptation significantly improves the results for semantic segmentation of urban scenes.

To summarize, feature-level domain adaptation provides interesting opportunities for synthetic-to-real adaptation, but these methods still mostly represent work in progress. In our experience, feature- and model-level DA is usually a simpler and more robust approach, easier to get to work, so we expect new exciting developments in this direction and recommend to try this family of methods for synthetic to-real DA (unless actual refined images are required).

10.6 Domain Adaptation for Control and Robotics

In the field of control, joint domain adaptation is usually intended to transfer control policies learned in a simulated environment to a real setting. As we have already discussed in Sections 7.3 and 7.2, simulated environments are almost inevitable in

(a)   (b)

Fig. 10.9 Real and simulated robots used by Rusu et al. [744]: (a) real camera frames; (b) synthetic images rendered by MuJoCo.

reinforcement learning for robotics, as they allow to scale the datasets up compared to real data and cover a much wider range of situations than real data that could be used for imitational learning (see also the survey [841]). In this setting, domain adaptation is performed either for the control itself or jointly for the control and synthetic data.

The field began even before deep learning; for instance, Saxena et al. [762] learned a model for estimating grasp locations for previously unseen objects on synthetic data. In Cutler et al. [173], the results of training on a simulator serve as a prior for subsequent learning in the real world. Moreover, in [174, 175] Cutler et al. proceed to multifidelity simulators, training reinforcement learning agents in a series of simulators with increasing realism; we note this idea as potentially fruitful for other domains as well.

DeepMind researchers Rusu et al. [744] studied the possibility for transfer learning from simulated environments to the real world in the context of end-to-end reinforcement learning for robotics. They use the general idea of *progressive networks* [743], an architecture designed for multitask learning and transfer where each subsequent column in the network solves a new task and receives as input the hidden activations from previous columns. Rusu et al. present a modification of this idea for robot transfer learning, then train the first column in the MuJoCo physics simulator [862], and then transfer to a real *Jaco* robotic arm, using the *Asynchronous Advantage Actor-Critic* (A3C) framework for reinforcement learning [606]. Even with a relatively simple rendering engine, illustrated in Fig. 10.9, the authors report improved results for progressive networks compared to simple transfer via fine-tuning.

There are plenty of works that consider similar kinds of transfer learning, known in robotics as closing the *reality gap*. In particular, Bousmalis et al. [86] use a simulated environment to learn robotic grasping with a domain adaptation model called Grasp-GAN that makes synthetic images more realistic with a refiner (see Section 10.3); they argue that the added realism improves the results for control transfer. Tzeng et al. [880] propose a framework that combines supervised domain adaptation (that requires paired images) and unsupervised DA (that aligns the domains on the level of distributions); to do that, they introduce the notion of a "weak pairing" between images in the source and target domains and learn to find matching synthetic images to produce aligned data. The resulting model is successfully applied to training a visuomotor policy for real robots.

Pan et al. [977] consider sim-to-real translation for autonomous driving; they convert synthetic images to a scene parsing representation and then generate a realistic image by a generator corresponding to this parsing representation; the reinforcement learning agent receives this image as part of its driving environment. An even simpler approach, taken, e.g., by Xu et al. [845], would be to directly use the segmentation masks as input for the RL agent.

Researchers from *Wayve* Bewley et al. [68] perform domain adaptation for learning to drive from simulation; they claim to present the first end-to-end driving policy transferred from an (obviously supervised) synthetic setting to the fully unsupervised real domain. Their model does image translation and control transfer at the same time, learning the control on a jointly learned latent embedding space.

The architecture consists of two encoders E_S^{WVE} and E_T^{WVE}, two generators G_S^{WVE} and G_T^{WVE}, two discriminators D_S^{WVE} and D_T^{WVE}, and a controller C^{WVE}. The image translator follows the MUNIT architecture [536], with two convolutional variational autoencoder networks that swap the latent embeddings to translate between domains, i.e., for $\mathbf{x}_S \sim \mathcal{X}_S, \mathbf{x}_T \sim \mathcal{X}_T$

$$
\mathbf{z}_S = E_S^{\text{WVE}}(\mathbf{x}_S) + \epsilon, \quad \hat{\mathbf{x}}_S = G_S^{\text{WVE}}(\mathbf{z}_S),
$$
$$
\mathbf{z}_T = E_T^{\text{WVE}}(\mathbf{x}_T) + \epsilon, \quad \hat{\mathbf{x}}_T = G_T^{\text{WVE}}(\mathbf{z}_T),
$$

and the translation is to compute $\mathbf{z}_S$ with E_S^{WVE} and then predict $\hat{\mathbf{x}}$ with G_T^{WVE} and vice versa. The overall generator loss function is

$$
\mathcal{L}^{\text{WVE}} = \lambda_0 \mathcal{L}_{\text{rec}}^{\text{WVE}} + \lambda_1 \mathcal{L}_{\text{cyc}}^{\text{WVE}} + \lambda_2 \mathcal{L}_{\text{ctrl}}^{\text{WVE}} + \lambda_3 \mathcal{L}_{\text{cyctrl}}^{\text{WVE}} +
$$
$$
+ \lambda_4 \mathcal{L}_{\text{LSGAN}}^{\text{WVE}} + \lambda_5 \mathcal{L}_{\text{perc}}^{\text{WVE}} + \lambda_6 \mathcal{L}_{\text{zrec}}^{\text{WVE}},
$$

where

- $\mathcal{L}_{\text{rec}}^{\text{WVE}}$ is the L_1 image reconstruction loss in both domains:

$$
\mathcal{L}_{\text{rec}}^{\text{WVE}}(\mathbf{x}_S) = \| G_S^{\text{WVE}}(E_S^{\text{WVE}}(\mathbf{x}_S)) - \mathbf{x}_S \|_1,
$$
$$
\mathcal{L}_{\text{rec}}^{\text{WVE}}(\mathbf{x}_T) = \| G_T^{\text{WVE}}(E_T^{\text{WVE}}(\mathbf{x}_T)) - \mathbf{x}_T \|_1;
$$

- $\mathcal{L}_{\text{cyc}}^{\text{WVE}}$ is the cycle consistency loss for both domains:

$$
\mathcal{L}_{\text{cyc}}^{\text{WVE}}(\mathbf{x}_S) = \| G_S^{\text{WVE}}(E_T^{\text{WVE}}(G_T^{\text{WVE}}(E_S^{\text{WVE}}(\mathbf{x}_S)))) - \mathbf{x}_S \|_1
$$

and similar for $\mathcal{X}_T$;
- $\mathcal{L}_{\text{ctrl}}^{\text{WVE}}$ is the control loss that compares the controls produced on the training set with the autopilot: $\mathcal{L}_{\text{ctrl}}^{\text{WVE}}(\mathbf{x}_S) = \| C^{\text{WVE}}(E_S^{\text{WVE}}(\mathbf{x}_S)) - \mathbf{c} \|_1$ for ground truth control $\mathbf{c}$, and similar for $\mathcal{X}_T$;

- $\mathcal{L}_{\text{cyctrl}}^{\text{WVE}}$ is the control cycle consistency loss that makes the controls similar for images translated to another domain:

$$\mathcal{L}_{\text{cyctrl}}^{\text{WVE}}(\mathbf{x}_S) = \|C^{\text{WVE}}(E_T^{\text{WVE}}(G_S^{\text{WVE}}(E_S^{\text{WVE}}(\mathbf{x}_S)))) - C^{\text{WVE}}(E_S^{\text{WVE}}(\mathbf{x}_S))\|_1,$$

and similar for $\mathcal{X}_T$;
- $\mathcal{L}_{\text{LSGAN}}^{\text{WVE}}$ is the LSGAN adversarial loss applied to both generator-discriminator pairs (see also Section 4.5 for a discussion of LSGAN);
- $\mathcal{L}_{\text{perc}}^{\text{WVE}}$ is the perceptual loss (see above) for both image translation directions with instance normalization applied before, as shown in [373];
- $\mathcal{L}_{\text{zrec}}^{\text{WVE}}$ is the latent reconstruction loss: $\mathcal{L}_{\text{zrec}}^{\text{WVE}}(\mathbf{z}_S) = \|E_T^{\text{WVE}}(G_T^{\text{WVE}}(\mathbf{z}_S)) - \mathbf{z}_S\|_1$ and similar for $\mathcal{X}_T$.

Bewley et al. compare their approach with a number of transfer learning baselines, show excellent results for end-to-end learning to drive, and even perform real-world experiments with the trained policy. Similar techniques have been used without synthetic data in the loop as well; e.g., Wulfmeier et al. [945] use a similar model for domain adaptation to handle appearance changes in outdoor robotics, i.e., changes in weather conditions, lighting, and the like.

We have already discussed the works of Inoue et al. [386], Zhang et al. [999], James et al. [396], and Yang et al. [969] who make real data more similar to synthetic for computer vision problems related to robotic grasping and visual navigation (see Section 10.4). Importantly, these models are not merely translating images but are also tested on real-world robots. Zhang et al. not only show improvements in semantic segmentation results but also conduct real-world robotic experiments for indoor and outdoor visual navigation tasks, first training a navigation policy in a simulated environment and then directly deploying it on a robot in a real environment, while James et al. test their solution on a real robotic hand, training the *QT-Opt* policy [422] to grasp from a simulation with 5000 additional real-life grasping episodes better than the same policy trained on 580,000 real episodes, a more than 99% reduction in required real-world input.

Another direction where synthetic data might be useful for learning control is to generate synthetic behaviours to improve imitation learning [647]. Bansal et al. [39] discuss the insufficient data problem in imitation learning: for learning to drive, even 30 million real-world expert driving examples that combine into more than 60 days of driving is not sufficient to train an end-to-end driving model. To alleviate this lack of data, they present their imitation learning framework *ChauffeurNet* with data where synthetic perturbations have been introduced to expert driving examples. This allows to cover corner cases such as collisions and off-road driving, i.e., bad examples that should be avoided but that are lacking in expert examples altogether. Interestingly, perturbations are introduced into intermediate representations rather than in raw sensor input or controller outputs.

To sum up, closing the reality gap is one of the most important problems in the field of control and robotics. Important breakthroughs in this direction appear constantly, but there is still some way to go before self-driving cars and robotic arms are able

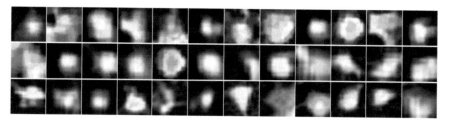

Fig. 10.10 Real and generated lung nodules from [158]: nodules 1–18 (numbered left to right) are synthetic, and 19–36 are real benign nodules.

to train in a simulated environment and then perfectly transfer these skills to the real world. On the other hand, the question of whether these techniques will actually be needed at the end also still remains open. Some interesting recent advances in robotics do use synthetic environments for training but do not use any explicit domain adaptation, using sufficiently varied domain randomization as the main tool; recall the *Dactyl* robotic arm we discussed in Section 7.4.

10.7   Case Study: GAN-Based Domain Adaptation for Medical Imaging

Medical imaging is a field where labeled data is especially hard to come by. First, while manual labeling is hard and expensive enough for regular computer vision problems, in medical imaging it is far more expensive still because it cannot be crowdsourced to anonymous annotators. For most problems, medical imaging data can only be reliably labeled by a trained professional, often with a medical degree. Second, for obvious privacy reasons it is very hard to arrange for publishing real datasets, and collecting a large enough labeled dataset to train a standard object detection or segmentation model would in many cases require a concerted effort from several different hospitals; thus, with the exception of some public competitions, most researchers in the field use private datasets and are not allowed to share their data. Third, some pathologies simply do not have sufficiently large and diverse datasets collected yet. At the same time, often there are relatively large generic datasets available, e.g., of healthy tissue but not of a specific pathology of interest.

While we emphasize GAN-based generation methods, we note that there have been successful attempts to use rendered synthetic data for medical imaging tasks that are based on recent developments in medical visualization and rendering tools. For example, Mahmood et al. [573] use the recently developed cinematic rendering technique for CT [223] (a photorealistic simulation of the propagation of light through tissue) to train a CNN for depth estimation in endoscopy data.

GANs have been widely applied to generating realistic medical images [45, 53, 469, 627, 967, 974]. Moreover, since the images are domain-specific and often low resolution, the quality of GAN-produced images has relatively quickly reached the level where it can in many applications pass the "visual Turing test", fooling even trained specialists. For a characteristic example, see the lung nodule samples generated in [158] and illustrated in Fig. 10.10. It is indeed very hard to distinguish real images from fake ones, but the general quality and resolution of these radiology images are low enough that even a simple GAN could do a very good job in this case. Another example is given by GAN-based generation of magnetic resonance (MR) images of the brain in [109, 314, 317]. Therefore, it is no wonder that GAN-based domain adaptation (DA) techniques, especially based on fusing and augmenting real images, are increasingly finding their way into medical imaging. In this section, we give a brief overview of recent work in this domain.

In some works, synthetic data is generated from scratch, i.e., GANs are trained to convert random noise into synthetic images (see also Section 9.4). Frid-Adar et al. [244, 245] used two standard GAN architectures, deep convolutional GAN (DCGAN) [696] and auxiliary classifier GAN (ACGAN) [637] with class label auxiliary information, to generate synthetic computed tomography (CT) images of liver lesions. They report significantly improved results in image classification with CNNs when training on synthetic data compared to standard augmentations of their highly limited dataset (182 two-dimensional scans divided into three types of lesions). Baur et al. [967] attempt high-resolution skin lesion synthesis, comparing several GAN architectures and obtaining highly realistic results even with a small training dataset.

Han et al. [316, 317] concentrate on brain magnetic resonance (MR) images. They use progressively growing GANs (PGGAN) [435] to generate 256×256 MR images and then compare two different refinement approaches: SimGAN [793] as discussed in Section 10.2 and UNIT [536], an unsupervised image-to-image translation architecture that maps each domain into a shared latent space with a VAE-GAN architecture [497] (we remark that the original paper [536] also applies UNIT, among other things, to synthetic-to-real translation; see also Section 4.7 where we discuss GAN-based style transfer architectures). Han et al. report improved results when combining GAN-based synthetic data with classic domain adaptation techniques.

Neff [621] uses a slightly different approach: to generate synthetic data for segmentation, he uses a standard WGAN-GP architecture [303] but generates image-segmentation pairs, i.e., images with an additional channel that shows the segmentation mask. Neff reports improved segmentation results with U-Net [730] after augmenting a real dataset with synthetic image-segmentation pairs. Mahmood et al. [572] show an interesting take on the problem by doing the reverse: they make real medical images look more like synthetic images in order to then apply a network trained on synthetic data (see also Section 10.4). With this approach, they improve state-of-the-art results in-depth estimation for endoscopy images.

In general, segmentation problems in medical imaging are especially hard to label, and segmentation data is especially lacking in many cases. In this context, recent works have often employed conditional GANs and *pix2pix* models to generate realistic images from randomized segmentation masks. For example, Bailo et al. [33] consider red blood cell image generation with the *pix2pixHD* model [909]. Namely, their conditional GAN optimizes

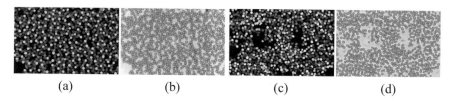

| (a) | (b) | (c) | (d) |

Fig. 10.11   Red blood cell images generated by Bailo et al. [33]: (a) real segmentation mask; (b) the corresponding real blood image; (c) synthetic segmentation mask; (d) the corresponding synthetic blood image.

$$\min_{G^{\text{P2P}}} \left[\max_{D_1^{\text{P2P}}, D_2^{\text{P2P}}} \left[\mathcal{L}_{\text{GAN}}^{\text{P2P}}(G^{\text{P2P}}, D_1^{\text{P2P}}) + \mathcal{L}_{\text{GAN}}^{\text{P2P}}(G^{\text{P2P}}, D_2^{\text{P2P}}) \right] + \right.$$

$$\left. + \lambda_1 \left(\mathcal{L}_{\text{FM}}^{\text{P2P}}(G^{\text{P2P}}, D_1^{\text{P2P}}) + \mathcal{L}_{\text{FM}}^{\text{P2P}}(G^{\text{P2P}}, D_2^{\text{P2P}}) \right) + \lambda_2 \mathcal{L}_{\text{PR}}^{\text{P2P}}(G^{\text{P2P}}(\mathbf{s}, E^{\text{P2P}}(\mathbf{x})), \mathbf{x}) \right],$$

where

- $\mathbf{x}$ is an input image, $\mathbf{s}$ is a segmentation mask (it serves as input to the generator), D_1^{P2P} and D_2^{P2P} are two discriminators that have the same architecture but operate on different image scales (original and 2x downsampled), $\mathcal{L}_{\text{GAN}}^{\text{P2P}}(G, D)$ is the regular GAN loss;
- $E^{\text{P2P}}(\mathbf{x})$ is the feature encoder network that encodes low-level features of the objects with instance-wise pooling; its output is fed to the generator $G^{\text{P2P}}(\mathbf{s}, E^{\text{P2P}}(\mathbf{x}))$ and can be used to manipulate object style in generated images (see [909] for more details);
- $\mathcal{L}_{\text{FM}}^{\text{P2P}}$ is the *feature matching loss* that makes features at different layers of the discriminators (we denote the input to the ith layer of D as $D^{(i)}$) match for $\mathbf{x}$ and $G(\mathbf{s})$:

$$\mathcal{L}_{\text{FM}}^{\text{P2P}}(G, D) = \mathbb{E}_{(\mathbf{s}, \mathbf{x})} \left[\sum_{i=1}^{L} \frac{1}{N_i} \left\| D^{(i)}(\mathbf{s}, \mathbf{x}) - D^{(i)}(\mathbf{s}, G(\mathbf{s}, E^{\text{P2P}}(\mathbf{x}))) \right\|_1 \right];$$

- $\mathcal{L}_{\text{PR}}^{\text{P2P}}$ is the *perceptual reconstruction loss* for some feature encoder F:

$$\mathcal{L}_{\text{PR}}^{\text{P2P}}(G, D) = \mathbb{E}_{(\mathbf{s}, \mathbf{x})} \left[\sum_{i=1}^{L'} \frac{1}{M_i} \left\| F^{(i)}(\mathbf{x}) - F^{(i)}(G(\mathbf{s}, E^{\text{P2P}}(\mathbf{x}))) \right\|_1 \right].$$

The real dataset in [33] consisted of only 60 manually annotated 1920×1200 RGB images (with another 40 images used for testing), albeit with a lot of annotated objects (669 blood cells per image on average). Bailo et al. also developed a scheme for sampling randomized but realistic segmentation masks to use for synthetic data generation. Sample images, shown in Fig. 10.11, again show very realistic images produced by this GAN-based architecture, to a large extent thanks to the regular and relatively simple structure of the images that need to be generated. Bailo et al. report improved segmentation results with FCN and improved detection with Faster R-CNN when trained on a combination of real and synthetic data.

Zhao et al. [1014] consider the problem of generating filamentary structured images, such as retinal fundus and neuronal images, from a ground truth segmentation map, with an emphasis on generating images in multiple different styles. Their FILA-sGAN approach is based on GAN-based image style transfer ideas [407, 886]. Its generator loss function is

$$\mathcal{L}^{\text{FIL}} = \mathcal{L}_{\text{GAN}}^{\text{FIL}}(G^{\text{FIL}}, D^{\text{FIL}}) + \lambda_1 \mathcal{L}_{\text{cont}}^{\text{FIL}}(G^{\text{FIL}}) + \lambda_2 \mathcal{L}_{\text{sty}}^{\text{FIL}}(G^{\text{FIL}}) + \lambda_3 \mathcal{L}_{\text{TV}}^{\text{FIL}}(G^{\text{FIL}}),$$

where

- $G^{\text{FIL}} : \mathbf{y} \rightarrow \hat{\mathbf{x}}$ is a generator that takes a binary image and produces a "phantom" $\hat{\mathbf{x}}$, D^{FIL} is a synthetic vs. real discriminator, and $\mathcal{L}_{\text{GAN}}^{\text{FIL}}$ is the standard GAN loss;
- $\mathcal{L}_{\text{cont}}^{\text{FIL}}$ is the *content loss* that makes the filamentary structure of a generated phantom $\hat{\mathbf{x}}$ match the real raw image $\mathbf{x}$, evidenced through the features $\phi^{(i)}$ for some standard CNN feature extractor such as VGG-19:

$$\mathcal{L}_{\text{cont}}^{\text{FIL}}(G^{\text{FIL}}) = \sum_l \frac{1}{W_l H_l} \left\| \phi^{(l)}(\mathbf{x}) - \phi^{(l)}(\hat{\mathbf{x}}) \right\|_F^2,$$

where $\mathbf{x}$ is the real raw image, l spans the CNN blocks and layers, W_l and H_l are the width and height of the corresponding feature maps, and $\| \cdot \|_F$ is the Frobenius matrix norm;
- $\mathcal{L}_{\text{sty}}^{\text{FIL}}$ is the *style loss* that minimizes the textural difference between $\hat{\mathbf{x}}$ and a style image $\mathbf{x}_s$:

$$\mathcal{L}_{\text{sty}}^{\text{FIL}}(G^{\text{FIL}}) = \sum_l \frac{\omega_l}{W_l H_l} \left\| \mathbb{G}^{(l)}(\mathbf{x}_s) - \mathbb{G}^{(l)}(\hat{\mathbf{x}}) \right\|_F^2,$$

where $\mathbf{x}_s$ is the style image, $\mathbb{G}^{(l)}$ is the Gram matrix of the features in CNN block l, and ω_l is its weight (a hyperparameter);
- $\mathcal{L}_{\text{TV}}^{\text{FIL}}$ is the *total variation loss* that serves as a regularizer and encourages $\hat{\mathbf{x}}$ to be smooth:

$$\mathcal{L}_{\text{TV}}^{\text{FIL}}(G^{\text{FIL}}) = \sum_{i,j} \left(\left\| \hat{\mathbf{x}}_{i,j+1} - \hat{\mathbf{x}}_{i,j} \right\|_2^2 + \left\| \hat{\mathbf{x}}_{i+1,j} - \hat{\mathbf{x}}_{i,j} \right\|_2^2 \right).$$

As a result, Zhao et al. report highly realistic filamentary structured images generated from a segmentation map and a single style image in a variety of different styles; some examples are shown in Fig. 10.12. Importantly for us, they also report improved segmentation results with state-of-the-art approaches to the corresponding segmentation task.

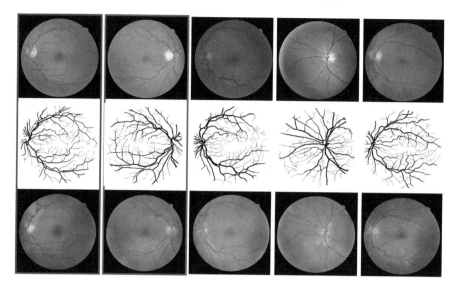

Fig. 10.12   Real and generated retinal images by Zhao et al. [1014]; in each column, the top image is a real image, the middle is the ground truth filament segmentation, and the bottom is a synthetic image generated from this segmentation map.

A similar architecture based on a conditional GAN has been used by Sirazitdinov et al. [804] to generate tube-like objects on X-ray images, such as puncture needles, wires, or catheters; finding such objects on X-rays is a very important task for interventional radiology but real data with such objects is (fortunately) hard to come by.

In other works, Hou et al. [357] use GANs to refine synthesized histopathology images (similar to SimGAN discussed in Section 10.2), with improved nucleus segmentation and glioma classification results. Tang et al. [850] use the *pix2pix* model [389] to generate realistic computed tomography (CT) images from customized lymph node masks, reporting improved lymph node segmentation with U-Net [730]. In [849], the same researchers use the multimodal image-to-image translation (MUNIT) model [373], which we discussed in Section 4.7, to generate realistic chest X-rays with custom abnormalities, reporting improved segmentation with both U-Net and their developed model XLSor. Han et al. [315] were the first to apply 3D GAN-based DA to produce data for 3D object detection, i.e., bounding boxes; they use it in the context of synthetizing CT images of lung nodules.

In a related approach, synthetic data can be generated from real data, but in a different domain. For example, Zhang et al. [1013] learn a CycleGAN-based architecture [1025] to learn volume-to-volume (i.e., 3D) translation for unpaired datasets of CT and MR images (domain A and domain B, respectively). Moreover, they augment the basic CycleGAN with segmentors S_A^{VOL} and S_B^{VOL} that help preserve segmentation mask consistency. In total, their model optimizes the loss function

$$\mathcal{L}^{\text{VOL}} = \mathcal{L}^{\text{VOL}}_{\text{GAN}}(G^{\text{VOL}}_A, D^{\text{VOL}}_A) + \mathcal{L}^{\text{VOL}}_{\text{GAN}}(G^{\text{VOL}}_B, D^{\text{VOL}}_B) + \lambda_1 \mathcal{L}^{\text{VOL}}_{\text{cyc}}(G^{\text{VOL}}_A, G^{\text{VOL}}_B) +$$
$$+ \lambda_2 \mathcal{L}^{\text{VOL}}_{\text{shape}}(S^{\text{VOL}}_A, S^{\text{VOL}}_B, G^{\text{VOL}}_A, G^{\text{VOL}}_B),$$

where

- $G^{\text{VOL}}_A : B \to A$ and $G^{\text{VOL}}_B : A \to B$ are CycleGAN generators, and D^{VOL}_A and D^{VOL}_B are discriminators in domains A and B, respectively, trained to distinguish between real and synthetic (generated by G^{VOL}) images by the standard GAN loss function $\mathcal{L}^{\text{VOL}}_{\text{GAN}}$;
- $\mathcal{L}^{\text{VOL}}_{\text{cyc}}$ is the cycle consistency loss

$$\mathcal{L}^{\text{VOL}}_{\text{cyc}}(G^{\text{VOL}}_A, G^{\text{VOL}}_B) = \mathbb{E}_{\mathbf{x}_A} \left[\left\| G^{\text{VOL}}_A(G^{\text{VOL}}_B(\mathbf{x}_A)) - \mathbf{x}_A \right\|_1 \right]$$
$$+ \mathbb{E}_{\mathbf{x}_B} \left[\left\| G^{\text{VOL}}_B(G^{\text{VOL}}_A(\mathbf{x}_B)) - \mathbf{x}_B \right\|_1 \right],$$

- $S^{\text{VOL}}_A : A \to Y$ and $S^{\text{VOL}}_B : B \to Y$ are segmentors that produce 3D segmentation masks, and $\mathcal{L}^{\text{VOL}}_{\text{shape}}$ is the shape consistency loss

$$\mathcal{L}^{\text{VOL}}_{\text{shape}}(S^{\text{VOL}}_A, S^{\text{VOL}}_B, G^{\text{VOL}}_A, G^{\text{VOL}}_B) =$$
$$\mathbb{E}_{\mathbf{x}_B} \left[-\frac{1}{N} \sum_i \mathbf{y}^i_B \log S^{\text{VOL}}_A(G^{\text{VOL}}_A(\mathbf{x}_B))_i \right]$$
$$+ \mathbb{E}_{\mathbf{x}_A} \left[-\frac{1}{N} \sum_i \mathbf{y}^i_B \log S^{\text{VOL}}_A(G^{\text{VOL}}_A(\mathbf{x}_B))_i \right],$$

where $\mathbf{y}_A$ and $\mathbf{y}_B$ are ground truth segmentation results for $\mathbf{x}_A$ and $\mathbf{x}_B$, respectively.

Zhang et al. report that 3D segmentation in their architecture improves not only over the baseline model trained only on real data but also over the standard approach of fine-tuning S^{VOL}_A and S^{VOL}_B separately on generated synthetic data. Ben-Cohen et al. [59] present a similar architecture for cross-modal synthetic data generation of PET scans from CT images, also with improved segmentation results for lesion detection. Similar image-to-image translation techniques have been applied to generating images from 2D MR brain images to CT and back [401, 628, 932], PET to CT [28], cardiac CT to MR [125], virtual H&E staining, including transformation from unstained to stained lung histology images [48] and stain style transfer [784], multi-contrast MRI (from contrast to contrast) [182], 3D cross-modality MRI [979], different styles of prostate histopathology [717], different datasets of chest X-rays [128], and others.

Model-based domain adaptation (that we discussed in Section 10.5) has also been applied in the context of medical imaging. Often it has been used to do domain transfer between different types of real images, e.g., between different parts of the brain [65] or from *in vitro* to *in vivo* images [166], but synthetic-to-real DA has also been a major topic. As early as 2013, Heimann et al. [333] generated synthetic training data in the form of digitally reconstructed radiographs for ultrasound transducer localization. To close the domain gap between synthetic and real images, they used standard instance weighting and found significant improvements in the resulting detections. Kamnitsas et al. [425] use unsupervised DA for brain lesion segmentation in 3D, switching from one type of MR images to another in domain adaptation. They use a state-of-the-art 3D multi-scale fully convolutional segmentation network [426] and a domain discriminator that makes intermediate feature representations of the segmentation networks indistinguishable across the domains.

In general, GAN-based architectures for medical imaging, either generating synthetic data or adapting real data from other domains, represent promising directions of further research and will, in my opinion, define state of the art in the field for years to come. However, at present the architectures used in different works differ a lot, and comparisons across different GAN-based architectures are usually lacking: each work compares their architecture only with the baselines. Further research and large-scale experimental studies are needed to determine which architectures work best for various domain adaptation problems related to medical imaging.

10.8 Conclusion

This chapter has been devoted to a survey of domain adaptation approaches to synthetic data applications. Virtually all applications of synthetic data face the same problem of synthetic-to-real transfer: we need to make a model trained on synthetic data work on real data that we will face in the actual application. The most straightforward way is, of course, not to do any transfer or do it with domain randomization (see Section 9.1) and pray that it works. But in this chapter, we have discussed ways to actively help the models transfer from synthetic to real data better.

We have seen that approaches to this regard can be broadly classified into two categories. In synthetic-to-real *refinement*, which we discussed in Sections 10.1, 10.2, and 10.3, domain adaptation changes the *data* itself, and it is possible to see a "refined" image, usually (but not always) made more realistic by the refinement procedure. There even exists the opposite approach that we have seen in Section 10.4: make real data look more "synthetic" so that the models trained on synthetic data will better recognize the refined images.

Model-based domain adaptation approaches, which we have considered in Section 10.5, leave the data in place but change the model itself or the training procedure. They usually train a joint embedding or feature space for both synthetic and real inputs, requiring that the model does not differentiate between the two domains. We have also seen some domain adaptation techniques in control and robotics (Section 10.6) and a case study of GAN-based domain adaptation for medical imaging (Section 10.7).

With this, we are nearing the end of the book. The next chapter is devoted to a field that looks completely orthogonal to what we have done before: differential privacy. It proves to be important for synthetic data because in certain applications, synthetic datasets are used to protect real data which is too sensitive to release. Differential privacy is a way (actually, *the* way) to provide guarantees that a released synthetic dataset, probably produced by some generative model, does not leak sensitive information about the real data that the model had been trained on. In the next chapter, we will see what kind of guarantees can exist in this regard.

Chapter 11
Privacy Guarantees in Synthetic Data

In this chapter, we discuss another important field of applications for synthetic data: ensuring privacy. In many real-world problems, real data is sensitive enough that it is impossible to release. One possible solution could be to train generative models that would produce new synthetic datasets based on real data, while the real data itself would remain secret. But how can we be sure that real data will not be inadvertently leaked? Guarantees in this regard can be provided by the framework of differential privacy. We give a brief introduction to differential privacy, its relation to machine learning, and the guarantees that it can provide for synthetic data generation.

11.1 Why is Privacy Important?

In many domains, real data is not only valuable, but also sensitive; it should be protected by law, commercial interest, and common decency. The unavailability of real data is exactly what makes synthetic data solutions attractive in these domains. But models for generating synthetic data have to train on real datasets anyway, so how do we know we are not revealing it? A number of famous examples show that naive attempts to anonymize data are often insufficient. Let me begin with a few illustrative examples that have become famous in the studies of privacy in computer science.

Probably the first such example dates back to 1997, when the Massachusetts Group Insurance Commission published a carefully anonymized dataset with the medical history of state employees. When it was published, a Ph.D. student from MIT, Latanya Sweeney, spent $20 on a list of all voters from Cambridge, MA (a perfectly legal operation), joined the two datasets according to zip code, birth date, and sex (the three fields that they have in common), and immediately identified William Weld, then the Governor of Massachusetts. She was able to confirm her findings because Weld had a recent public medical incident, but she did not use

© The Author(s), under exclusive license to Springer Nature Switzerland AG 2021
S. I. Nikolenko, *Synthetic Data for Deep Learning*, Springer Optimization
and Its Applications 174, https://doi.org/10.1007/978-3-030-75178-4_11

that information in the deanonymization procedure. This was one of the first such incidents that attracted public attention, and Dr. Sweeney's works dating back to the 1990s were among the first not only to raise concerns, but also present specific algorithms with which privacy could be violated even in anonymized data and suggest some ways to efficiently preserve privacy in practice [832, 833].

The second famous example came on August 4, 2006, when *AOL Research*, a division of the internet company AOL that was at the time still a huge internet and email provider (#1 in the United States) and a very popular web portal, published anonymized search queries for over 650,000 users over a 3-month period. Naturally, AOL research did not mean to do any harm: data was released purely for research purposes, and the dataset did not contain anything other than search queries grouped only by user id. But search queries often spoke for themselves: in five days, on August 9, *The New York Times* ran a profile on one of the searchers whom they were able to identify personally from the queries. This profile did not contain anything incriminating, but other search histories were less than innocent: some suggested that the user might be getting ready to commit murder (here's an ethical question for you: should AOL or *Google* be doing anything about this?), and the vagaries of "User 927" even became the basis of an experimental play staged in a Philadelphia theatre in 2008. No, I am not going to cite what this user searched for, but, fortunately and somewhat ironically, this information is just a couple of search queries away...

Search queries are, of course, an easy suspect for revealing sensitive information. *AOL Research* quickly admitted that they made a mistake, removed the dataset in three days (not that it helped, of course), and as a result of the scandal Maureen Govern, the CTO of AOL, resigned in a couple of weeks. But what could go wrong if we publish a much more restricted data type? Say, only the fields where the users do not type in any odd thing? Say, their ratings for products in a recommender engine?..

Alas, the third example is exactly this. Every researcher working in the field of recommender systems knows about the Netflix Prize, a competition held from 2006 to 2009 with a grand prize of one million dollars [64]. This was way before *Kaggle* was a thing, and one could say that *Kaggle* was founded in 2010, in part, as a result of the resounding success of the Netflix Prize. This competition brought to light several new ideas about recommender systems that blossomed into whole directions of research and for a long time defined state of the art in recommender systems [54, 468]. The Netflix Prize dataset only contained the identifiers of movies (names of the movies were known) and ratings that a given user has given them; the users only had numerical ids, there was no personal information disclosed.

However, even this kind of dataset proved to be dangerous: in 2008, researchers from the University of Texas discovered that they could match IMDB user profiles (which are public) with their anonymized Netflix profiles from the published dataset with very high confidence [619]. This means that they could mine information about movie preferences that the users chose not to disclose to their public IMDB profile; needless to say, some of this information might, again, be rather sensitive. As a result, a class action lawsuit was filed against Netflix based on the arguments from [619]; the company settled with the plaintiffs but had to cancel the second Netflix Prize, which had already been announced at the time.

These three examples show that the computational privacy is a very fragile thing. The adversary might have additional information, such as a list of voters or public IMDB profiles. The adversary does not need to attack a large fraction of the dataset because a successful attack on even a small portion of the data might be damaging; after all, the vast majority of people couldn't care less about who knows their movie ratings. A sparse dataset with high-dimensional information about the users helps the adversary: high confidence in the case of the Netflix Prize became possible precisely because there were a lot of movies in the dataset to mine for correlations (about 20,000). And finally, in all three cases, the datasets were not published by malicious hackers or people who didn't know any better: they were published by experienced researchers, in case of AOL and Netflix by researchers who worked in computer science. Still, the adversaries proved to be more resourceful: in these cases, finding a crack in the wall is a much easier job than building a perfect all-encompassing barrier.

When researchers recognized the preservation of privacy as a computer science problem, formal negative results also followed quickly. A famous paper by Dinur and Nissim [199] showed that a few database queries (e.g., taking sums or averages of subsets) suffice to bring about strong violations of privacy even if the database attempts to preserve privacy by introducing noise. Formally, one of their results was that if a database of n private bits $d_1, \ldots, d_n$ responds to queries defined by subsets of bits $S \subseteq \{1, \ldots, n\}$ by specifying the sums $q_S(D) = \sum_{i \in S} d_i$ approximately, and the error in the database's answers is on the order of $o(\sqrt{n})$ (it is hard to imagine a useful database that responds to queries with an error of $\sqrt{n}$ or more!), then a polynomial number of queries suffices to reconstruct *almost all*, i.e., $n - o(n)$ private bits in D. Subsequent results made this even stronger.

This problem also pertains to machine learning. If a machine learning model has trained on a dataset with a few outliers, how do we know it does not "memorize" these outliers directly and will not divulge them to an adversary? For a sufficiently expressive model, such memorization is quite possible, and note that the outliers are usually the most sensitive data points. For instance, Carlini et al. [114] show that state of the art language models do memorize specific sequences of symbols, and one can extract, e.g., a secret string of numbers from the original dataset with a reasonably high success rate.

How does all this relate to synthetic data? Machine learning on privacy-sensitive datasets might be an important field of application for synthetic data: wouldn't it be great if AOL or Netflix didn't have to publish their *real* datasets but would publish information about *synthetic* users instead? This would immediately alleviate all privacy concerns. On the other hand, choosing a uniform distribution for the ratings or random character strings for search queries would render such synthetic datasets completely useless: naturally, to be useful the distribution of synthetic data must resemble the distribution of real data. But then wouldn't we be divulging private information? Looks like we need to dig a little deeper.

11.2   Introduction to Differential Privacy

The field of *differential privacy*, pioneered by Dwork et al. [214, 216, 218], was largely motivated by considerations such as the ones we saw in the previous section. The works of Cynthia Dwork have been widely recognized as some of the most novel and interesting advances in modern computer science: Prof. Dwork received the Dijkstra Prize in 2007, the Gödel Prize in 2017 for the work [216], the Hamming Medal and the Knuth Prize in 2020.

In the main definition of the field, a mechanism (randomized algorithm) M is called (ϵ, δ)-*differentially private* for some positive real parameters ϵ and δ if for any two databases D and D' that differ in only a single point x, $D \setminus \{x\} = D' \setminus \{x\}$, and any subset of outputs S

$$p\left(M(D) \in S\right) \le e^{\epsilon} p\left(M(D') \in S\right) + \delta,$$

or, equivalently, for every point s in the output range of M

$$\left| \ln \frac{p\left(M(D) = s\right)}{p\left(M(D') = s\right)} \right| \le \epsilon \quad \text{with probability } 1 - \delta.$$

The ratio $\ln \frac{p(M(D)=s)}{p(M(D')=s)}$ is an important quantity called *privacy loss* that needs to be bounded in absolute value.

The intuition here is that an adversary who receives only the outputs of M should have a hard time learning anything about any single point in D. The same intuition could be reformulated in terms of a Bayesian update of beliefs (recall Section 2.2, where we discussed how the Bayes theorem is the foundation of all machine learning): an adversary, after learning the result $M(D) = s$, updates their beliefs about the two databases (that is, about the question whether the dataset contains some specific point x) as

$$\frac{p(D \mid M(D) = s)}{p(D' \mid M(D) = s)} = \frac{p(D)}{p(D')} \frac{p(M(D) = s \mid D)}{p(M(D') = s \mid D')},$$

and the latter ratio on the right-hand side is precisely the privacy loss whose logarithm's absolute value is bounded by ϵ in the definition.

This definition has a number of important desirable qualities. First, it is robust to the introduction of *additional information*, that is, knowledge of some events available to an adversary: naturally, additional information regarding the database will help the adversary, but the definition still remains in place: an (ϵ, δ)-differentially private mechanism will remain (ϵ, δ)-differentially private and will not help the adversary further. Second, it is immune to *postprocessing*: an adversary cannot compute some function of the private mechanism's result $M(D)$ and compromise the privacy, i.e., the privacy loss cannot be increased by thinking hard about the results of M. Third, it is *composable*: if an adversary has access to two mechanisms, M_1 which is (ϵ_1, δ_1)-differentially private and M_2 with parameters (ϵ_2, δ_2), any composition of them will have parameters not exceeding $(\epsilon_1 + \epsilon_2, \delta_1 + \delta_2)$ regardless of whether M_1 and M_2 know about each other; this allows for modular design of private architectures. Fourth, it allows for *group privacy*, that is, when the databases differ by k elements an $(\epsilon, 0)$-differentially private mechanism will become at most $(k\epsilon, 0)$-differentially private.

Unfortunately, this definition conceals an unpleasant tradeoff. If we set $\delta = 0$ the definition becomes too strong: for example, it is too pessimistic for repeated applications of M (the exponent grows linearly). But if not, δ may hide a complete failure of privacy preservation: for instance, an (ϵ, δ)-differentially private mechanism may reveal the entire database with probability δ or reveal the δ share of data with probability 1. Therefore, in practice, it should hold that $\delta \ll \frac{1}{n}$.

I do not want to get to a much deeper discussion of differential privacy than the definitions, so I will conclude this brief intro with an example of the *Laplace mechanism*, probably the simplest and most classical example of a differentially private mechanism. Suppose that we are sending numerical queries to a database of n integer numbers, that is, a query is a function $f : \mathbb{N}^n \to \mathbb{R}^k$. An important property of such functions f in this case is their L_1-*sensitivity*, a measure of how much changing a single element in the database can change the function:

$$\Delta f = \max_{D, D': D \text{ and } D' \text{ differ in one point}} \left\| f(D) - f(D') \right\|_1 .$$

The Laplace mechanism works as follows: when someone asks to compute $f(D)$, it computes the correct answer and gives out a version of it perturbed by the Laplace distribution (hence the name):

$$M_L(D, f, \epsilon) = f(D) + \left(Y_1\ Y_2 \ldots Y_k \right), \text{ where } Y_i \sim \text{Lap}\left(\frac{\Delta f}{\epsilon} \right)$$

for some constant ϵ and for the Laplace distribution

$$\text{Lap}(x \mid b) = \frac{1}{2b} e^{-\frac{1}{b}|x|}.$$

Let us now compare the distributions of M_L results on two databases D and D' that differ at a single point. For some point $\mathbf{z} \in \mathbb{R}^k$,

$$\frac{p(M_L(D, f, \epsilon) = \mathbf{z})}{p(M_L(D', f, \epsilon) = \mathbf{z})} = \prod_{i=1}^{k} \frac{e^{-\frac{\epsilon}{\Delta f}|f(D)_i\ z_i|}}{e^{-\frac{\epsilon}{\Delta f}|f(D')_i - z_i|}} =$$

$$= \prod_{i=1}^{k} e^{\frac{\epsilon(|f(D')_i - z_i| - |f(D)_i - z_i|)}{\Delta f}} \leq \prod_{i=1}^{k} e^{\frac{\epsilon(|f(D')_i - f(D)_i|)}{\Delta f}} = e^{\frac{\epsilon\|f(D) - f(D')\|_1}{\Delta f}} \leq e^{\epsilon},$$

where the first inequality is the triangle inequality and the second is by our assumption on the L_1-sensitivity of f. Similarly, $\frac{p(M_L(D, f, \epsilon)=\mathbf{z})}{p(M_L(D', f, \epsilon)=\mathbf{z})} \geq e^{-\epsilon}$, and we have proved that the Laplace mechanism is $(\epsilon, 0)$-differentially private.

A similar (but much more involved and cumbersome) argument shows that the same can be achieved with L_2-sensitivity and Gaussian noise. In other words, if we define L_2-sensitivity as

$$\Delta_2 f = \max_{D, D':D \text{ and } D' \text{ differ in one point}} \left\| f(D) - f(D') \right\|_2$$

and define the Gaussian mechanism as

$$M_{\mathcal{N}}(D, f, \epsilon) = f(D) + \left(Y_1 \; Y_2 \; \dots \; Y_k \right), \quad \text{where } Y_i \sim \mathcal{N}\left(0, \sigma^2\right),$$

then $M_{\mathcal{N}}$ will be (ϵ, δ)-differentially private if we let

$$\sigma \geq c\frac{\Delta_2 f}{\epsilon}, \quad \text{where } c^2 > 2\ln\frac{1.25}{\delta};$$

see, e.g., [218] for details.

11.3   Differential Privacy in Deep Learning

We are most interested in machine learning applications for privacy: how can we give access to the results of learning without giving access to the training data? Before we proceed to applying differential privacy to machine learning, we should define a formal setting for such considerations. What should we allow the adversary to do? One might think that we can hide the model from the adversary, but both theory and practice show that if we provide an interface for running inference on the model (which we definitely have to assume), a smart adversary can learn so much about the model that it doesn't make much sense to distinguish these two cases. Therefore, research in this field mostly concentrates on how to keep training data private while giving the model and its weights to the adversary (the "white box" scenario).

Note that a model that generalizes well does not necessarily preserve privacy. Generalization is an average-case notion, and it characterizes how well the model's accuracy (or another objective function) transfers to new data. Privacy, on the other hand, is a worst-case notion, and it deals with the corner cases and the information that the entire model provides, not just its performance. For example, if you train an SVM for classification, it might generalize very well, but the model will explicitly contain (and thus provide to any adversary) full and unperturbed information about its support vectors, which can hardly be called privacy-preserving. And let's not even get started on nearest neighbors...

Using a "standard model" that has been tried and tested also doesn't really help. For example, Zhang et al. [992], in a very important paper that has already become a classic of deep learning research, studied standard models such as *AlexNet* on standard datasets such as *ImageNet* (we discussed *AlexNet* in Section 3.2). Their experiment was to introduce a *random permutation* of the labels, that is, assign labels from 1000 *ImageNet* classes completely at random, thus making the dataset entirely unlearnable. After training *AlexNet*, they indeed saw purely random-looking accuracy on the test set (about 0.1% top-1 accuracy and about 0.5% top-5 accuracy), but on the training set the model actually achieved more than 90% top-1 accuracy, not much worse than after training on original labels with the exact same learning parameters! This means that even in the absence of any possibility for generalization and extracting useful features, modern deep learning models can learn quite a lot by simply memorizing the data; note also that *AlexNet* is by modern standards a pretty small and weak network...

Therefore, if we want to be able to train on real data, we need to somehow introduce privacy-preserving transformations into the model training process. Since, in this book, we are mainly interested in deep learning, I will not go into preserving privacy with other machine learning models; there is a growing body of research in this field, and I can refer, e.g., to the surveys [238, 287, 312, 400] and references therein. Our focus in this section is on how to make complex high-dimensional optimization, such as training deep neural networks with stochastic gradient descent (recall Sections 2.4 and 2.5), respect privacy constraints. As we have already seen, the basic approach to achieving differential privacy is to add noise to the output of M, just like the Laplace and Gaussian mechanisms do. Many classical works on the subject focus on estimating and reducing the amount of noise necessary to ensure privacy under various assumptions [214–217, 520]. However, it is not immediately obvious how to apply this idea to a deep neural network. There are two major approaches to achieving differential privacy in deep learning, that is, in stochastic gradient descent.

The most important advance in this field came from Abadi et al. [1], who suggested a method for controlling the influence of the training data during stochastic gradient descent called *Differentially Private SGD* (DP-SGD). They use the Gaussian mechanism that we introduced in the previous section, so the basic idea is to add Gaussian noise to the gradients on every step of the SGD. But in order to estimate the variance σ for the necessary noise, the Gaussian mechanism needs to know an estimate on the influence that each individual example can have on the gradient $\mathbf{g}_k$ computed on the minibatch at step k. How can we get such an estimate when we do not have any prior bound on the gradients? We have to bound them ourselves!

Specifically, Abadi et al. clip the gradients on each SGD iteration to a predefined value of the L_2-norm and add Gaussian noise to the resulting gradient value. The entire DP-SGD scheme is presented in Algorithm 9. By careful analysis of the privacy loss variable, i.e., $\log \frac{p(A(D)=s)}{p(A(D')=s)}$ above, Abadi et al. show that the resulting algorithm preserves differential privacy under reasonable choices of the clipping and random noise parameters. Moreover, this is a general approach that is agnostic to the network architecture and can be extended to various first-order optimization algorithms based on SGD.

A year later, Papernot et al. [653] (actually, mostly the same group of researchers from *Google*) presented the *Private Aggregation of Teacher Ensembles* (PATE) approach. In PATE, the final "student" model is trained from an ensemble of "teacher" models that have access to sensitive data, while the "student" model only has access

---

Algorithm 9: Differentially private stochastic gradient descent

---

Initialize $\mathbf{w}_0, k := 0$;

repeat

    $D_k := \text{Sample}(D)$;

    for $d \in D_k$ **do**

        $\mathbf{g}_k(d) := \nabla_{\mathbf{w}} f(\mathbf{w}_k, d)$;

        $\bar{\mathbf{g}}_k(d) := \mathbf{g}_k(d) / \max\left(1, \frac{1}{C} \|\mathbf{g}_k(d)\|_2\right)$;

    end

    $\mathbf{g}_k := \frac{1}{|D_k|}\left(\sum_{d \in D_k} \bar{\mathbf{g}}_k(d) + \mathcal{N}\left(0, \sigma^2 C^2 \mathbf{I}\right)\right)$;

    $\mathbf{w}_{k+1} := \mathbf{w}_k - \alpha_k \mathbf{g}_k$;

    $k := k + 1$;

until *a stopping condition is met*;

---

to (noisy) aggregated results of "teacher" models, which allows to control disclosure and preserve privacy. A big advantage of this approach is that "teacher" models can be treated as black box while still providing rigorous differential privacy guarantees based on the same moments accounting technique from [1]. Incidentally, the best results were obtained with adversarial training for the "student" in a semi-supervised fashion, where the entire dataset is available for the "student" but labels are only provided for a subset of it, preserving privacy.

In conclusion, I think it is important to note the practical side of things. Differential privacy is a worst-case theoretical concept, and definitions of an (ϵ, δ)-differentially private mechanism might have reminded the reader of definitions from theoretical cryptography, where usually nothing is possible to actually achieve and even the best results are often either negative or impossible to apply in practice. But differential privacy for deep learning is a field that has actual implementations. The original paper by Abadi et al. was already accompanied by a repository that added differentially private variations of *Tensorflow* optimizers[1]. And the latest news is the release of *Opacus*, a library developed by *Facebook* researchers Davide Testuggine and Ilya Mironov that enables differential privacy for *PyTorch* models[2].

Thus, deep learning with differential privacy guarantees may eventually provide a good answer to the problem of preserving information regarding the datasets. But if you want to release a *dataset* for the general public, say organize a *Kaggle* competition, rather than just publish your model while keeping the original dataset private, you still cannot avoid the generation of synthetic data with privacy guarantees. This is exactly what we will discuss in the next section.

---

[1] At the time of writing (late 2020), the *Tensorflow Privacy* library is alive and well supported: https://github.com/tensorflow/privacy.

[2] At the time of writing (late 2020), this library has been very recently released, so it obviously also does not lack support: https://github.com/pytorch/opacus

11.4 Differential Privacy Guarantees for Synthetic Data Generation

In this section, we review the applications of differential privacy and related concepts to synthetic data generation. The purpose is similar: the release of a synthetic dataset generated by some model trained on real data should not disclose information regarding the individual points in this real dataset. Our review is slanted towards deep learning; for a more complete picture of the field, we refer to the surveys in [71, 214]. However, we do note the efforts devoted to generating differentially private synthetic datasets in classical machine learning. In particular, Lu et al. [559] develop a model for making sensitive databases private by fixing a set of queries to the database and perturbing the outputs to ensure differential privacy. Zhang et al. present the *PrivBayes* approach [997]: construct a Bayesian network that captures the correlations and dependencies between data attributes, inject noise into the marginals that constitute this network, and then sample from the perturbed network to produce the private synthetic dataset. In a similar effort, the *DataSynthetizer* model by Ping et al. [673] is able to take a sensitive dataset as input and generate a synthetic dataset that has the same statistics and structure but at the same time provides differential privacy guarantees.

We also note some privacy-related applications of synthetic data that are not about differential privacy. For example, Ren et al. [721] present an adversarial architecture for video face anonymization; their model learns to modify the original real video to remove private information while at the same time still maximizing the performance of action recognition models (see also Section 6.6).

The general approaches we have discussed in the previous section have been modified and applied for producing synthetic data with generative models, mostly, of course, with generative adversarial networks. Although the methods are similar, we note an important conceptual difference that synthetic data brings in this case. *Model release* approaches in the previous section assumed access to and full control of model training. *Data release* approaches (here we use the terminology from [871]) that perform synthetic data generation have the following advantages:

- they can provide private data to third parties to construct better models and develop new techniques or use computational resources that might be unavailable to the holders of sensitive data;
- moreover, these third parties are able to pool synthetic data from different sources, while in the model release framework this would require a transfer of sensitive data;
- synthetic data can be either traded of freely made public, which is an important step towards reproducibility of research, especially in such fields as bioinformatics and healthcare, where reproducibility is an, especially, important problem and where, at the same time, sensitive data abounds.

In this section, we discuss existing constructions of GANs that provide rigorous privacy guarantees for the resulting generated data. Basically, in the ideal case, a differentially private GAN has to generate an artificial dataset that would be sampled from the same distribution p_{data} but with differential privacy guarantees as discussed above. One general remark that is used in most of these works is that in a GAN-based architecture, it suffices to have privacy guarantees or additional privacy-preserving modifications (such as adding noise) only in the discriminator since gradient updates for the generator are functions of discriminator updates. Another important remark is that in cases when we generate differentially private synthetic data, a drop in quality for subsequent "student" models trained on synthetic data is expected in nearly all cases, not because of any deficiencies of synthetic data vs. real in general but because the nature of differential privacy requires adding random noise to the generative model training.

Xie et al. [955] present the differentially private GAN (DPGAN) model, which is basically the already classical Wasserstein GAN [27, 303] but with additional noise on the gradient of the Wasserstein distance, in a fashion following the DP-SGD approach (Section 11.3). They apply DPGAN to generate electronic health records, showing that classifiers trained on synthetic records have accuracy approaching that of classifiers trained on real data, while guaranteeing differential privacy. This was further developed by Zhang et al. [1002], who used the Improved WGAN framework [303] and obtained excellent results on the synthetic data generated from various subsets of the LSUN dataset [1002], which is already a full-scale image dataset, albeit at low resolution (64×64).

Beaulieu-Jones et al. [50] apply the same idea to generating electronic health records, specifically training on the data of the Systolic Blood Pressure Trial (SPRINT) data analysis challenge [205, 854], which are in nature low-dimensional time series. They used the DP-SGD approach for the Auxiliary Classifier GAN (AC-GAN) architecture [637] and studied how the accuracy of various classifiers drops when passing to synthetic data. Triastcyn and Faltings [871] continue this line of work and show that differential privacy guarantees can be obtained by adding a special Gaussian noise layer to the discriminator network. They show good results for "student" models trained on synthetically generated data for MNIST, but already at the SVHN dataset the performance degrades more severely.

Bayesian methods are a natural fit for differential privacy since they deal with entire distributions of parameters and lend themselves easily to adding extra noise needed for DP guarantees. In a combination of generative models and Bayesian methods, a Bayesian variant of the GAN framework, which provides representations of full posterior distributions over the parameters, was provided by Saatchi and Wilson [747]. The idea of their *Bayesian GAN* is to introduce prior distributions on generator parameters θ_g and discriminator parameters θ_d, $p\left(\theta_g \mid \alpha_g\right)$, and $p\left(\theta_d \mid \alpha_d\right)$, respectively, and infer posteriors over θ_g and θ_d

$$p\left(\boldsymbol{\theta}_g \mid Z, \boldsymbol{\theta}_d\right) \propto p\left(\boldsymbol{\theta}_g \mid \boldsymbol{\alpha}_g\right) \prod_{n=1}^{N_g} D\left(G\left(\mathbf{z}_n; \boldsymbol{\theta}_g\right); \boldsymbol{\theta}_d\right),$$

$$p\left(\boldsymbol{\theta}_d \mid Z, X, \boldsymbol{\theta}_g\right) \propto p\left(\boldsymbol{\theta}_d \mid \boldsymbol{\alpha}_d\right) \prod_{n=1}^{N_d} D\left(\mathbf{x}_n; \boldsymbol{\theta}_d\right) \prod_{n=1}^{N_g}\left(1 - D\left(G\left(\mathbf{z}_n; \boldsymbol{\theta}_g\right); \boldsymbol{\theta}_d\right)\right),$$

where $\mathbf{x}_n$ are real inputs, $\mathbf{z}_n$ are random noise samples, and N_d and N_g are the numbers of real and fake samples, respectively.

Saatchi and Wilson provide learning algorithms in this setting, marginalizing the above posteriors over random noise by Monte Carlo integration and sampling from posterior distributions with stochastic gradient Hamiltonian Monte Carlo [136, 1033]. Arnold et al. [30] adapted the BayesGAN framework for differential privacy by injecting noise into the gradients during training, which was shown by Wang et al. [918] to lead to DP guarantees. They apply the resulting *DP-BayesGAN* framework to microdata, i.e., medium-dimensional samples of 40 explanatory variables of different nature and one dependent variable.

As for the PATE framework, it cannot be directly applied to GANs since noisy aggregation of a PATE ensemble is not a differentiable function that could serve as part of a GAN discriminator. Ács et al. [6] proposed to use a differentially private clustering method to split the data into k clusters, then train a separate generative models (the authors tried VAE) on their own clusters, and then create a mixture of the resulting models that would inherit differential privacy properties as well. A recent work by Jordon et al. [976] circumvents the non-differentiability problem by training a "student-discriminator" on already differentially private synthetic data produced by the generator. The learning procedure alternates between updating "teacher" classifiers for a fixed generator on real samples and updating the "student-discriminator" classifier and the generator for fixed "teachers". PATE-GAN works well on low-dimensional data but begins to lose ground on high-dimensional datasets such as, e.g., the UCI Epileptic Seizure Recognition dataset (with 184 features).

However, these results are still underwhelming; it has proven very difficult to stabilize GAN training with the additional noise necessary for differential privacy guarantees, which has not allowed researchers to progress to, say, higher resolution images so far. In a later work, Triastcyn and Faltings [870] consider a different approach: they use the *empirical DP* framework [3, 124, 168, 768], an approach that empirically estimates the privacy of a posterior distribution, and the modification that ensures privacy is usually a sufficiently diffuse prior. In this framework, evaluating the privacy would reduce to training a GAN on the original dataset D, removing one sample from D to obtain D', retraining the GAN and comparing the probabilities of all outcomes, and so on, repeating these experiments enough times to obtain empirical estimates for ϵ and δ. For realistic GANs, a large number of retrainings is impractical, so Triastcyn and Faltings modify this procedure to make it operate directly on the generated set $\tilde{D}$ rather than the original dataset D. They study the tradeoff of privacy vs. accuracy of the "student" models trained on synthetic data and show that GANs can fall into the region of practical values for both privacy and accuracy. Their proposed modification of the architecture (a single randomizing layer close to the end of the discriminator) strengthens DP guarantees while preserving good generation quality for datasets up to *CelebA* [542]; in fact, it appears to serve as a regularizer and improve generation.

Frigerio et al. [246] extend the DPGAN framework to continuous, categorical, and time series data. They use the Wasserstein GAN loss function [303], extending the moment accountant to this case. To handle discrete variables, the generator produces an output for every possible value with a softmax layer on top, and its results are sent to the discriminator. Bindschadler [71] presents a *seedbased* modification of synthetic data generation: an algorithm that produces data records through a generative model conditioned on some seed real data record; this significantly improves quality but introduces correlations between real and synthetic data. To avoid correlations, Bindschadler introduces privacy tests that reject unsuitable synthetic data points. The approach can be used in complex models based on encoder-decoder architecture by adding noise to a seed in the latent space; it has been evaluated across different domains from census data to celebrity face images, the latter through a VAE/GAN architecture [497].

Finally, we note that synthetic data produced with differential privacy guarantees is also starting to gain legal status; in a technical report [58], Bellovin et al. from the Stanford Law School discuss various definitions of privacy from the point of view of what kind of data can be released. They conclude: "...as we recommend, synthetic data may be combined with differential privacy to achieve a best-of-both-worlds scenario", i.e., combining added utility of synthetic data produced by generative models with formal privacy guarantees.

11.5   Case Study: Synthetic Data in Economics, Healthcare, and Social Sciences

Synthetic data is increasingly finding its way into economics, healthcare, and social sciences in a variety of applications. We discuss this set of models and applications here since often the main concern that drives researchers in these fields to synthetic data is not lack of data *per se* but rather privacy issues. A number of models that guarantee differential privacy have already been discussed above, so in this section, we concentrate on other approaches and applications.

As long ago as 1993, Rubin [740] discussed the dangers of releasing microdata (i.e., information about individual transactions) and the extremely complicated legal status of data releases, as the released data might be used to derive protected information even if it had been masked by standard techniques. To avoid these complications, Rubin proposed to use imputed synthetic data instead: given a dataset with confidential information, "forget" and impute confidential values for a sample from this dataset, using the same background variables but drawing confidential data from the predictions of some kind of imputation model. Repeating the process for several samples, we get a multiply-imputed population that can then be released. In the same year (actually, the same special issue of the *Journal of Official Statistics*), Little [530] suggested to also keep the non-confidential part of the information to improve imputation. By now, synthetic datasets produced by multiple imputation are a well-established field of statistics, with applications to finance and economics [195, 713], healthcare [18], social sciences [102], survey statistics [4, 13], and other domains. Since the main emphasis of the present survey is on synthetic data for deep learning, we do not go into details about multiple imputation and refer to the book [207] and the main recent sources in the field [206, 700, 712, 714].

In a very recent work, Heaton and Witte [332] propose another interesting take on synthetic data in finance. They begin with the well-known problem of overfitting during backtesting: since there is a very large number of financial products and relatively short time series available for them, one can always find a portfolio (subset of products) that works great during backtesting, but it does not necessarily reflect future performance. The authors suggest to use synthetic data not to train financial strategies (they regard it as infeasible), but rather to *evaluate* developed strategies, generating synthetic data with a different distribution of abnormalities and testing strategies for robustness in these altered circumstances. Interestingly, the motivation here is not to improve or choose the best strategies, but to obtain evidence of their robustness that could be used for regulatory purposes. As a specific application, the authors use existing fraud detection algorithms to find anomalies in the Kaggle Credit Card Fraud Detection Dataset [180] and generate synthetic data that balances the found abnormalities.

However, at present, I know of no direct applications where synthetically generated financial time series that would lead to improved results in financial forecasting, developing financial strategies, and the like. In general, financial time series are notorious for not being amenable to either prediction or accurate modeling, and even with current state-of-the-art economic models, it looks like generating useful synthetic financial time series is still in the future. Moreover, the reasons we discussed in Section 8.4 regarding why synthetic data is unpopular in natural language processing apply here as well.

As for healthcare, this is again a field where the need for synthetic data was understood very early, and this need was mostly caused by privacy concerns: hospitals are required to protect the confidentiality of their patients. Ever since the first works in this direction, dating back to early 1990s, researchers mostly concentrated on generating synthetic electronic medical records (EMR) in order to preserve privacy [44]. Among more recent works, MDClone [594] is a system that samples synthetic EMRs from the distributions learned on existing cohorts, without actually reusing original data points. Walonoski et al. [898] present the *Synthea* software suite designed to simulate the lifespans of synthetic patients and produce realistic synthetic EMRs. McLaghlan [591] discusses realism in synthetic EMR generation and methods for its validation, and in another work presents a state transition machine that incorporates domain knowledge and clinical practice guidelines to generate realistic synthetic EMRs [592].

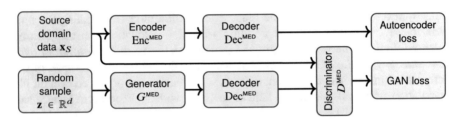

Fig. 11.1 The architecture of *medGAN* [150]. Blocks with identical labels have shared weights.

Another related direction of research concentrates not on individual EMRs, but on modeling entire populations of potential patients. Synthetic micro-populations produced by Smith et al. [805] are intended to match various sociodemographic conditions found in real cities and use them in imitational modeling to estimate the effect of interventions. Moniz et al. [610, 746] create synthetic EMRs made available on the CDC Public Health grid for imitational modeling. Buczak et al. [98] generate synthetic EMRs for an outbreak of a certain disease (together with background records). Kartoun [441] progressed from individual EMRs to entire virtual patient repositories, concentrating on preserving the correct general statistics while using simulated individual records. However, most of this work does not make use of modern formalizations of differential privacy or recent developments in generative models, and only very recently researchers have attempted to bring those into the healthcare microdata domain as well.

A direct application of GANs for synthetic EMR generation was presented by Choi et al. [150]. Their *medGAN* model consists of a generator G^{MED}, discriminator D^{MED}, and an autoencoder with encoder Enc^{MED} and decoder Dec^{MED}; the architecture is shown in Fig. 11.1. The autoencoder is trained to reconstruct real data $\mathbf{x}_T \sim \mathcal{X}_T$, while G^{MED} learns to generate latent representations $G^{\text{MED}}(\mathbf{z})$ from a random seed $\mathbf{z}$ such that D^{MED} will not be able to differentiate between $\text{Dec}^{\text{MED}}(G^{\text{MED}}(\mathbf{z}))$ and a real sample $\mathbf{x}_T \sim \mathcal{X}_T$. Privacy in the *medGAN* model is established empirically, and the main justification for privacy is the fact that *medGAN* uses real data only for the discriminator and never trains the generator on any real samples. We note, however, that in terms of generation *medGAN* is not perfect: for example, Patel et al. [659] present the *CorrGAN* model for correlated discrete data generation (with no regard for privacy) and show improvements over *medGAN* with a relatively straightforward architecture.

The DP-SGD framework has also been applied to GANs in the context of medical data. We have discussed Beaulieu-Jones et al. [50] above. Another important application for synthetic data across many domains, including but not limited to finance, would be to generate synthetic time series. This, however, has proven to be a more difficult problem, and solutions are only starting to appear. In particular, Hyland et al. [228] present the Recurrent GAN (RGAN) and Recurrent Conditional GAN architectures designed to generate realistic real-valued multi-dimensional time series. They applied the architecture to generating medical time series (vitals measured for

ICU patients) and reported successful generation and ability to train classifiers on synthetic data, although there was a significant drop in quality when testing on real data. Hyland et al. also discuss the possibility to use a differentially private training procedure, applying the DP-SGD framework to the discriminator and thus achieving differential privacy for the RGAN training. The authors report that after this procedure, synthetic-to-real test results deteriorate significantly but remain reasonable in classification tasks on ICU patient vitals.

Finally, we note another emerging field of research related to generating synthetic EMRs for the sake of privacy: generating clinical notes and free-text fields in EMRs with neural language models (see also Section 8.4). Latest advances in deep learning for natural language processing have led to breakthroughs in large-scale language modeling [193, 359, 697], and it has been applied to smaller datasets of clinical notes as well. Lee [505] uses an encoder-decoder architecture to generate chief complaint texts for EMRs. Guan et al. [302] propose a GAN architecture called *mtGAN* (medical text GAN) for the generation of synthetic EMR text. It is based on the SeqGAN architecture [982] and is trained with the REINFORCE algorithm; the primary difference is a condition added by Guan et al. to be able to generate EMRs for a specific disease or other features. Melamud and Shivade [597] compare LSTM-based language models for generating synthetic clinical notes, suggesting a new privacy measure and showing promising results. Further advances in this direction may be related to the recently developed differentially private language models [593].

11.6 Conclusion

In this chapter, we have investigated a motivation for synthetic data very different from the rest of the book. Instead of making synthetic data that alleviates the hardships of collecting and labeling real data, here we have been making synthetic data because real data, while available, cannot be published for legal or ethical reasons. This motivation has led to very different methods: now the main problem is to guarantee that real data is not released by publishing the synthetic dataset. Therefore, in this chapter, we have been mostly talking about the main (and probably the only) approach that can provide these guarantees: differential privacy.

Next, we come to the final chapter of the book. In it, we will try to highlight the most promising directions for further research, ideas that I believe will bring interesting advances in the nearest future. Who knows, maybe my readers will be among those who take up these directions and bring machine learning to new heights with synthetic data. Let's find out together!

Chapter 12
Promising Directions for Future Work

In this concluding chapter, we discuss the next steps that we can expect from the field of synthetic data for deep learning. We consider four different ideas that are starting to gain traction in this field. First, procedural generation of synthetic data can allow for much larger synthetic datasets or datasets generated on the fly. Second, recent works try to make the shift from domain randomization to the generation feedback loop, adapting synthetic data generation to the model and problem at hand. Third, we discuss how to best incorporate additional knowledge into the domain adaptation architectures, and fourth, show examples of introducing extra modalities into synthetic datasets with the purpose to improve downstream tasks that formally might not even use these modalities.

12.1 Procedural Generation of Synthetic Data

The first direction that we highlight as important for further study in the field of synthetic data is *procedural generation*. Take, for instance, synthetic indoor scenes that we discussed in Section 7.3. Note that in the main synthetic dataset for indoor scenes, SUNCG, the 3D scenes, and their layouts were created manually. While we have seen that this approach has an advantage of several orders of magnitude in terms of labeled images over real datasets, it still cannot scale up to millions of 3D scenes. The only way to achieve that would be to learn a model that can generate the *contents* of a 3D scene (in this case, an indoor environment with furniture and supported objects), varying it stochastically according to some random input seed. This is a much harder problem than it might seem: e.g., a bedroom is much more than just a predefined set of objects placed at random positions.

There is a large related field of procedural content generation for video games [336, 786, 863], but we highlight a recent work by Qi et al. [687] as representative for state of the art and more directly related to synthetic data. They propose a human-centric

© The Author(s), under exclusive license to Springer Nature Switzerland AG 2021
S. I. Nikolenko, *Synthetic Data for Deep Learning*, Springer Optimization
and Its Applications 174, https://doi.org/10.1007/978-3-030-75178-4_12

(a)   (b)   (c)   (d)

Fig. 12.1 Sample interior scenes generated procedurally by Qi et al. [687], with two rendered views and a heatmap: (a) bathroom; (b) bedroom; (c) dining room; (d) garage.

approach to modeling indoor scene layout, learning a stochastic scene grammar based on an attributed spatial AND-OR graph [1026] that relates scene components, objects, and corresponding human activities. A scene configuration is represented by a parse graph whose probability is modeled with potential functions corresponding to functional grouping relations between furniture, relations between a supported object and supporting furniture, and human-centric relations between the furniture based on the map of sampled human trajectories in the scene. After learning the weights of potential functions corresponding to the relations between objects, MCMC sampling can be used to generate new indoor environments. Qi et al. train their model on the very same SUNCG dataset and show that the resulting layouts are hard to distinguish (by an automated state-of-the-art classifier trained on layout segmentation maps) from the original SUNCG data. The sample pictures—some of them are reproduced in Fig. 12.1—also look quite convincing.

In Section 7.2, we have already discussed *ProcSy* by Khan et al. [449]. In addition to randomizing weather and lighting conditions, another interesting part of their work is the procedural generation of cities and outdoor scenes. They base this procedural generation on the method of Parish and Müller [656], which is, in turn, based on the notion of Lindenmayer systems (L-systems) [686] and embodied in the *CityEngine* tool [954] (see Section 7.2). They use a part of real *OpenStreetMaps* data for the road network and buildings, but we hope that future work based on the same ideas can offer fully procedural modeling of cities and road networks.

We believe that procedural generation can lead to an even larger scale of synthetic data adoption in the future, covering cases when simply placing the objects at random is not enough. The nature of the model used for procedural generation may differ depending on the specific field of application, but, in general, for procedural generation, one needs to train probabilistic generative models that allow for both learning from real or manually prepared synthetic data and then generating new samples.

Algorithm 10: Learning in the *SimOpt* framework

Initialize $p_\phi(\xi)$, ϵ;
repeat
  | Env := Simulation (p_ϕ);
  | π_{θ,p_ϕ} := RL(Env);
  | $\tau_{\text{real}}^{\text{ob}} \sim$ RealRollout (π_{θ,p_ϕ});
  | $\xi \sim p_\phi$;
  | $\tau_\xi^{\text{ob}} \sim$ SimRollout $(\pi_{\theta,p_\phi}, \xi)$;
  | $c(\xi) := D\left(\tau_{\text{real}}^{\text{ob}}, \tau_\xi^{\text{ob}}\right)$;
  | $p_\phi :=$ UpdateDistribution $(p_\phi, \xi, c(\xi), \epsilon)$;
until *a stopping condition is met*;

12.2 From Domain Randomization to the Generation Feedback Loop

Chebotar et al. [126] make one of the first steps in a very interesting direction by introducing the *SimOpt* framework. They are also working on domain transfer, transferring continuous control policies for robotic arms from synthetic-to-real domain. But importantly, they attempt to close the feedback loop between synthetic data generation and domain transfer via domain randomization. Previous works on domain randomization (see Section 9.1) manually tuned the distribution of simulation parameters $p_\phi(\xi)$ such that a policy trained on $D_{\xi \sim p_\phi}$ would perform well. On the contrary, in [126] the parameters of $p_\phi(\xi)$ are learned automatically via a feedback loop from the results of real observations.

The *SimOpt* framework is presented in Algorithm 10 and illustrated in Figure 12.2. The idea is to create an environment from the simulation with a given distribution of parameters p_ϕ, then train a policy π_{θ,p_ϕ} with some reinforcement learning algorithm,

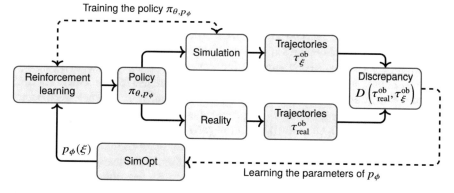

Fig. 12.2 The structure of the *SimOpt* framework [126].

and then sample two sets of trajectories with the trained policy π_{θ,p_ϕ}: real-world observation trajectories τ_{real}^{ob} and simulated observation trajectories τ_ξ^{ob}. The idea is to update the distribution p_ϕ in such a way that the two sets of trajectories will become closer to each other, which means that training in simulation will produce agents that perform in similar ways in the simulation and in the real world.

A similar approach on the level of data augmentation was presented in [706]. As we discussed in Section 3.4, data augmentation differs from synthetic data in that it modifies real data rather than creates new; the modifications are usually done with predefined transformation functions (TFs) that do not change the target labels. This assumption is somewhat unrealistic: e.g., if we augment by shifting or cropping the image for image classification, we might crop out exactly the object that determines the class label. The work [706] relaxes this assumption, treating TFs as black boxes that might move the data point out of all necessary classes, into the "null" class, but cannot mix up different classes of objects. The authors train a generative sequence model with an adversarial objective that learns a sequence of TFs that would not move data points into the "null" class by training a null class discriminator $D_\phi^\emptyset$ and a generator G_θ for sequences of TFs $h_L \circ \ldots \circ h_1$:

$$\min_\theta \max_\phi \mathbb{E}_{\tau \sim G_\theta} \mathbb{E}_{x \sim \mathcal{U}} \left[\log(1 - D_\phi^\emptyset(h_{\tau_L} \circ \ldots \circ h_{\tau_1}(x))) \right] + \mathbb{E}_{x' \sim \mathcal{U}} \left[\log(D_\phi^\emptyset(x')) \right],$$

where $\mathcal{U}$ is some distribution of (possibly unlabeled) data. Since TFs are not necessarily differentiable or deterministic, learning G_θ is defined in the syntax of reinforcement learning.

Pashevich et al. [658] note that the space of augmentation functions is very large (for 8 different transformations they estimate to have $\approx 3.6 \cdot 10^{14}$ augmentation functions), and propose to use Monte Carlo Tree Search (MCTS) [169, 463] to find the best augmentations by automatic exploration. They apply this idea to augmenting synthetic images for sim-to-real policy transfer for robotic manipulation and report improved results in real-world tasks such as cube stacking or cup placing.

The next step was taken by *Google Brain* researchers Cubuk et al. [172], who continued this work by presenting a framework called *AutoAugment* for learning augmentation strategies from data. Their approach is modeled after recent advances in *neural architecture search* [35, 57, 1031, 1032] that have recently yielded new families of neural architectures for feature extraction from images and image classification (*EfficientNet*, see Section 3.2 for more details), object detection (the *EfficientDet* family, see Section 3.3), and more. Cubuk et al. use reinforcement learning to find the best augmentation policy composed of a number of parameterized operations. As a result, they significantly improve state-of-the-art results on such classical datasets as CIFAR-10, CIFAR-100, and ImageNet.

In [172], the controller is trained by proximal policy optimization, a rather involved reinforcement learning algorithm [772]. Also, reinforcement learning on this scale definitely takes up a *lot* of computational resources. So, can this research help us when we are not *Google Brain* and cannot run this pipeline for our problem? Cubuk et al. note that the resulting augmentation strategies can indeed transfer across a wide variety of datasets and network architectures. On the other hand, this transferability is far from perfect, and so far the results of *AutoAugment* pop up in other works as often as the authors would probably like.

Zakharov et al. [986] look at a similar idea from the point of view of domain randomization (see also Section 9.1). Their framework consists of a recognition network that does the basic task (say, object detection and pose estimation) and a deception network that transforms the synthetic input with an encoder-decoder architecture. Training is done in an adversarial way, alternating between two phases:

- fixing the weights of the deception network, perform updates of the recognition network as usual, serving synthetic images transformed by the deception network as inputs;
- fixing the weights of the recognition network, perform updates of the deception network with the same objective but with reversed gradients, updating the deception network so as to make the inputs hard for the recognition network.

The deception network is organized and constrained in such a way that its transformations do not change the ground truth labels or change them in predictable ways. This adversarial framework exhibits performance comparable to state-of-the-art domain adaptation frameworks and shows superior generalization capabilities (better avoiding overfitting).

There are two recent works that represent important steps towards closing this feedback loop. First, the *Meta-Sim* framework [433], which we discussed in Section 9.2 in the context of high-level procedural scene generation, also makes inroads in this direction: the distribution parameters for synthetic data generation are tuned not only to bring the synthetic data distribution closer to the real one but also to improve the performance on downstream tasks such as object detection. The difference here is that instead of low-level parameters of image augmentation functions *Meta-Sim* adapts high-level parameters such as the synthetic scene structure captured as a scene graph.

Second, the *Visual Adversarial Domain Randomization and Augmentation* (VADRA) model by Khirodkar et al. [450] makes the next step in developing domain randomization ideas: instead of simply randomizing synthetic data or making it similar to real, let us learn a policy π_ω that generates rendering parameters in such a way that the downstream model learns best. They use the REINFORCE algorithm to obtain stochastic gradients for the objective $J(\omega)$ which consists of the downstream model performance (for supervised data) and the errors of a domain classifier (for unsupervised data); this is the "adversarial" part of VADRA.

The structure of learning in the VADRA framework, shown in Fig. 12.3, is a good illustration to the general idea of closing the feedback loop. In VADRA, the policy π_ω produces rendering parameters θ that are used by the synthetic data generator to produce a data sample $(\mathbf{x}, y)$. This data sample is used to produce a hypothesis (train a model for the downstream task), and the error of this hypothesis (performance of this model) is used as the reward for learning π_ω. As a result, VADRA works much better for synthetic-to-real transfer on problems such as object detection and segmentation than regular domain randomization.

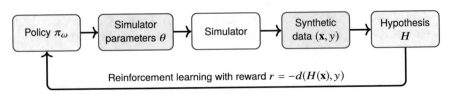

Fig. 12.3 The structure of the VADRA framework [450].

Similar ideas have been recently explored by Mehta et al. [595], who present *Active Domain Randomization*, again learning a policy for generating better simulated instance, but this time in the context of generating Markov decision processes for reinforcement learning, Ruiz et al. [741], who also learn a policy π_ω that outputs the parameters for a simulator, learning to generate data to maximize validation accuracy, with reinforcement learning techniques, and Louppe et al. [558], who provide an inference algorithm based on variational approximations for fitting the parameters of a domain-specific non-differentiable simulator.

We believe that this meta-approach to automatically learning the best way to generate synthetic data, both high-level and low-level, is an important new direction that might work well for other applications too. In my opinion, this idea might be further improved by methods such as the SPIRAL framework by Ganin et al. [258] or neural painter models [375, 617, 1021] that train adversarial architectures to generate images in the form of sequences of brushstrokes or higher-level image synthesis programs with instructions such as "place object X at location Y"; these or other kinds of high-level descriptions for images might be more convenient for the generation feedback loop. Whether with these ideas or others, I expect further developments in this direction in the nearest future.

12.3   Improving Domain Adaptation with Domain Knowledge

To showcase this direction, let us consider one more work on gaze estimation (see Section 10.2) that presents a successful application of a hybrid approach to image refinement. Namely, Wang et al. [905] propose a very different approach to generating synthetic data for gaze estimation: a hierarchical generative model (HGM) that is able to operate both top-down, generating new synthetic images of eyes, and bottom-up, performing Bayesian inference to estimate the gaze in a given new image.

The general structure of their approach is shown in Figure 12.4. Specifically, Wang et al. design a probabilistic hierarchical generative shape model (HGSM) based on 27 eye-related landmarks that together represent the shape of a human eye. The model connects personal parameters that define variation between humans, visual axis parameters that define eye gaze, and eye shape parameters. The structure of HGSM is based on anatomical studies, and its parameters are learned from the

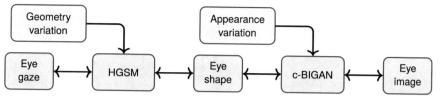

Fig. 12.4 General structure of the hierarchical generative model for eye image synthesis and gaze estimation [905]. Left to right: top-down image synthesis pipeline; right to left: bottom-up eye gaze estimation pipeline.

UnityEyes dataset [934]. During generation, HGSM generates eye shape parameters (positions of the 27 landmarks in the eyeball's spherical coordinate system) based on the given gaze direction.

The second part of the pipeline generates the actual images with a conditional BiGAN (bidirectional GAN) architecture [201]. Bidirectional GAN is an architecture that learns to transform data in both directions, from latent representations to the objects and back, while regular GANs learn to generate only the objects from latent representations. The conditional BiGAN (c-BiGAN) modification developed in [905] does the same with a condition, which in this case are the eye shape parameters produced by HGSM. As a result, the model by Wang et al. can work in both directions:

- generate eye images by sampling the gaze from a prior distribution, sampling 2D eye shape parameters from HGSM, and then using the generator G of c-BiGAN to generate a refined image;
- infer gaze parameters from an eye image by first estimating the eye shape through the encoder E of c-BIGAN and then performing Bayesian inference in the HGSM to find the posterior distribution of gaze parameters.

Wang et al. report performance improvements of the model itself applied to gaze estimation over [793] for sufficiently large training sets, and also show that the synthetic data generated by the model improves the results of standard gaze estimators (*LeNet*, as used in [1003]).

In general, the approach of [905] to combining probabilistic generative models that incorporate domain knowledge and GAN-based architectures appears to be a very interesting direction for further research. I believe this approach can be suitable for applications other than gaze estimation.

12.4 Additional Modalities for Domain Adaptation Architectures

Another natural idea that has not been used too widely yet is to use the additional data modalities such as depth maps or 3D volumetric data, which are available for free in synthetic datasets but usually unavailable in real ones, to improve the tasks that might, on the surface, not require these additional modalities at all.

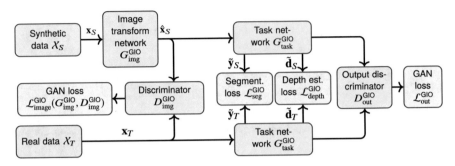

Fig. 12.5 General structure of the GIO-Ada model with input- and output-level domain adaptation [142].

Pioneering work in this direction has been recently done by Chen et al. [142]. They note that depth estimation and segmentation are related tasks that are increasingly learned together with multitask learning architectures [444, 665, 961] and then propose to use this idea to improve synthetic-to-real domain adaptation. Their model, called *Geometrically Guided Input-Output Adaptation* (GIO-Ada), is based on the PatchGAN architecture [389] intended for image translation.

We have illustrated the GIO-Ada model on Figure 12.5. Similar to the refiners considered in Section 10.3, they train a GAN generator to refine the synthetic image, but the refiner takes as input not just a synthetic image $\mathbf{x}_S$ but an input triple $(\mathbf{x}_S, \mathbf{y}, \mathbf{d})$, where $\mathbf{x}_S$ is a synthetic image, $\mathbf{y}$ is its segmentation map, and $\mathbf{d}$ is its depth map. Moreover, they also incorporate *output-level adaptation*, where a separate generator predicts segmentation and depth maps, and a discriminator tries to distinguish whether these maps came from a synthetic transformed image or from a real one. In this way, the model can use the depth information to obtain additional cues to further improve segmentation on real images, and the output-level adaptation brings segmentation results on synthetic and real domains closer together.

Specifically, the GIO-Ada model optimizes, in an adversarial way, the following objective:

$$\min_{G_{\text{img}}^{\text{GIO}}, G_{\text{task}}^{\text{GIO}}} \max_{D_{\text{img}}^{\text{GIO}}, D_{\text{out}}^{\text{GIO}}} \left[\mathcal{L}_{\text{seg}}^{\text{GIO}} + \lambda_1 \mathcal{L}_{\text{depth}}^{\text{GIO}} + \lambda_2 \mathcal{L}_{\text{image}}^{\text{GIO}} + \lambda_3 \mathcal{L}_{\text{out}}^{\text{GIO}} \right],$$

where

- $G_{\text{img}}^{\text{GIO}}$ is an image transformation network that performs input-level adaptation, i.e., produces a transformed image $\hat{\mathbf{x}} = G_{\text{img}}^{\text{GIO}}(\mathbf{x}, \mathbf{y}, \mathbf{d})$;
- $G_{\text{task}}^{\text{GIO}}$ is the output-level adaptation network that predicts the segmentation and depth maps $(\tilde{\mathbf{y}}, \tilde{\mathbf{d}}) = G_{\text{task}}^{\text{GIO}}(\mathbf{x})$;
- $D_{\text{img}}^{\text{GIO}}$ is the image discriminator that tries to distinguish between $\hat{\mathbf{x}}_S$ and $\mathbf{x}_T$, and $D_{\text{out}}^{\text{GIO}}$ is the output discriminator that distinguishes between $(\tilde{\mathbf{y}}_S, \tilde{\mathbf{d}}_S) = G_{\text{task}}^{\text{GIO}}(\hat{\mathbf{x}}_S)$ and $(\tilde{\mathbf{y}}_T, \tilde{\mathbf{d}}_T) = G_{\text{task}}^{\text{GIO}}(\mathbf{x}_T)$;

- $\mathcal{L}_{\mathrm{seg}}^{\mathrm{GIO}}$ is the segmentation cross-entropy loss $\mathcal{L}_{\mathrm{seg}}^{\mathrm{GIO}} = \mathbb{E}_{\mathbf{x}_S \sim p_{\mathrm{syn}}} \left[\mathrm{CE}\left(\mathbf{y}_S, \tilde{\mathbf{y}}_S\right) \right]$;
- $\mathcal{L}_{\mathrm{depth}}^{\mathrm{GIO}}$ is the depth estimation L_1-loss $\mathcal{L}_{\mathrm{depth}}^{\mathrm{GIO}} = \mathbb{E}_{\mathbf{x}_S \sim p_{\mathrm{syn}}} \left[\left\| \mathbf{d}_S - \tilde{\mathbf{d}}_S \right\|_1 \right]$;
- $\mathcal{L}_{\mathrm{image}}^{\mathrm{GIO}}$ is the GAN loss for $D_{\mathrm{img}}^{\mathrm{GIO}}$:

$$\mathcal{L}_{\mathrm{image}}^{\mathrm{GIO}} = \mathbb{E}_{\mathbf{x}_T \sim p_{\mathrm{real}}} \left[\log D_{\mathrm{img}}^{\mathrm{GIO}}(\mathbf{x}_T) \right] + \mathbb{E}_{\mathbf{x}_S \sim p_{\mathrm{syn}}} \left[\log(1 - D_{\mathrm{img}}^{\mathrm{GIO}}(\hat{\mathbf{x}}_S)) \right];$$

- $\mathcal{L}_{\mathrm{out}}^{\mathrm{GIO}}$ is the GAN loss for $D_{\mathrm{out}}^{\mathrm{GIO}}$:

$$\mathcal{L}_{\mathrm{out}}^{\mathrm{GIO}} = \mathbb{E}_{\mathbf{x}_T \sim p_{\mathrm{real}}} \left[\log D_{\mathrm{out}}^{\mathrm{GIO}}(\tilde{\mathbf{y}}_T, \tilde{\mathbf{d}}_T) \right] + \mathbb{E}_{\mathbf{x}_S \sim p_{\mathrm{syn}}} \left[\log(1 - D_{\mathrm{out}}^{\mathrm{GIO}}(\tilde{\mathbf{y}}_S, \tilde{\mathbf{d}}_S)) \right].$$

Chen et al. report promising results on standard synthetic-to-real adaptations: Virtual KITTI to KITTI and SYNTHIA to Cityscapes (see Section 7.2). This, however, looks to us as merely a first step in the very interesting direction of using additional data modalities easily provided by synthetic data generators to further improve domain adaptation.

12.5   Conclusion

And with this, we come to the last section of the whole book. In this work, we have attempted a survey of one of the most promising general techniques on the rise in modern deep learning, especially computer vision: synthetic data. This source of virtually limitless perfectly labeled data has been explored in many problems, but we believe that many more potential use cases still remain.

In direct applications of synthetic data, we have discussed many different domains and use cases, from basic computer vision tasks such as stereo disparity estimation or semantic segmentation to full-scale simulated environments for autonomous driving, unmanned aerial vehicles, and robotics. In the domain adaptation part, we have surveyed a wide variety of generative models for synthetic-to-real refinement and for feature-level domain adaptation. As another important field of synthetic data applications, we have considered data generation with differential privacy guarantees. We have also reviewed the works dedicated to improving synthetic data generation and outlined potential promising directions for further research.

In general, throughout this book, we have seen synthetic data work well across a wide variety of tasks and domains. I believe that synthetic data is essential for further development of deep learning: many applications require labeling which is expensive or impossible to do by hand, other applications have a wide underlying data distribution that real datasets do not or cannot fully cover, yet other applications may benefit from additional modalities unavailable in real datasets, and so on. Moreover, I believe that synthetic data will find new applications in the near future. For example, while this book does not yet have a section devoted to sound and speech processing, works that use synthetic data in this domain are already beginning to appear [510, 745]. As synthetic data becomes more realistic (where necessary) and encompasses more use cases and modalities, I expect it to play an increasingly important role in deep learning.

I strongly believe that synthetic data will be indispensable in the future of deep learning. Let us build this future together!

References

1. Abadi, M., Chu, A., Goodfellow, I., McMahan, H.B., Mironov, I., Talwar, K., Zhang, L.: Deep learning with differential privacy. In: Proceedings of the 2016 ACM SIGSAC Conference on Computer and Communications Security, CCS '16, pp. 308–318. ACM, New York, NY, USA (2016). https://doi.org/10.1145/2976749.2978318
2. Abbasi, A., Kalkan, S., Sahillioglu, Y.: Deep 3d semantic scene extrapolation. Vis. Comput. **35**, 271–279 (2018)
3. Abowd, J., Schneider, M., Vilhuber, L.: Differential privacy applications to bayesian and linear mixed model estimation. J. Priv. Confid. **5**(1), 185–205 (2013)
4. Abowd, J., Stinson, M., Benedetto, G.: Final report to the social security administration on the sipp/ssa/irs public use file project. Technical report, Longitudinal Employer—Household Dynamics Program, U.S. Bureau of the Census, Washington, DC (2006) https://hdl.handle.net/1813/43929
5. Abu Alhaija, H., Mustikovela, S.K., Mescheder, L., Geiger, A., Rother, C.: Augmented reality meets computer vision: efficient data generation for urban driving scenes. Int. J. Comput. Vis. **126**(9), 961–972 (2018). https://doi.org/10.1007/s11263-018-1070-x
6. Ács, G., Melis, L., Castelluccia, C., Cristofaro, E.D.: Differentially private mixture of generative neural networks. CoRR (2017). arXiv e-prints abs:1709.04514
7. Adiprawita, W., Ahmad, A.S., Semibiring, J.: Hardware in the loop simulator in UAV rapid development life cycle. CoRR (2008). arXiv e-prints abs:0804.3874
8. Agrawal, A., Lu, J., Antol, S., Mitchell, M., Zitnick, C.L., Parikh, D., Batra, D.: VQA: visual question answering. Int. J. Comput. Vis. **123**(1), 4–31 (2017). https://doi.org/10.1007/s11263-016-0966-6
9. Aguero, C., Koenig, N., Chen, I., Boyer, H., Peters, S., Hsu, J., Gerkey, B., Paepcke, S., Rivero, J., Manzo, J., Krotkov, E., Pratt, G.: Inside the virtual robotics challenge: simulating real-time robotic disaster response. IEEE Trans. Autom. Sci. Eng. **12**(2), 494–506 (2015). https://doi.org/10.1109/TASE.2014.2368997
10. Akbani, O., Gokrani, A., Quresh, M., Khan, F.M., Behlim, S.I., Syed, T.Q.: Character recognition in natural scene images. In: 2015 International Conference on Information and Communication Technologies (ICICT), pp. 1–6 (2015). https://doi.org/10.1109/ICICT.2015.7469575
11. Al-Kaff, A., Martín, D., García, F., de la Escalera, A., Armingol, J.M.: Survey of computer vision algorithms and applications for unmanned aerial vehicles. Expert Syst. Appl. **92**, 447–463 (2018). https://doi.org/10.1016/j.eswa.2017.09.033

12. Al-Musawi, N.J., Hasson, S.T.: Improving video streams summarization using synthetic noisy video data. Int. J. Adv. Comput. Sci. Appl. **6**(12) (2015). https://doi.org/10.14569/IJACSA.2015.061233

13. Alfons, A., Kraft, S., Templ, M., Filzmoser, P.: Simulation of close-to-reality population data for household surveys with application to eu-silc. Stat. Methods Appl. **20**(3), 383–407 (2011). https://doi.org/10.1007/s10260-011-0163-2

14. Alonso, E., Moysset, B., Messina, R.O.: Adversarial generation of handwritten text images conditioned on sequences. CoRR (2019). arXiv e-prints abs:1903.00277

15. Alvernaz, S., Togelius, J.: Autoencoder-augmented neuroevolution for visual doom playing. In: 2017 IEEE Conference on Computational Intelligence and Games (CIG) pp. 1–8 (2017)

16. Amodei, D., Hernandez, D., Sastry, G., Clark, J., Brockman, G., Sutskever, I.: Ai and compute (2018). https://openai.com/blog/ai-and-compute/

17. Amos, B., Ludwiczuk, B., Satyanarayanan, M.: Openface: A general-purpose face recognition library with mobile applications. Technical Report, CMU-CS-16-118, CMU School of Computer Science (2016)

18. An, D., Little, R.J.A., McNally, J.W.: A multiple imputation approach to disclosure limitation for high-age individuals in longitudinal studies. Stat. Med. **29**(17), 1769–1778 (2010). https://doi.org/10.1002/sim.3974

19. Andersen, P., Goodwin, M., Granmo, O.: Deep rts: A game environment for deep reinforcement learning in real-time strategy games. In: 2018 IEEE Conference on Computational Intelligence and Games (CIG), pp. 1–8 (2018). https://doi.org/10.1109/CIG.2018.8490409

20. Anderson, C., Du, X., Vasudevan, R., Johnson-Roberson, M.: Stochastic sampling simulation for pedestrian trajectory prediction (2019). arXiv e-prints abs:1903.01860

21. Ando, H., Lubashevsky, I., Zgonnikov, A., Saito, Y.: Statistical properties of car following: theory and driving simulator experiments (2015). arXiv e-prints arXiv:1511.04640

22. Andreas, J., Rohrbach, M., Darrell, T., Klein, D.: Deep compositional question answering with neural module networks. CoRR (2015). arXiv e-prints abs:1511.02799

23. Andreas, J., Rohrbach, M., Darrell, T., Klein, D.: Learning to compose neural networks for question answering. CoRR (2016). arXiv e-prints abs:1601.01705

24. Andriluka, M., Pishchulin, L., Gehler, P., Schiele, B.: 2d human pose estimation: New benchmark and state of the art analysis. In: IEEE Conference on Computer Vision and Pattern Recognition (CVPR) (2014)

25. Andriluka, M., Pishchulin, L., Gehler, P., Schiele, B.: 2d human pose estimation: New benchmark and state of the art analysis. In: 2014 IEEE Conference on Computer Vision and Pattern Recognition, pp. 3686–3693 (2014). https://doi.org/10.1109/CVPR.2014.471

26. Angelino, C.V., Baraniello, V.R., Cicala, L.: High altitude UAV navigation using imu, GPS and camera. In: Proceedings of the 16th International Conference on Information Fusion, FUSION 2013, Istanbul, Turkey, 9–12 July 2013, pp. 647–654 (2013)

27. Arjovsky, M., Chintala, S., Bottou, L.: Wasserstein generative adversarial networks. In: D. Precup, Y.W. Teh (eds.) Proceedings of the 34th International Conference on Machine Learning. Proceedings of Machine Learning Research, vol. 70, pp. 214–223. PMLR, International Convention Centre, Sydney, Australia (2017)

28. Armanious, K., Yang, C., Fischer, M., Küstner, T., Nikolaou, K., Gatidis, S., Yang, B.: Medgan: Medical image translation using gans. CoRR (2018). arXiv e-prints abs:1806.06397

29. Arnab, A., Zheng, S., Jayasumana, S., Romera-Paredes, B., Larsson, M., Kirillov, A., Savchynskyy, B., Rother, C., Kahl, F., Torr, P.H.S.: Conditional random fields meet deep neural networks for semantic segmentation: Combining probabilistic graphical models with deep learning for structured prediction. IEEE Signal Process. Mag. **35**(1), 37–52 (2018). https://doi.org/10.1109/MSP.2017.2762355

30. Arnold, C., Neunhoeffer, M., Sternberg, S.: Releasing differentially private synthetic microdata with bayesian gans (2019). http://www.marcel-neunhoeffer.com/pdf/papers/dp_gan.pdf

31. Aubry, M., Maturana, D., Efros, A., Russell, B., Sivic, J.: Seeing 3d chairs: exemplar part-based 2d-3d alignment using a large dataset of cad models. In: CVPR (2014)

32. Aubry, M., Russell, B.C.: Understanding deep features with computer-generated imagery. In: Proceedings of the 2015 IEEE International Conference on Computer Vision (ICCV), ICCV '15, pp. 2875–2883. IEEE Computer Society, Washington, DC, USA (2015). https://doi.org/10.1109/ICCV.2015.329

33. Bailo, O., Ham, D., Shin, Y.M.: Red blood cell image generation for data augmentation using conditional generative adversarial networks. CoRR (2019). arXiv e-prints abs:1901.06219

34. Bak, S., Carr, P., Lalonde, J.F.: Domain adaptation through synthesis for unsupervised person re-identification. In: ECCV (2018)

35. Baker, B., Gupta, O., Naik, N., Raskar, R.: Designing neural network architectures using reinforcement learning. CoRR (2016). arXiv e-prints abs:1611.02167

36. Baker, S., Scharstein, D., Lewis, J.P., Roth, S., Black, M.J., Szeliski, R.: A database and evaluation methodology for optical flow. Int. J. Comput. Vis. **92**(1), 1–31 (2011). https://doi.org/10.1007/s11263-010-0390-2

37. Bansal, A., Castillo, C., Ranjan, R., Chellappa, R.: The do's and don'ts for cnn-based face verification (2017). arXiv preprint arXiv:1705.07426

38. Bansal, A., Nanduri, A., Castillo, C.D., Ranjan, R., Chellappa, R.: Umdfaces: An annotated face dataset for training deep networks (2016). arXiv preprint arXiv:1611.01484v2

39. Bansal, M., Krizhevsky, A., Ogale, A.S.: Chauffeurnet: Learning to drive by imitating the best and synthesizing the worst (2018). arXiv preprint arXiv:1812.03079

40. Bao, Y., Zhou, H., Huang, S., Li, L., Mou, L., Vechtomova, O., Dai, X.Y., Chen, J.: Generating sentences from disentangled syntactic and semantic spaces. In: Proceedings of the 57th Annual Meeting of the Association for Computational Linguistics, pp. 6008–6019. Association for Computational Linguistics, Florence, Italy (2019)

41. Barati, E., Chen, X., Zhong, Z.: Attention-based deep reinforcement learning for multi-view environments. CoRR (2019). arXiv preprint arXiv:1905.03985

42. Barron, J.L., Fleet, D.J., Beauchemin, S.S.: Performance of optical flow techniques. Int. J. Comput. Vis. **12**(1), 43–77 (1994). https://doi.org/10.1007/BF01420984

43. Barron, J.L., Fleet, D.J., Beauchemin, S.S.: Performance of optical flow techniques. Int. J. Comput. Vis. **12**(1), 43–77 (1994). https://doi.org/10.1007/BF01420984

44. Barrows, R.C., Clayton, P.D.: Privacy, confidentiality, and electronic medical records. J. Am. Med. Inform. Assoc. (JAMIA) **3**(2), 139–48 (1996)

45. Baur, C., Albarqouni, S., Navab, N.: Melanogans: High resolution skin lesion synthesis with gans. CoRR (2018). arXiv preprint arXiv:1804.04338

46. Bayraktar, E., Yigit, C.B., Boyraz, P.: A hybrid image dataset toward bridging the gap between real and simulation environments for robotics. Mach. Vis. Appl. **30**, 23–40 (2018)

47. Bayraktar, E., Yigit, C.B., Boyraz, P.: A hybrid image dataset toward bridging the gap between real and simulation environments for robotics. Mach. Vis. Appl. **30**(1), 23–40 (2019). https://doi.org/10.1007/s00138-018-0966-3

48. Bayramoglu, N., Kaakinen, M., Eklund, L., Heikkilä, J.: Towards virtual h e staining of hyperspectral lung histology images using conditional generative adversarial networks. In: 2017 IEEE International Conference on Computer Vision Workshops (ICCVW), pp. 64–71 (2017). https://doi.org/10.1109/ICCVW.2017.15

49. Beattie, C., Leibo, J.Z., Teplyashin, D., Ward, T., Wainwright, M., Küttler, H., Lefrancq, A., Green, S., Valdés, V., Sadik, A., Schrittwieser, J., Anderson, K., York, S., Cant, M., Cain, A., Bolton, A., Gaffney, S., King, H., Hassabis, D., Legg, S., Petersen, S.: Deepmind lab. CoRR (2016). arXiv preprint arXiv:1612.03801

50. Beaulieu-Jones, B.K., Wu, Z.S., Williams, C., Lee, R., Bhavnani, S.P., Byrd, J.B., Greene, C.S.: Privacy-preserving generative deep neural networks support clinical data sharing. bioRxiv (2018). https://doi.org/10.1101/159756

51. Bechtel, W.: Connectionism and the philosophy of mind: an overview*. South. J. Philos. **26**(S1), 17–41 (1988). https://doi.org/10.1111/j.2041-6962.1988.tb00461.x

52. Bechtel, W., Abrahamsen, A.: Connectionism and the Mind: An Introduction to Parallel Processing in Networks. Blackwell, Cambridge, MA (1990)

53. Beers, A., Brown, J.M., Chang, K., Campbell, J.P., Ostmo, S., Chiang, M.F., Kalpathy-Cramer, J.: High-resolution medical image synthesis using progressively grown generative adversarial networks. CoRR (2018). arXiv preprint arXiv:1805.03144

54. Bell, R.M., Koren, Y.: Lessons from the netflix prize challenge. SIGKDD Explor. Newsl. **9**(2), 75–79 (2007). https://doi.org/10.1145/1345448.1345465

55. Bellemare, M.G., Naddaf, Y., Veness, J., Bowling, M.: The arcade learning environment: an evaluation platform for general agents. J. Artif. Int. Res. **47**(1), 253–279 (2013)

56. Bellifemine, F.L., Caire, G., Greenwood, D.: Developing Multi-Agent Systems with JADE (Wiley Series in Agent Technology). John Wiley & Sons Inc, USA (2007)

57. Bello, I., Zoph, B., Vasudevan, V., Le, Q.V.: Neural optimizer search with reinforcement learning. In: D. Precup, Y.W. Teh (eds.) Proceedings of the 34th International Conference on Machine Learning. Proceedings of Machine Learning Research, vol. 70, pp. 459–468. PMLR, International Convention Centre, Sydney, Australia (2017)

58. Bellovin, S.M., Dutta, P.K., Reitinger, N.: Privacy and synthetic datasets. Technical Report, SSRN (2018)

59. Ben-Cohen, A., Klang, E., Raskin, S.P., Soffer, S., Ben-Haim, S., Konen, E., Amitai, M.M., Greenspan, H.: Cross-modality synthesis from CT to PET using FCN and GAN networks for improved automated lesion detection. CoRR (2018). arXiv preprint arXiv:1802.07846

60. Benaim, S., Wolf, L.: One-sided unsupervised domain mapping. In: Proceedings of the 31st International Conference on Neural Information Processing Systems, NIPS'17, pp. 752–762. Curran Associates Inc., USA (2017)

61. Benenson, R., Popov, S., Ferrari, V.: Large-scale interactive object segmentation with human annotators. In: CVPR (2019)

62. Bengio, Y., Lamblin, P., Popovici, D., Larochelle, H.: Greedy layer-wise training of deep networks. In: Neural Information Processing Systems (NIPS) (2007)

63. Bengio, Y., Louradour, J., Collobert, R., Weston, J.: Curriculum learning. In: Proceedings of the 26th Annual International Conference on Machine Learning, ICML '09, p. 41–48. Association for Computing Machinery, New York, NY, USA (2009). https://doi.org/10.1145/1553374.1553380

64. Bennett, J., Lanning, S.: The netflix prize. In: Proceedings of the KDD Cup Workshop 2007, pp. 3–6. ACM, New York (2007)

65. Bermúdez-Chacón, R., Becker, C., Salzmann, M., Fua, P.: Scalable unsupervised domain adaptation for electron microscopy. In: Ourselin, S., Joskowicz, L., Sabuncu, M.R., Unal, G., Wells, W. (eds.) Medical Image Computing and Computer-Assisted Intervention–MICCAI 2016, pp. 326–334. Springer International Publishing, Cham (2016)

66. Berthelot, D., Schumm, T., Metz, L.: BEGAN: boundary equilibrium generative adversarial networks. CoRR (2017). arXiv preprint abs:1703.10717

67. Bertiche, H., Madadi, M., Escalera, S.: CLOTH3D: clothed 3d humans. CoRR (2019). arXiv preprint abs:1912.02792

68. Bewley, A., Rigley, J., Liu, Y., Hawke, J., Shen, R., Lam, V.D., Kendall, A.: Learning to drive from simulation without real world labels (2018). arXiv preprint abs:1812.03823

69. Bhatti, S., Desmaison, A., Miksik, O., Nardelli, N., Siddharth, N., Torr, P.H.S.: Playing doom with slam-augmented deep reinforcement learning. CoRR (2016). arXiv preprint abs:1612.00380

70. Bielski, A., Favaro, P.: Emergence of object segmentation in perturbed generative models (2019). arXiv preprint abs:1905.12663

71. Bindschadler, V.: Privacy-preserving seedbased data synthesis. Ph.D. thesis, University of Illinois at Urbana-Champaign (2018)

72. Bisagno, N., Conci, N.: Virtual camera modeling for multi-view simulation of surveillance scenes. In: 2018 26th European Signal Processing Conference (EUSIPCO), pp. 2170–2174 (2018). https://doi.org/10.23919/EUSIPCO.2018.8553409

73. Bittar, A., de Oliveira, N.M.F., de Figueiredo, H.V.: Hardware-in-the-loop simulation with x-plane of attitude control of a suav exploring atmospheric conditions. J. Intell. Robot. Syst. **73**(1), 271–287 (2014). https://doi.org/10.1007/s10846-013-9905-8

74. Blanz, V., Scherbaum, K., Seidel, H.: Fitting a morphable model to 3d scans of faces. In: 2007 IEEE 11th International Conference on Computer Vision, pp. 1–8 (2007). https://doi. org/10.1109/ICCV.2007.4409029

75. Blanz, V., Vetter, T.: A morphable model for the synthesis of 3d faces. In: Proceedings of the 26th Annual Conference on Computer Graphics and Interactive Techniques, SIGGRAPH '99, pp. 187–194. ACM Press/Addison-Wesley Publishing Co., New York, NY, USA (1999). https://doi.org/10.1145/311535.311556

76. Blanz, V., Vetter, T.: Face recognition based on fitting a 3d morphable model. IEEE Trans. Pattern Anal. Mach. Intell. **25**(9), 1063–1074 (2003). https://doi.org/10.1109/TPAMI.2003. 1227983

77. Bochinski, E., Eiselein, V., Sikora, T.: Training a convolutional neural network for multi-class object detection using solely virtual world data. In: 2016 13th IEEE International Conference on Advanced Video and Signal Based Surveillance (AVSS), pp. 278–285 (2016). https://doi. org/10.1109/AVSS.2016.7738056

78. Bochkovskiy, A., Wang, C.Y., Liao, H.Y.M.: YOLOv4: optimal speed and accuracy of object detection (2020). arXiv e-prints arXiv:2004.10934

79. Bojanowski, P., Grave, E., Joulin, A., Mikolov, T.: Enriching word vectors with subword information. CoRR (2016). arXiv e-prints abs:1607.04606

80. Bolón-Canedo, V., Sánchez-Maroño, N., Alonso-Betanzos, A.: A review of feature selection methods on synthetic data. Knowl. Inf. Syst. **34**(3), 483–519 (2013). https://doi.org/10.1007/ s10115-012-0487-8

81. Bondi, E., Dey, D., Kapoor, A., Piavis, J., Shah, S., Fang, F., Dilkina, B.N., Hannaford, R., Iyer, A., Joppa, L., Tambe, M.: Airsim-w: A simulation environment for wildlife conservation with uavs. In: COMPASS (2018)

82. Bonetto, M., Korshunov, P., Ramponi, G., Ebrahimi, T.: Privacy in mini-drone based video surveillance. In: 2015 11th IEEE International Conference and Workshops on Automatic Face and Gesture Recognition (FG) **04**, 1–6 (2015)

83. Bonin-Font, F., Ortiz, A., Oliver, G.: Visual navigation for mobile robots: A survey. J. Intell. Rob. Syst. **53**(3), 263 (2008). https://doi.org/10.1007/s10846-008-9235-4

84. Bonnans, J.F., Gilbert, J.C., Lemarechal, C., Sagastizábal, C.A.: Numerical Optimization: Theoretical and Practical Aspects. Springer, Berlin Heidelberg (2013)

85. Borrego, J., Dehban, A., Figueiredo, R., Moreno, P., Bernardino, A., Santos-Victor, J.: Applying domain randomization to synthetic data for object category detection. CoRR (2018). arXiv e-prints abs:1807.09834

86. Bousmalis, K., Irpan, A., Wohlhart, P., Bai, Y., Kelcey, M., Kalakrishnan, M., Downs, L., Ibarz, J., Pastor, P., Konolige, K., Levine, S., Vanhoucke, V.: Using simulation and domain adaptation to improve efficiency of deep robotic grasping. 2018 IEEE International Conference on Robotics and Automation (ICRA) pp. 4243–4250 (2018)

87. Bousmalis, K., Silberman, N., Dohan, D., Erhan, D., Krishnan, D.: Unsupervised pixel-level domain adaptation with generative adversarial networks. 2017 IEEE Conference on Computer Vision and Pattern Recognition (CVPR) pp. 95–104 (2017)

88. Bousmalis, K., Trigeorgis, G., Silberman, N., Krishnan, D., Erhan, D.: Domain separation networks. In: Proceedings of the 30th International Conference on Neural Information Processing Systems, NIPS'16, pp. 343–351. Curran Associates Inc., USA (2016)

89. Brock, A., Donahue, J., Simonyan, K.: Large scale GAN training for high fidelity natural image synthesis. CoRR (2018). arXiv e-prints abs:1809.11096

90. Brock, A., Donahue, J., Simonyan, K.: Large scale GAN training for high fidelity natural image synthesis. In: International Conference on Learning Representations (2019)

91. Brooks, R.A.: Intelligence without reason. In: Proceedings of the 12th International Joint Conference on Artificial Intelligence, Vol. 1, IJCAI'91, p. 569–595. Morgan Kaufmann Publishers Inc., San Francisco, CA, USA (1991)

92. Brooks, R.A.: Intelligence without representation. Artif. Intell. **47**(1–3), 139–159 (1991). https://doi.org/10.1016/0004-3702(91)90053-M

93. Brooks, R.A., Mataric, M.J.: Real Robots, Real Learning Problems, pp. 193–213. Springer US, Boston, MA (1993). https://doi.org/10.1007/978-1-4615-3184-5_8
94. Brown, T.B., Mann, B., Ryder, N., Subbiah, M., Kaplan, J., Dhariwal, P., Neelakantan, A., Shyam, P., Sastry, G., Askell, A., Agarwal, S., Herbert-Voss, A., Krueger, G., Henighan, T., Child, R., Ramesh, A., Ziegler, D.M., Wu, J., Winter, C., Hesse, C., Chen, M., Sigler, E., Litwin, M., Gray, S., Chess, B., Clark, J., Berner, C., McCandlish, S., Radford, A., Sutskever, I., Amodei, D.: Language models are few-shot learners (2020)
95. Bruls, T., Porav, H., Kunze, L., Newman, P.: Generating all the roads to rome: Road layout randomization for improved road marking segmentation (2019). arXiv e-prints abs:1907.04569
96. Bubeck, S.: Convex optimization: algorithms and complexity. Found. Trends Mach. Learn. **8**(3–4), 231–357 (2015). https://doi.org/10.1561/2200000050
97. Buchberger, M., Jorg, K.W., von Puttkamer, E.: Laserradar and sonar based world modeling and motion control for fast obstacle avoidance of the autonomous mobile robot mobot-iv. In: [1993] Proceedings IEEE International Conference on Robotics and Automation, pp. 534–540 vol.1 (1993). https://doi.org/10.1109/ROBOT.1993.292034
98. Buczak, A.L., Babin, S., Moniz, L.: Data-driven approach for creating synthetic electronic medical records. BMC Med. Inform. Decis. Mak. **10**(1), 59 (2010). https://doi.org/10.1186/1472-6947-10-59
99. Van den Bulcke, T., Van Leemput, K., Naudts, B., van Remortel, P., Ma, H., Verschoren, A., De Moor, B., Marchal, K.: Syntren: a generator of synthetic gene expression data for design and analysis of structure learning algorithms. BMC Bioinform. **7**(1), 43 (2006). https://doi.org/10.1186/1471-2105-7-43
100. Buls, E., Kadikis, R., Cacurs, R., Arents, J.: Generation of synthetic training data for object detection in piles. In: International Conference on Machine Vision (2019)
101. Bunel, R., Hausknecht, M.J., Devlin, J., Singh, R., Kohli, P.: Leveraging grammar and reinforcement learning for neural program synthesis. CoRR (2018). arXiv e-prints abs:1805.04276
102. Burgard, J.P., Kolb, J.P., Merkle, H., Münnich, R.: Synthetic data for open and reproducible methodological research in social sciences and official statistics. AStA Wirtschafts- und Sozialstatistisches Archiv **11**(3), 233–244 (2017). https://doi.org/10.1007/s11943-017-0214-8
103. Buslaev, A., Iglovikov, V.I., Khvedchenya, E., Parinov, A., Druzhinin, M., Kalinin, A.A.: Albumentations: Fast and flexible image augmentations. Information **11**(2) (2020). https://doi.org/10.3390/info11020125
104. Buslaev, A.V., Parinov, A., Khvedchenya, E., Iglovikov, V.I., Kalinin, A.A.: Albumentations: fast and flexible image augmentations. CoRR (2018). arXiv e-prints abs:1809.06839
105. Butler, D.J., Wulff, J., Stanley, G.B., Black, M.J.: A naturalistic open source movie for optical flow evaluation. In: Fitzgibbon, A., Lazebnik, S., Perona, P., Sato, Y., Schmid, C. (eds.) Computer Vision–ECCV 2012, pp. 611–625. Springer, Berlin, Heidelberg (2012)
106. Byrne, J., Taylor, C.J.: Expansion segmentation for visual collision detection and estimation. In: 2009 IEEE International Conference on Robotics and Automation, pp. 875–882 (2009). https://doi.org/10.1109/ROBOT.2009.5152487
107. Caesar, H., Bankiti, V., Lang, A.H., Vora, S., Liong, V.E., Xu, Q., Krishnan, A., Pan, Y., Baldan, G., Beijbom, O.: nuscenes: A multimodal dataset for autonomous driving. CoRR (2019). arXiv e-prints abs:1903.11027
108. Cai, Z., Fan, Q., Feris, R.S., Vasconcelos, N.: A unified multi-scale deep convolutional neural network for fast object detection. CoRR (2016). arXiv e-prints abs:1607.07155
109. Calimeri, F., Marzullo, A., Stamile, C., Terracina, G.: Biomedical data augmentation using generative adversarial neural networks. In: Lintas, A., Rovetta, S., Verschure, P.F., Villa, A.E. (eds.) Artificial Neural Networks and Machine Learning–ICANN 2017, pp. 626–634. Springer International Publishing, Cham (2017)
110. Calli, B., Singh, A., Walsman, A., Srinivasa, S., Abbeel, P., Dollar, A.M.: The ycb object and model set: Towards common benchmarks for manipulation research. In: 2015 International

Conference on Advanced Robotics (ICAR), pp. 510–517 (2015). https://doi.org/10.1109/ICAR.2015.7251504

111. Emídio de Campos, T., Rakesh Babu, B., Varma, M.: Character recognition in natural images. In: VISAPP 2009—Proceedings of the 4th International Conference on Computer Vision Theory and Applications, vol. 2, pp. 273–280 (2009)

112. Cao, Q., Shen, L., Xie, W., Parkhi, O.M., Zisserman, A.: Vggface2: A dataset for recognising faces across pose and age. In: International Conference on Automatic Face and Gesture Recognition (2018)

113. Capo, E.: State of the art on: Deep learning for video games ai development (2018). http://www.honours-programme.deib.polimi.it/2018-I/Deliverable1/CSE_CAPO_SOTA.pdf

114. Carlini, N., Liu, C., Kos, J., Erlingsson, Ú., Song, D.: The secret sharer: Measuring unintended neural network memorization & extracting secrets. CoRR (2018). arXiv e-prints abs:1802.08232

115. Carlucci, F.M., Russo, P., Caputo, B.: A deep representation for depth images from synthetic data. In: 2017 IEEE International Conference on Robotics and Automation (ICRA), pp. 1362–1369 (2017). https://doi.org/10.1109/ICRA.2017.7989162

116. Castagno, J.D., Yao, Y., Atkins, E.M.: Realtime rooftop landing site identification and selection in urban city simulation (2019). arXiv preprint abs:1903.03829

117. Castelle, M.: Deep learning as an epistemic ensemble (2018). https://castelle.org/pages/deep-learning-as-an-epistemic-ensemble.html

118. Chan, A.B., Zhang-Sheng John Liang, Vasconcelos, N.: Privacy preserving crowd monitoring: Counting people without people models or tracking. In: 2008 IEEE Conference on Computer Vision and Pattern Recognition, pp. 1–7 (2008). https://doi.org/10.1109/CVPR.2008.4587569

119. Chan, C., Ginosar, S., Zhou, T., Efros, A.A.: Everybody dance now (2018). arXiv preprint abs:1808.07371

120. Chandu, K.R., Pyreddy, M.A., Felix, M., Joshi, N.N.: Textually enriched neural module networks for visual question answering. CoRR (2018). arXiv preprint abs:1809.08697

121. Chang, A.X., Dai, A., Funkhouser, T.A., Halber, M., Nießner, M., Savva, M., Song, S., Zeng, A., Zhang, Y.: Matterport3d: Learning from RGB-D data in indoor environments. CoRR (2017). arXiv preprint abs:1709.06158

122. Chang, A.X., Funkhouser, T.A., Guibas, L.J., Hanrahan, P., Huang, Q., Li, Z., Savarese, S., Savva, M., Song, S., Su, H., Xiao, J., Yi, L., Yu, F.: Shapenet: An information-rich 3d model repository. CoRR (2015). arXiv preprint abs:1512.03012

123. Chao, Q., Bi, H., Li, W., Mao, T., Wang, Z., Lin, M.C., Deng, Z.: A survey on visual traffic simulation: models, evaluations, and applications in autonomous driving. Comput. Graph. Forum 39(1), 287–308 (2020)

124. Charest, A.S., Hou, Y.: On the meaning and limits of empirical differential privacy. J. Priv. Confident. 7(3), 185–205 (2017)

125. Chartsias, A., Joyce, T., Dharmakumar, R., Tsaftaris, S.A.: Adversarial image synthesis for unpaired multi-modal cardiac data. In: SASHIMIMICCAI (2017)

126. Chebotar, Y., Handa, A., Makoviychuk, V., Macklin, M., Issac, J., Ratliff, N.D., Fox, D.: Closing the sim-to-real loop: Adapting simulation randomization with real world experience. CoRR (2018). arXiv preprint abs:1810.05687

127. Chellapilla, K., Puri, S., Simard, P.: High performance convolutional neural networks for document processing. In: International Workshop on Frontiers in Handwriting Recognition (2006)

128. Chen, C., Dou, Q., Chen, H., Heng, P.: Semantic-aware generative adversarial nets for unsupervised domain adaptation in chest x-ray segmentation. CoRR (2018). arXiv preprint abs:1806.00600

129. Chen, G., Choi, W., Yu, X., Han, T., Chandraker, M.: Learning efficient object detection models with knowledge distillation. In: I. Guyon, U.V. Luxburg, S. Bengio, H. Wallach, R. Fergus, S. Vishwanathan, R. Garnett (eds.) Advances in Neural Information Processing Systems, vol. 30. Curran Associates, Inc. (2017)

130. Chen, H., Engkvist, O., Wang, Y., Olivecrona, M., Blaschke, T.: The rise of deep learning in drug discovery. Drug Discovery Today **23**(6), 1241–1250 (2018). https://doi.org/10.1016/j.drudis.2018.01.039

131. Chen, K., Loy, C.C., Gong, S., Xiang, T.: Feature mining for localised crowd counting. In: BMVC (2012)

132. Chen, L., Jin, L., Du, X., Li, S., Liu, M.: Deforming the Loss Surface to Affect the Behaviour of the Optimizer. arXiv e-prints arXiv:2009.08274 (2020)

133. Chen, L., Papandreou, G., Kokkinos, I., Murphy, K., Yuille, A.L.: Deeplab: Semantic image segmentation with deep convolutional nets, atrous convolution, and fully connected crfs. IEEE Trans. Pattern Anal. Mach. Intell. **40**(4), 834–848 (2018). https://doi.org/10.1109/TPAMI.2017.2699184

134. Chen, L., Wang, W., Zhu, J.: Learning transferable uav for forest visual perception (2018). arXiv preprint abs:1806.03626

135. Chen, L.C., Zhu, Y., Papandreou, G., Schroff, F., Adam, H.: Encoder-decoder with atrous separable convolution for semantic image segmentation. In: Ferrari, V., Hebert, M., Sminchisescu, C., Weiss, Y. (eds.) Computer Vision–ECCV 2018, pp. 833–851. Springer International Publishing, Cham (2018)

136. Chen, T., Fox, E., Guestrin, C.: Stochastic gradient hamiltonian monte carlo. In: E.P. Xing, T. Jebara (eds.) Proceedings of the 31st International Conference on Machine Learning, *Proceedings of Machine Learning Research*, vol. 32(2), pp. 1683–1691. PMLR, Bejing, China (2014). http://proceedings.mlr.press/v32/cheni14.html

137. Chen, X., Golovinskiy, A., Funkhouser, T.: A benchmark for 3d mesh segmentation. ACM Trans. Graph. **28**(3), 73:1–73:12 (2009). https://doi.org/10.1145/1531326.1531379

138. Chen, X., Liu, C., Song, D.: Execution-guided neural program synthesis. In: International Conference on Learning Representations (2019)

139. Chen, X., Xu, C., Yang, X., Tao, D.: Attention-gan for object transfiguration in wild images. In: Ferrari, V., Hebert, M., Sminchisescu, C., Weiss, Y. (eds.) Computer Vision–ECCV 2018, pp. 167–184. Springer International Publishing, Cham (2018)

140. Chen, Y., Assael, Y., Shillingford, B., Budden, D., Reed, S., Zen, H., Wang, Q., Cobo, L.C., Trask, A., Laurie, B., Gulcehre, C., van den Oord, A., Vinyals, O., de Freitas, N.: Sample efficient adaptive text-to-speech (2019)

141. Chen, Y., Chen, W., Chen, Y., Tsai, B., Wang, Y.F., Sun, M.: No more discrimination: Cross city adaptation of road scene segmenters. CoRR (2017). arXiv e-prints abs:1704.08509

142. Chen, Y., Li, W., Chen, X., Van Gool, L.: Learning semantic segmentation from synthetic data: A geometrically guided input-output adaptation approach. In: The IEEE Conference on Computer Vision and Pattern Recognition (CVPR) (2019)

143. Chen, Y.T., Garbade, M., Gall, J.: 3d semantic scene completion from a single depth image using adversarial training (2019). arXiv e-prints arXiv:1905.06231

144. Chernodub, A., Oliynyk, O., Heidenreich, P., Bondarenko, A., Hagen, M., Biemann, C., Panchenko, A.: Targer: Neural argument mining at your fingertips. In: Proceedings of the 57th Annual Meeting of the Association of Computational Linguistics (ACL'2019). Florence, Italy (2019)

145. Cheung, E., Wong, T.K., Bera, A., Wang, X., Manocha, D.: Lcrowdv: Generating labeled videos for simulation-based crowd behavior learning. In: Hua, G., Jégou, H. (eds.) Computer Vision–ECCV 2016 Workshops, pp. 709–727. Springer International Publishing, Cham (2016)

146. Ching, T., Himmelstein, D.S., Beaulieu-Jones, B.K., Kalinin, A.A., Do, B.T., Way, G.P., Ferrero, E., Agapow, P.M., Zietz, M., Hoffman, M.M., Xie, W., Rosen, G.L., Lengerich, B.J., Israeli, J., Lanchantin, J., Woloszynek, S., Carpenter, A.E., Shrikumar, A., Xu, J., Cofer, E.M., Lavender, C.A., Turaga, S.C., Alexandari, A.M., Lu, Z., Harris, D.J., DeCaprio, D., Qi, Y., Kundaje, A., Peng, Y., Wiley, L.K., Segler, M.H.S., Boca, S.M., Swamidass, S.J., Huang, A., Gitter, A., Greene, C.S.: Opportunities and obstacles for deep learning in biology and medicine. J. R. Soc. Interface **15**(141), 20170387 (2018). https://doi.org/10.1098/rsif.2017.0387

147. Chociej, M., Welinder, P., Weng, L.: ORRB—openai remote rendering backend. CoRR (2019). arXiv e-prints abs:1906.11633

148. Choi, C., Christensen, H.I.: Rgb-d object tracking: A particle filter approach on gpu. In: 2013 IEEE/RSJ International Conference on Intelligent Robots and Systems, pp. 1084–1091 (2013). https://doi.org/10.1109/IROS.2013.6696485

149. Choi, D., An, T., Ahn, K., Choi, J.: Driving experience transfer method for end-to-end control of self-driving cars. CoRR (2018). arXiv e-prints abs:1809.01822

150. Choi, E., Biswal, S., Malin, B., Duke, J., Stewart, W.F., Sun, J.: Generating multi-label discrete patient records using generative adversarial networks. In: F. Doshi-Velez, J. Fackler, D. Kale, R. Ranganath, B. Wallace, J. Wiens (eds.) Proceedings of the 2nd Machine Learning for Healthcare Conference, Proceedings of Machine Learning Research, vol. 68, pp. 286–305. PMLR, Boston, Massachusetts (2017)

151. Choi, S., Zhou, Q.Y., Miller, S., Koltun, V.: A large dataset of object scans (2016). arXiv e-prints arXiv:1602.02481

152. Chollet, F.: Xception: Deep learning with depthwise separable convolutions. CoRR (2016). arXiv e-prints abs:1610.02357

153. Chollet, F.: Deep Learning with Python, 1st edn. Manning Publications Co., USA (2017)

154. Chollet, F.: On the measure of intelligence (2019). arXiv e-prints arXiv:1911.01547

155. Chopra, S., Hadsell, R., LeCun, Y.: Learning a similarity metric discriminatively, with application to face verification. In: 2005 IEEE Computer Society Conference on Computer Vision and Pattern Recognition (CVPR'05), vol. 1, pp. 539–546 vol. 1 (2005). https://doi.org/10.1109/CVPR.2005.202

156. Chorowski, J., Weiss, R., Bengio, S., van den Oord, A.: Unsupervised speech representation learning using wavenet autoencoders. IEEE Trans. Audio Speech Lang. Process. (2019)

157. Chum, L., Subramanian, A., Balasubramanian, V.N., Jawahar, C.V.: Beyond supervised learning: A computer vision perspective. J. Indian Inst. Sci. 99(2), 177–199 (2019). https://doi.org/10.1007/s41745-019-0099-3

158. Chuquicusma, M.J.M., Hussein, S., Burt, J., Bagci, U.: How to fool radiologists with generative adversarial networks? a visual turing test for lung cancer diagnosis. In: 2018 IEEE 15th International Symposium on Biomedical Imaging (ISBI 2018), pp. 240–244 (2018). https://doi.org/10.1109/ISBI.2018.8363564

159. Ciresan, D.C., Meier, U., Masci, J., Gambardella, L.M., Schmidhuber, J.: Flexible, high performance convolutional neural networks for image classification. In: International Joint Conference on Artificial Intelligence IJCAI, pp. 1237–1242 (2011)

160. Ciresan, D.C., Meier, U., Masci, J., Schmidhuber, J.: Multi-column deep neural network for traffic sign classification. Neural Netw. 32, 333–338 (2012)

161. Clevert, D.A., Unterthiner, T., Hochreiter, S.: Fast and Accurate Deep Network Learning by Exponential Linear Units (ELUs). arXiv e-prints arXiv:1511.07289 (2015)

162. Clowes, M.: On seeing things. Artif. Intell. 2(1), 79–116 (1971). https://doi.org/10.1016/0004-3702(71)90005-1

163. Cohen, T.S., Geiger, M., Köhler, J., Welling, M.: Spherical cnns. CoRR (2018). arXiv e-prints abs:1801.10130

164. Cohen, T.S., Welling, M.: Steerable cnns. CoRR (2016). arXiv e-prints abs:1612.08498

165. Cohn, D., Atlas, L., Ladner, R.: Improving generalization with active learning. Mach. Learn. 15(2), 201–221 (1994). https://doi.org/10.1007/BF00993277

166. Conjeti, S., Katouzian, A., Roy, A.G., Peter, L., Sheet, D., Carlier, S., Laine, A., Navab, N.: Supervised domain adaptation of decision forests: Transfer of models trained in vitro for in vivo intravascular ultrasound tissue characterization. Med. Image Anal. 32, 1–17 (2016). https://doi.org/10.1016/j.media.2016.02.005

167. Cordts, M., Omran, M., Ramos, S., Rehfeld, T., Enzweiler, M., Benenson, R., Franke, U., Roth, S., Schiele, B.: The cityscapes dataset for semantic urban scene understanding. In: Proceedings of the IEEE Conference on Computer Vision and Pattern Recognition (CVPR) (2016)

168. Cormode, G., Procopiuc, C.M., Shen, E., Srivastava, D., Yu, T.: Empirical privacy and empirical utility of anonymized data. In: 2013 IEEE 29th International Conference on Data Engineering Workshops (ICDEW), pp. 77–82 (2013). https://doi.org/10.1109/ICDEW.2013. 6547431

169. Coulom, R.: Efficient selectivity and backup operators in monte-carlo tree search. In: H.J. van den Herik, P. Ciancarini, H.H.L.M.J. Donkers (eds.) Computers and Games, pp. 72–83. Springer Berlin Heidelberg, Berlin, Heidelberg (2007)

170. Courbon, J., Mezouar, Y., Guénard, N., Martinet, P.: Vision-based navigation of unmanned aerial vehicles. Control Eng. Pract. **18**(7), 789–799 (2010). https://doi.org/10.1016/j. conengprac.2010.03.004 (Special Issue on Aerial Robotics)

171. Courty, N., Allain, P., Creusot, C., Corpetti, T.: Using the agoraset dataset: assessing for the quality of crowd video analysis methods. Pattern Recogn. Lett. **44**, 161–170 (2014). https:// doi.org/10.1016/j.patrec.2014.01.004 (Pattern Recognition and Crowd Analysis)

172. Cubuk, E.D., Zoph, B., Mané, D., Vasudevan, V., Le, Q.V.: Autoaugment: Learning augmentation policies from data. CoRR (2018). arXiv e-prints abs:1805.09501

173. Cutler, M., How, J.P.: Efficient reinforcement learning for robots using informative simulated priors. In: 2015 IEEE International Conference on Robotics and Automation (ICRA), pp. 2605–2612 (2015). https://doi.org/10.1109/ICRA.2015.7139550

174. Cutler, M., Walsh, T.J., How, J.P.: Reinforcement learning with multi-fidelity simulators. In: IEEE International Conference on Robotics and Automation (ICRA), pp. 3888–3895. Hong Kong (2014). http://markjcutler.com/papers/Cutler14_ICRA.pdf

175. Cutler, M., Walsh, T.J., How, J.P.: Real-world reinforcement learning via multifidelity simulators. IEEE Transactions on Robotics **31**(3), 655–671 (2015). http://markjcutler.com/papers/ Cutler15_TRO.pdf

176. Dai, A., Chang, A.X., Savva, M., Halber, M., Funkhouser, T., Nießner, M.: Scannet: Richly-annotated 3d reconstructions of indoor scenes. In: Proc. Computer Vision and Pattern Recognition (CVPR), IEEE (2017)

177. Dai, J., Li, Y., He, K., Sun, J.: R-fcn: Object detection via region-based fully convolutional networks. In: D.D. Lee, M. Sugiyama, U.V. Luxburg, I. Guyon, R. Garnett (eds.) Advances in Neural Information Processing Systems 29, pp. 379–387. Curran Associates, Inc. (2016)

178. Dai, J., Li, Y., He, K., Sun, J.: R-FCN: object detection via region-based fully convolutional networks. CoRR (2016). arXiv e-prints abs:1605.06409

179. Dai, Z., Yang, Z., Yang, Y., Carbonell, J., Le, Q., Salakhutdinov, R.: Transformer-XL: Attentive language models beyond a fixed-length context. In: Proceedings of the 57th Annual Meeting of the Association for Computational Linguistics, pp. 2978–2988. Association for Computational Linguistics, Florence, Italy (2019)

180. Dal Pozzolo, A., Caelen, O., Le Borgne, Y.A., Waterschoot, S., Bontempi, G.: Learned lessons in credit card fraud detection from a practitioner perspective. Expert Syst. Appl. **41**, 4915–4928 (2014). https://doi.org/10.1016/j.eswa.2014.02.026

181. Dang, Q., Yin, J., Wang, B., Zheng, W.: Deep learning based 2d human pose estimation: A survey. Tsinghua Sci. Technol. **24**(6), 663–676 (2019)

182. Dar, S.U.H., Yurt, M., Karacan, L., Erdem, A., Erdem, E., Çukur, T.: Image synthesis in multi-contrast MRI with conditional generative adversarial networks. CoRR (2018). arXiv e-prints abs:1802.01221

183. Das, A., Datta, S., Gkioxari, G., Lee, S., Parikh, D., Batra, D.: Embodied question answering. In: 2018 IEEE/CVF Conference on Computer Vision and Pattern Recognition Workshops (CVPRW), pp. 2135–213509 (2018). https://doi.org/10.1109/CVPRW.2018.00279

184. Davydow, A., Nikolenko, S.I.: Land cover classification with superpixels and jaccard index post-optimization. In: Proceedings of the IEEE Conference on Computer Vision and Pattern Recognition 2018 Workshops, pp. 280–284 (2018)

185. Dawson, M., Zisserman, A., Nellaker, C.: Mining faces from biomedical literature using deep learning. In: Proceedings of the 8th ACM International Conference on Bioinformatics, Computational Biology, and Health Informatics, ACM-BCB '17, pp. 562–567. ACM, New York, NY, USA (2017). https://doi.org/10.1145/3107411.3107476

186. Debar, H., Dacier, M., Lampart, S.: An experimentation workbench for intrusion detection systems. IBM TJ Watson Research Center (1999)

187. Deng, J., Dong, W., Socher, R., Li, L.J., Li, K., Fei-Fei, L.: Imagenet: A large-scale hierarchical image database. In: 2009 IEEE conference on computer vision and pattern recognition, pp. 248–255. Ieee (2009)

188. Deng, J., Guo, J., Xue, N., Zafeiriou, S.: Arcface: Additive angular margin loss for deep face recognition. In: IEEE Conference on Computer Vision and Pattern Recognition, CVPR 2019, Long Beach, CA, USA, 16–20 June 2019, pp. 4690–4699. Computer Vision Foundation. IEEE (2019). https://doi.org/10.1109/CVPR.2019.00482

189. Deng, J., Guo, J., Zhou, Y., Yu, J., Kotsia, I., Zafeiriou, S.: Retinaface: Single-stage dense face localisation in the wild. CoRR (2019). arXiv e-prints abs:1905.00641

190. Desouza, G.N., Kak, A.C.: Vision for mobile robot navigation: a survey. IEEE Trans. Pattern Anal. Mach. Intell. **24**(2), 237–267 (2002). https://doi.org/10.1109/34.982903

191. Devlin, J., Bunel, R.R., Singh, R., Hausknecht, M., Kohli, P.: Neural program meta-induction. In: I. Guyon, U.V. Luxburg, S. Bengio, H. Wallach, R. Fergus, S. Vishwanathan, R. Garnett (eds.) Advances in Neural Information Processing Systems 30, pp. 2080–2088. Curran Associates, Inc. (2017)

192. Devlin, J., Chang, M., Lee, K., Toutanova, K.: BERT: pre-training of deep bidirectional transformers for language understanding. In: Proceedings of the 2019 Conference of the North American Chapter of the Association for Computational Linguistics: Human Language Technologies, NAACL-HLT 2019, Minneapolis, MN, USA, 2–7 June 2019, Vol. 1 (Long and Short Papers), pp. 4171–4186 (2019)

193. Devlin, J., Chang, M.W., Lee, K., Toutanova, K.: Bert: Pre-training of deep bidirectional transformers for language understanding. arXiv preprint arXiv:1810.04805 (2018)

194. Devlin, J., Uesato, J., Bhupatiraju, S., Singh, R., Mohamed, A., Kohli, P.: Robustfill: Neural program learning under noisy I/O. CoRR (2017). arXiv e-prints abs:1703.07469

195. DiCesare, G.: Imputation, estimation and missing data in finance. Ph.D. thesis, University of Waterloo (2006)

196. Dilipkumar, D., Póczos, B.: Generative adversarial image refinement for handwriting recognition (2017). https://www.ml.cmu.edu/research/dap-papers/F17/dap-dilipkumar-deepak.pdf

197. Ding, C., Tao, D.: A comprehensive survey on pose-invariant face recognition. ACM Trans. Intell. Syst. Technol. **7**(3), 37:1–37:42 (2016). https://doi.org/10.1145/2845089

198. Dinh, L., Sohl-Dickstein, J., Bengio, S.: Density estimation using real NVP. CoRR (2016). arXiv e-prints abs:1605.08803

199. Dinur, I., Nissim, K.: Revealing information while preserving privacy. In: Proceedings of the Twenty-second ACM SIGMOD-SIGACT-SIGART Symposium on Principles of Database Systems, PODS '03, pp. 202–210. ACM, New York, NY, USA (2003). https://doi.org/10.1145/773153.773173

200. Doersch, C.: Tutorial on variational autoencoders (2021)

201. Donahue, J., Krähenbühl, P., Darrell, T.: Adversarial feature learning. CoRR (2016). arXiv e-prints abs:1605.09782

202. Dosovitskiy, A., Fischer, P., Ilg, E., Häusser, P., Hazirbas, C., Golkov, V., v. d. Smagt, P., Cremers, D., Brox, T.: Flownet: Learning optical flow with convolutional networks. In: 2015 IEEE International Conference on Computer Vision (ICCV), pp. 2758–2766 (2015). https://doi.org/10.1109/ICCV.2015.316

203. Dosovitskiy, A., Ros, G., Codevilla, F., Lopez, A., Koltun, V.: CARLA: An open urban driving simulator. In: Proceedings of the 1st Annual Conference on Robot Learning, pp. 1–16 (2017)

204. Dozat, T.: Incorporating nesterov momentum into adam. In: ICLR Workshop, pp. 2013–2016 (2016)

205. Drazen, J.M., Morrissey, S., Campion, E.W., Jarcho, J.A.: A sprint to the finish. N. Engl. J. Med. **373**(22), 2174–2175 (2015). https://doi.org/10.1056/NEJMe1513991. PMID: 26551058

206. Drechsler, J.: Generating multiply imputed synthetic datasets: Theory and implementation. Ph.D. thesis, Otto-Friedrich-Universität Bamberg (2010)

207. Drechsler, J.: Synthetic Datasets for Statistical Disclosure Control. Lecture Notes in Statistics, vol. 201. Springer (2011)

208. Driemeyer, T.: Rendering with Mental Ray. Springer, New York (2001)

209. Duchi, J., Hazan, E., Singer, Y.: Adaptive subgradient methods for online learning and stochastic optimization. J. Mach. Learn. Res. **12**(61), 2121–2159 (2011). http://jmlr.org/papers/v12/duchi11a.html

210. Dumoulin, V., Shlens, J., Kudlur, M.: A learned representation for artistic style. ICLR (2017)

211. Dumoulin, V., Visin, F.: A guide to convolution arithmetic for deep learning (2016). ArXiv e-prints

212. Dvornik, N., Mairal, J., Schmid, C.: Modeling visual context is key to augmenting object detection datasets. CoRR (2018). arXiv e-prints abs:1807.07428

213. Dwibedi, D., Misra, I., Hebert, M.: Cut, paste and learn: Surprisingly easy synthesis for instance detection. 2017 IEEE International Conference on Computer Vision (ICCV) pp. 1310–1319 (2017)

214. Dwork, C.: Differential privacy: A survey of results. In: Agrawal, M., Du, D., Duan, Z., Li, A. (eds.) Theory and Applications of Models of Computation, pp. 1–19. Springer, Berlin Heidelberg, Berlin, Heidelberg (2008)

215. Dwork, C., Kenthapadi, K., McSherry, F., Mironov, I., Naor, M.: Our data, ourselves: Privacy via distributed noise generation. In: Vaudenay, S. (ed.) Advances in Cryptology–EUROCRYPT 2006, pp. 486–503. Springer, Berlin, Heidelberg (2006)

216. Dwork, C., McSherry, F., Nissim, K., Smith, A.: Calibrating noise to sensitivity in private data analysis. In: Proceedings of the Third Conference on Theory of Cryptography, TCC'06, pp. 265–284. Springer-Verlag, Berlin, Heidelberg (2006). https://doi.org/10.1007/11681878_14

217. Dwork, C., Nissim, K.: Privacy-preserving datamining on vertically partitioned databases. In: Franklin, M. (ed.) Advances in Cryptology–CRYPTO 2004, pp. 528–544. Springer, Berlin, Heidelberg (2004)

218. Dwork, C., Roth, A.: The algorithmic foundations of differential privacy. Found. Trends Theor. Comput. Sci. **9**(3–4), 211–407 (2014). https://doi.org/10.1561/0400000042

219. Edlinger, T., von Puttkamer, E.: Exploration of an indoor-environment by an autonomous mobile robot. In: Proceedings of IEEE/RSJ International Conference on Intelligent Robots and Systems (IROS'94), vol. 2, pp. 1278–1284 (1994). https://doi.org/10.1109/IROS.1994.407463

220. Efimova, V., Filchenkov, A.: Generative models for placing text on images. In: IFMO Young Researchers Congress (2019)

221. Eggert, C., Winschel, A., Lienhart, R.: On the benefit of synthetic data for company logo detection. In: Proceedings of the 23rd ACM International Conference on Multimedia, MM '15, pp. 1283–1286. ACM, New York, NY, USA (2015). https://doi.org/10.1145/2733373.2806407

222. Eggert, C., Zecha, D., Brehm, S., Lienhart, R.: Improving small object proposals for company logo detection. CoRR (2017). arXiv e-prints abs:1704.08881

223. Eid, M.H., Cecco, C.N.D., Nance, J.W., Caruso, D., Albrecht, M.H., Spandorfer, A.J., Santis, D.D., Varga-Szemes, A., Schoepf, U.J.: Cinematic rendering in ct: A novel, lifelike 3d visualization technique. AJR Am. J. Roentgenol. **209**(2), 370–379 (2017)

224. Ekbatani., H.K., Pujol., O., Segui., S.: Synthetic data generation for deep learning in counting pedestrians. In: Proceedings of the 6th International Conference on Pattern Recognition Applications and Methods—Vol. 1: ICPRAM, pp. 318–323. INSTICC, SciTePress (2017). https://doi.org/10.5220/0006119203180323

225. Elgendy, M.: Deep Learning for Vision Systems. Manning (2019)

226. Erhan, D., Bengio, Y., Courville, A., Manzagol, P.A., Vincent, P., Bengio, S.: Why does unsupervised pre-training help deep learning? J. Mach. Learn. Res. **11**, 625–660 (2010)

227. Ertugrul, I.O., Jeni, L.A., Cohn, J.F.: Facscaps: Pose-independent facial action coding with capsules. In: 2018 IEEE Conference on Computer Vision and Pattern Recognition Workshops, CVPR Workshops 2018, Salt Lake City, UT, USA, 18–22 June 2018, pp. 2130–2139. IEEE Computer Society (2018). https://doi.org/10.1109/CVPRW.2018.00287

228. Esteban, C., Hyland, S.L., Rätsch, G.: Real-valued (Medical) Time Series Generation with Recurrent Conditional GANs. arXiv e-prints arXiv:1706.02633 (2017)

229. Face recognition prize challenge (2017). https://www.nist.gov/programs-projects/face-recognition-prize-challenge-2017

230. Fadaee, M., Bisazza, A., Monz, C.: Data augmentation for low-resource neural machine translation. In: Proceedings of the 55th Annual Meeting of the Association for Computational Linguistics (Volume 2: Short Papers), pp. 567–573. Association for Computational Linguistics, Vancouver, Canada (2017). https://doi.org/10.18653/v1/P17-2090

231. Fang, J., Yan, F., Zhao, T., Zhang, F., Zhou, D., Yang, R., Ma, Y., Wang, L.: Simulating lidar point cloud for autonomous driving using real-world scenes and traffic flows (2018). arXiv e-prints abs:1811.07112

232. Farnebäck, G.: Two-frame motion estimation based on polynomial expansion. In: Bigun, J., Gustavsson, T. (eds.) Image Analysis, pp. 363–370. Springer, Berlin, Heidelberg (2003)

233. Fedus, W., Goodfellow, I., Dai, A.M.: MaskGAN: Better text generation via filling in the——. In: International Conference on Learning Representations (2018)

234. Felzenszwalb, P.F., Huttenlocher, D.P.: Efficient graph-based image segmentation. Int. J. Comput. Vision 59(2), 167–181 (2004). https://doi.org/10.1023/B:VISI.0000022288.19776.77

235. Fernández, C., Baiget, P., Roca, F., Gonzàlez, J.: Augmenting video surveillance footage with virtual agents for incremental event evaluation. Pattern Recogn. Lett. 32(6), 878–889 (2011). https://doi.org/10.1016/j.patrec.2010.09.027

236. Firman, M.: Rgbd datasets: Past, present and future. In: 2016 IEEE Conference on Computer Vision and Pattern Recognition Workshops (CVPRW), pp. 661–673 (2016). https://doi.org/10.1109/CVPRW.2016.88

237. Fletcher, R.: Practical Methods of Optimization. John Wiley & Sons, New York (1987)

238. Fletcher, S.C.: Data mining and privacy: Modeling sensitive data with differential privacy. Ph.D. thesis, Charles Sturt University (2017)

239. Fossen, T.I., Pettersen, K.Y., Nijmeijer, H.: Sensing and Control for Autonomous Vehicles: Applications to Land, Water and Air Vehicles. Lecture Notes in Control and Information Sciences, vol. 474. Springer (2017)

240. Freeman, W.T., Pasztor, E.C.: Learning low-level vision. In: Proceedings of the Seventh IEEE International Conference on Computer Vision, vol. 2, pp. 1182–1189 vol.2 (1999). https://doi.org/10.1109/ICCV.1999.790414

241. Freivalds, K., Liepins, R.: Improving the neural gpu architecture for algorithm learning. CoRR (2017). arXiv e-prints abs:1702.08727

242. Frey, B.J.: Graphical Models for Machine Learning and Digital Communication. MIT Press (1998)

243. Frey, B.J., Hinton, G.E., Dayan, P.: Does the wake-sleep algorithm produce good density estimators? In: Proceedings of the 8th International Conference on Neural Information Processing Systems, NIPS'95, p. 661–667. MIT Press, Cambridge, MA, USA (1995)

244. Frid-Adar, M., Diamant, I., Klang, E., Amitai, M., Goldberger, J., Greenspan, H.: Gan-based synthetic medical image augmentation for increased cnn performance in liver lesion classification. Neurocomputing 321, 321–331 (2018). https://doi.org/10.1016/j.neucom.2018.09.013

245. Frid-Adar, M., Diamant, I., Klang, E., Amitai, M., Goldberger, J., Greenspan, H.: Gan-based synthetic medical image augmentation for increased CNN performance in liver lesion classification. CoRR (2018). arXiv e-prints abs:1803.01229

246. Frigerio, L., de Oliveira, A.S., Gomez, L., Duverger, P.: Differentially private generative adversarial networks for time series, continuous, and discrete open data. In: Dhillon, G., Karlsson, F., Hedström, K., Zúquete, A. (eds.) ICT Systems Security and Privacy Protection, pp. 151–164. Springer International Publishing, Cham (2019)

247. Frosst, N., Sabour, S., Hinton, G.E.: DARCCC: detecting adversaries by reconstruction from class conditional capsules. CoRR (2018). arXiv e-prints abs:1811.06969

248. Fukushima, K.: Neural network model for a mechanism of pattern recognition unaffected by shift in position—Neocognitron. Trans. IECE **J62-A(10)**, 658–665 (1979)

249. Fukushima, K.: Artificial vision by multi-layered neural networks: Neocognitron and its advances. Neural Netw. **37**, 103–119 (2013)

250. Fukushima, K., Miyake, S.: Neocognitron: A self-organizing neural network model for a mechanism of visual pattern recognition. In: S.i. Amari, M.A. Arbib (eds.) Competition and Cooperation in Neural Nets, pp. 267–285. Springer, Berlin, Heidelberg (1982)

251. Furrer, F., Burri, M., Achtelik, M., Siegwart, R.: RotorS—A Modular Gazebo MAV Simulator Framework, pp. 595–625. Springer International Publishing, Cham (2016). https://doi.org/10.1007/978-3-319-26054-9_23

252. Gaber, C., Hemery, B., Achemlal, M., Pasquet, M., Urien, P.: Synthetic logs generator for fraud detection in mobile transfer services. In: 2013 International Conference on Collaboration Technologies and Systems (CTS), pp. 174–179 (2013). https://doi.org/10.1109/CTS.2013.6567225

253. Gaidon, A., Lopez, A., Perronnin, F.: The reasonable effectiveness of synthetic visual data. Int. J. Comput. Vision **126**(9), 899–901 (2018). https://doi.org/10.1007/s11263-018-1108-0

254. Gaidon, A., Wang, Q., Cabon, Y., Vig, E.: Virtual worlds as proxy for multi-object tracking analysis. In: CVPR (2016)

255. Galama, Y., Mensink, T.: Itergans: Iterative gans to learn and control 3d object transformation (2018). arXiv e-prints abs:1804.05651

256. Galinsky, R., Alekseyev, A., Nikolenko, S.I.: Improving neural network models for natural language processing in russian with synonyms. In: Proceedings of the 5th conference on Artificial Intelligence and Natural Language, pp. 45–51 (2016)

257. Galland, S., Gaud, N., Demange, J., Koukam, A.: Environment model for multiagent-based simulation of 3d urban systems. In: Proceedings of the 7th European Workshop on Multiagent Systems (EUMAS09) (2009)

258. Ganin, Y., Kulkarni, T., Babuschkin, I., Eslami, S.M.A., Vinyals, O.: Synthesizing programs for images using reinforced adversarial learning. CoRR (2018). arXiv e-prints abs:1804.01118

259. Ganin, Y., Lempitsky, V.: Unsupervised domain adaptation by backpropagation. In: Proceedings of the 32Nd International Conference on International Conference on Machine Learning—Volume 37, ICML'15, pp. 1180–1189. JMLR.org (2015)

260. Ganin, Y., Ustinova, E., Ajakan, H., Germain, P., Larochelle, H., Laviolette, F., Marchand, M., Lempitsky, V.: Domain-adversarial training of neural networks. J. Mach. Learn. Res. **17**(1), 2096–2030 (2016)

261. Gao, X., Gong, R., Shu, T., Xie, X., Wang, S., Zhu, S.: Vrkitchen: an interactive 3d virtual environment for task-oriented learning. CoRR (2019). arXiv e-prints abs:1903.05757

262. Garcia, R., Barnes, L.: Multi-uav simulator utilizing x-plane. J. Intell. Rob. Syst. **57**(1), 393 (2009). https://doi.org/10.1007/s10846-009-9372-4

263. Garipov, T., Izmailov, P., Podoprikhin, D., Vetrov, D.P., Wilson, A.G.: Loss surfaces, mode connectivity, and fast ensembling of dnns. In: S. Bengio, H. Wallach, H. Larochelle, K. Grauman, N. Cesa-Bianchi, R. Garnett (eds.) Advances in Neural Information Processing Systems 31, pp. 8789–8798. Curran Associates, Inc. (2018)

264. Gatys, L.A., Ecker, A.S., Bethge, M.: A neural algorithm of artistic style. CoRR (2015). arXiv e-prints abs:1508.06576

265. Gatys, L.A., Ecker, A.S., Bethge, M.: Texture synthesis and the controlled generation of natural stimuli using convolutional neural networks. CoRR (2015). arXiv e-prints abs:1505.07376

266. Gaud, N., Galland, S., Hilaire, V., Koukam, A.: An organisational platform for holonic and multiagent systems. In: Hindriks, K.V., Pokahr, A., Sardina, S. (eds.) Programming Multi-Agent Systems, pp. 104–119. Springer, Berlin, Heidelberg (2009)

267. Gaulton, A., Bellis, L., Chambers, J., Davies, M., Hersey, A., Light, Y., McGlinchey, S., Akhtar, R., Atkinson, F., Bento, A., Al-Lazikani, B., Michalovic, D., Overington, J.: Chembl: A large-scale bioactivity database for chemical biology and drug discovery. Nucleic Acids Res. **40**, 1100-D1107 (2011)

268. Gawehn, E., Hiss, J., Schneider, G.: Deep learning in drug discovery. Mol. Inf. **35**, (2015). https://doi.org/10.1002/minf.201501008

269. Geiger, A., Lenz, P., Stiller, C., Urtasun, R.: Vision meets robotics: the kitti dataset. Int. J. Robot. Res. (IJRR) (2013)

270. Georgakis, G., Mousavian, A., Berg, A.C., Kosecka, J.: Synthesizing training data for object detection in indoor scenes (2017). arXiv e-prints abs:1702.07836

271. Gerig, T., Morel-Forster, A., Blumer, C., Egger, B., Luthi, M., Schoenborn, S., Vetter, T.: Morphable face models—an open framework. In: 2018 13th IEEE International Conference on Automatic Face Gesture Recognition (FG 2018), pp. 75–82 (2018). https://doi.org/10.1109/FG.2018.00021

272. Germain, M., Gregor, K., Murray, I., Larochelle, H.: Made: Masked autoencoder for distribution estimation. In: F. Bach, D. Blei (eds.) Proceedings of the 32nd International Conference on Machine Learning, *Proceedings of Machine Learning Research*, vol. 37, pp. 881–889. PMLR, Lille, France (2015). http://proceedings.mlr.press/v37/germain15.html

273. Gers, F.A., Schmidhuber, E.: Lstm recurrent networks learn simple context-free and context-sensitive languages. IEEE Trans. Neural Netw. **12**(6), 1333–1340 (2001). https://doi.org/10.1109/72.963769

274. Ghiasi, G., Lin, T., Le, Q.V.: NAS-FPN: learning scalable feature pyramid architecture for object detection. In: IEEE Conference on Computer Vision and Pattern Recognition, CVPR 2019, Long Beach, CA, USA, 16–20 June 2019, pp. 7036–7045. Computer Vision Foundation, IEEE (2019). https://doi.org/10.1109/CVPR.2019.00720

275. Girin, L., Leglaive, S., Bie, X., Diard, J., Hueber, T., Alameda-Pineda, X.: Dynamical variational autoencoders: a comprehensive review (2020)

276. Girshick, R., Donahue, J., Darrell, T., Malik, J.: Rich feature hierarchies for accurate object detection and semantic segmentation. In: 2014 IEEE Conference on Computer Vision and Pattern Recognition, pp. 580–587 (2014)

277. Girshick, R.B.: Fast R-CNN. CoRR (2015). arXiv e-prints abs:1504.08083

278. Giuffrida, M.V., Scharr, H., Tsaftaris, S.A.: ARIGAN: synthetic arabidopsis plants using generative adversarial network. CoRR (2017). arXiv e-prints abs:1709.00938

279. Glassner, Y., Gispan, L., Ayash, A., Shohet, T.F.: Closing the gap towards end-to-end autonomous vehicle system. CoRR (2019). arXiv e-prints abs:1901.00114

280. Glorot, X., Bengio, Y.: Understanding the difficulty of training deep feedforward neural networks. In: Y.W. Teh, M. Titterington (eds.) Proceedings of the Thirteenth International Conference on Artificial Intelligence and Statistics, Proceedings of Machine Learning Research, vol. 9, pp. 249–256. PMLR, Chia Laguna Resort, Sardinia, Italy (2010)

281. Glorot, X., Bordes, A., Bengio, Y.: Deep sparse rectifier networks. In: AISTATS **15**, 315–323 (2011)

282. Glorot, X., Bordes, A., Bengio, Y.: Deep sparse rectifier neural networks. Proc. Mach. Learn. Res. **15**, 315 323 (2011)

283. Glorot, X., Bordes, A., Bengio, Y.: Domain adaptation for large-scale sentiment classification: A deep learning approach. In: Proceedings of the 28th International Conference on International Conference on Machine Learning, ICML'11, pp. 513–520. Omnipress, USA (2011)

284. Godard, C., Mac Aodha, O., Brostow, G.J.: Unsupervised monocular depth estimation with left-right consistency. In: CVPR (2017). http://visual.cs.ucl.ac.uk/pubs/monoDepth/

285. Goldman, E., Herzig, R., Eisenschtat, A., Ratzon, O., Levi, I., Goldberger, J., Hassner, T.: Precise detection in densely packed scenes. CoRR (2019). arXiv e-prints abs:1904.00853

286. Gómez-Bombarelli, R., Duvenaud, D.K., Hernández-Lobato, J.M., Aguilera-Iparraguirre, J., Hirzel, T.D., Adams, R.P., Aspuru-Guzik, A.: Automatic chemical design using a data-driven continuous representation of molecules. CoRR (2016). arXiv e-prints abs:1610.02415

287. Gong, M., Xie, Y., Pan, K., Feng, K., Qin, A.K.: A survey on differentially private machine learning [review article]. IEEE Comput. Intell. Mag. **15**(2), 49–64 (2020)

288. Gonzalez-Garcia, A., Weijer, J.v.d., Bengio, Y.: Image-to-image translation for cross-domain disentanglement. In: Proceedings of the 32Nd International Conference on Neural Information Processing Systems, NIPS'18, pp. 1294–1305. Curran Associates Inc., USA (2018)

289. Goodfellow, I., Bengio, Y., Courville, A.: Deep Learning. MIT Press (2016). http://www.deeplearningbook.org

290. Goodfellow, I., Pouget-Abadie, J., Mirza, M., Xu, B., Warde-Farley, D., Ozair, S., Courville, A., Bengio, Y.: Generative adversarial nets. In: Z. Ghahramani, M. Welling, C. Cortes, N.D. Lawrence, K.Q. Weinberger (eds.) Advances in Neural Information Processing Systems 27, pp. 2672–2680. Curran Associates, Inc. (2014)

291. Goodfellow, I.J.: NIPS 2016 tutorial: generative adversarial networks. CoRR (2017). arXiv e-prints abs:1701.00160

292. Goodfellow, I.J., Shlens, J., Szegedy, C.: Explaining and harnessing adversarial examples (2015)

293. Gou, J., Yu, B., Maybank, S.J., Tao, D.: Knowledge distillation: a survey (2021)

294. Goyal, Y., Khot, T., Agrawal, A., Summers-Stay, D., Batra, D., Parikh, D.: Making the v in vqa matter: Elevating the role of image understanding in visual question answering. Int. J. Comput. Vision **127**(4), 398–414 (2019). https://doi.org/10.1007/s11263-018-1116-0

295. Grard, M., Brégier, R., Sella, F., Dellandréa, E., Chen, L.: Object segmentation in depth maps with one user click and a synthetically trained fully convolutional network. In: Ficuciello, F., Ruggiero, F., Finzi, A. (eds.) Human Friendly Robotics, pp. 207–221. Springer International Publishing, Cham (2019)

296. Graves, A., Wayne, G., Danihelka, I.: Neural turing machines. CoRR (2014). arXiv e-prints abs:1410.5401

297. Gretton, A., Borgwardt, K.M., Rasch, M.J., Schölkopf, B., Smola, A.: A kernel two-sample test. J. Mach. Learn. Res. **13**(1), 723–773 (2012)

298. Griffiths, D., Boehm, J.: SynthCity: A large scale synthetic point cloud. arXiv e-prints arXiv:1907.04758 (2019)

299. Grosicki, E., Abed, H.E.: Icdar 2009 handwriting recognition competition. In: 2009 10th International Conference on Document Analysis and Recognition, pp. 1398–1402 (2009). https://doi.org/10.1109/ICDAR.2009.184

300. Gschwandtner, M., Kwitt, R., Uhl, A., Pree, W.: Blensor: Blender sensor simulation toolbox. In: Bebis, G., Boyle, R., Parvin, B., Koracin, D., Wang, S., Kyungnam, K., Benes, B., Moreland, K., Borst, C., DiVerdi, S., Yi-Jen, C., Ming, J. (eds.) Advances in Visual Computing, pp. 199–208. Springer, Berlin, Heidelberg (2011)

301. Gu, X., Cho, K., Ha, J.W., Kim, S.: DialogWAE: Multimodal response generation with conditional wasserstein auto-encoder. In: International Conference on Learning Representations (2019)

302. Guan, J., Li, R., Yu, S., Zhang, X.: Generation of synthetic electronic medical record text. In: 2018 IEEE International Conference on Bioinformatics and Biomedicine (BIBM), pp. 374–380 (2018). https://doi.org/10.1109/BIBM.2018.8621223

303. Gulrajani, I., Ahmed, F., Arjovsky, M., Dumoulin, V., Courville, A.C.: Improved training of wasserstein gans. In: I. Guyon, U.V. Luxburg, S. Bengio, H. Wallach, R. Fergus, S. Vishwanathan, R. Garnett (eds.) Advances in Neural Information Processing Systems 30, pp. 5767–5777. Curran Associates, Inc. (2017)

304. Gulwani, S.: Automating string processing in spreadsheets using input-output examples. SIGPLAN Not. **46**(1), 317–330 (2011). https://doi.org/10.1145/1925844.1926423

305. Guo, J., Lu, S., Cai, H., Zhang, W., Yu, Y., Wang, J.: Long text generation via adversarial training with leaked information. CoRR (2017). arXiv e-prints abs:1709.08624

306. Guo, Y., Zhang, L., Hu, Y., He, X., Gao, J.: Ms-celeb-1m: A dataset and benchmark for large-scale face recognition. CoRR (2016). arXiv e-prints abs:1607.08221

307. Gupta, A., Johnson, J., Fei-Fei, L., Savarese, S., Alahi, A.: Social GAN: socially acceptable trajectories with generative adversarial networks. CoRR (2018). arXiv e-prints abs:1803.10892

308. Gupta, A., Vedaldi, A., Zisserman, A.: Synthetic data for text localisation in natural images. CoRR (2016). arXiv e-prints abs:1604.06646

309. Gupta, O.K., Jarvis, R.A.: Using a virtual world to design a simulation platform for vision and robotic systems. In: Bebis, G., Boyle, R., Parvin, B., Koracin, D., Kuno, Y., Wang, J., Wang, J.X., Wang, J., Pajarola, R., Lindstrom, P., Hinkenjann, A., Encarnação, M.L., Silva, C.T., Coming, D. (eds.) Advances in Visual Computing, pp. 233–242. Springer, Berlin, Heidelberg (2009)

310. Gupta, S., Arbeláez, P.A., Girshick, R.B., Malik, J.: Inferring 3d object pose in RGB-D images. CoRR (2015). arXiv e-prints abs:1502.04652

311. Gupta, S., Arbeláez, P., Girshick, R., Malik, J.: Aligning 3d models to rgb-d images of cluttered scenes. In: 2015 IEEE Conference on Computer Vision and Pattern Recognition (CVPR), pp. 4731–4740 (2015). https://doi.org/10.1109/CVPR.2015.7299105

312. Ha, T., Dang, T.K., Dang, T.T., Truong, T.A., Nguyen, M.T.: Differential privacy in deep learning: An overview. In: 2019 International Conference on Advanced Computing and Applications (ACOMP), pp. 97–102 (2019)

313. Haidar, M.A., Rezagholizadeh, M.: Textkd-gan: Text generation using knowledgedistillation and generative adversarial networks. CoRR (2019). arXiv e-prints abs:1905.01976

314. Han, C., Hayashi, H., Rundo, L., Araki, R., Shimoda, W., Muramatsu, S., Furukawa, Y., Mauri, G., Nakayama, H.: Gan-based synthetic brain mr image generation. In: 2018 IEEE 15th International Symposium on Biomedical Imaging (ISBI 2018), pp. 734–738 (2018). https://doi.org/10.1109/ISBI.2018.8363678

315. Han, C., Kitamura, Y., Kudo, A., Ichinose, A., Rundo, L., Furukawa, Y., Umemoto, K., Nakayama, H., Li, Y.: Synthesizing Diverse Lung Nodules Wherever Massively: 3D Multi-Conditional GAN-based CT Image Augmentation for Object Detection. arXiv e-prints arXiv:1906.04962 (2019)

316. Han, C., Rundo, L., Araki, R., Furukawa, Y., Mauri, G., Nakayama, H., Hayashi, H.: Infinite brain MR images: Pggan-based data augmentation for tumor detection. CoRR (2019). arXiv e-prints abs:1903.12564

317. Han, C., Rundo, L., Araki, R., Nagano, Y., Furukawa, Y., Mauri, G., Nakayama, H., Hayashi, H.: Combining Noise-to-Image and Image-to-Image GANs: Brain MR Image Augmentation for Tumor Detection (2019). arXiv e-prints arXiv:1905.13456

318. Han, X., Zhang, Z., Du, D., Yang, M., Yu, J., Pan, P., Yang, X., Liu, L., Xiong, Z., Cui, S.: Deep reinforcement learning of volume-guided progressive view inpainting for 3d point scene completion from a single depth image (2019). arXiv e-prints abs:1903.04019

319. Handa, A., Patraucean, V., Badrinarayanan, V., Stent, S., Cipolla, R.: Scenenet: Understanding real world indoor scenes with synthetic data. CoRR (2015). arXiv e-prints abs:1511.07041

320. Handa, A., Patraucean, V., Stent, S., Cipolla, R.: Scenenet: An annotated model generator for indoor scene understanding. In: 2016 IEEE International Conference on Robotics and Automation (ICRA), pp. 5737–5743 (2016). https://doi.org/10.1109/ICRA.2016.7487797

321. Handa, A., Whelan, T., McDonald, J., Davison, A.J.: A benchmark for rgb-d visual odometry, 3d reconstruction and slam. In: 2014 IEEE International Conference on Robotics and Automation (ICRA), pp. 1524–1531 (2014). https://doi.org/10.1109/ICRA.2014.6907054

322. Hanson, S.J., Pratt, L.Y.: Comparing biases for minimal network construction with backpropagation. In: D.S. Touretzky (ed.) Advances in Neural Information Processing Systems 1, pp. 177–185. Morgan-Kaufmann (1989)

323. Hart, P.E., Nilsson, N.J., Raphael, B.: A formal basis for the heuristic determination of minimum cost paths. IEEE Trans. Syst. Sci. Cybern. **4**(2), 100–107 (1968). https://doi.org/10.1109/TSSC.1968.300136

324. Hartenfeller, M., Schneider, G.: Enabling future drug discovery by de novo design. Wiley Interdisc. Rev. Comput. Mol. Sci. **1**(5), 742–759 (2011). https://doi.org/10.1002/wcms.49

325. Hattori, H., Boddeti, V.N., Kitani, K., Kanade, T.: Learning scene-specific pedestrian detectors without real data. In: 2015 IEEE Conference on Computer Vision and Pattern Recognition (CVPR), pp. 3819–3827 (2015). https://doi.org/10.1109/CVPR.2015.7299006

326. He, H., Huang, G., Yuan, Y.: Asymmetric valleys: Beyond sharp and flat local minima. CoRR (2019). arXiv e-prints abs:1902.00744
327. He, K., Gkioxari, G., Dollár, P., Girshick, R.: Mask r-cnn. In: 2017 IEEE International Conference on Computer Vision (ICCV), pp. 2980–2988 (2017). https://doi.org/10.1109/ICCV.2017.322
328. He, K., Zhang, X., Ren, S., Sun, J.: Deep residual learning for image recognition. CoRR (2015). arXiv e-prints abs:1512.03385
329. He, K., Zhang, X., Ren, S., Sun, J.: Delving deep into rectifiers: Surpassing human-level performance on imagenet classification. CoRR (2015). arXiv e-prints abs:1502.01852
330. He, K., Zhang, X., Ren, S., Sun, J.: Deep residual learning for image recognition. In: Proc. 2016 CVPR, pp. 770–778 (2016). https://doi.org/10.1109/CVPR.2016.90
331. He, K., Zhang, X., Ren, S., Sun, J.: Identity mappings in deep residual networks. CoRR (2016). arXiv e-prints abs:1603.05027
332. Heaton, J., Witte, J.: Generating synthetic data to test financial strategies and investment products for regulatory compliance. Technical Report, SSRN (2019)
333. Heimann, T., Mountney, P., John, M., Ionasec, R.: Learning without labeling: Domain adaptation for ultrasound transducer localization. In: Mori, K., Sakuma, I., Sato, Y., Barillot, C., Navab, N. (eds.) Medical Image Computing and Computer-Assisted Intervention–MICCAI 2013, pp. 49–56. Springer, Berlin, Heidelberg (2013)
334. Heitz, G., Koller, D.: Learning spatial context: Using stuff to find things. In: Forsyth, D., Torr, P., Zisserman, A. (eds.) Computer Vision–ECCV 2008, pp. 30–43. Springer, Berlin, Heidelberg (2008)
335. Helbing, D., Molnár, P.: Social force model for pedestrian dynamics. Phys. Rev. E **51**, 4282–4286 (1995). https://doi.org/10.1103/PhysRevE.51.4282
336. Hendrikx, M., Meijer, S., Van Der Velden, J., Iosup, A.: Procedural content generation for games: a survey. ACM Trans. Multimedia Comput. Commun. Appl. **9**(1), 1–22 (2013). https://doi.org/10.1145/2422956.2422957
337. Hentati, A.I., Krichen, L., Fourati, M., Fourati, L.C.: Simulation tools, environments and frameworks for uav systems performance analysis. In: 2018 14th International Wireless Communications Mobile Computing Conference (IWCMC), pp. 1495–1500 (2018). https://doi.org/10.1109/IWCMC.2018.8450505
338. Hernandez-Juarez, D., Schneider, L., Espinosa, A., Vazquez, D., Lopez, A.M., Franke, U., Pollefeys, M., Moure, J.C.: Slanted stixels: Representing san francisco's steepest streets. In: British Machine Vision Conference (BMVC), 2017 (2017)
339. Hertz, J., Palmer, R.G., Krogh, A.S.: Introduction to the Theory of Neural Computation, 1st edn. Perseus Publishing (1991)
340. Heusel, M., Ramsauer, H., Unterthiner, T., Nessler, B., Klambauer, G., Hochreiter, S.: Gans trained by a two time-scale update rule converge to a nash equilibrium. CoRR (2017). arXiv e-prints abs:1706.08500
341. Hinterstoisser, S., Lepetit, V., Wohlhart, P., Konolige, K.: On pre-trained image features and synthetic images for deep learning. In: Leal-Taixé, L., Roth, S. (eds.) Computer Vision–ECCV 2018 Workshops, pp. 682–697. Springer International Publishing, Cham (2019)
342. Hinterstoisser, S., Pauly, O., Heibel, H., Marek, M., Bokeloh, M.: An Annotation Saved is an Annotation Earned: Using Fully Synthetic Training for Object Instance Detection (2019). arXiv e-prints arXiv:1902.09967
343. Hinton, G.: What is wrong with convolutional neural nets? Talk at the Brain & Cognitive Sciences—Fall Colloquium Series (2014). https://www.youtube.com/watch?v=rTawFwUvnLE
344. Hinton, G., Salakhutdinov, R.: Reducing the dimensionality of data with neural networks. Science **313**(5786), 504–507 (2006)
345. Hinton, G., Vinyals, O., Dean, J.: Distilling the knowledge in a neural network. In: NIPS Deep Learning and Representation Learning Workshop (2015)
346. Hinton, G.E., Deng, L., Yu, D., Dahl, G.E., Mohamed, A., Jaitly, N., Senior, A., Vanhoucke, V., Nguyen, P., Sainath, T.N., Kingsbury, B.: Deep neural networks for acoustic modeling in

speech recognition: The shared views of four research groups. IEEE Signal Process. Mag. **29**(6), 82–97 (2012)

347. Hinton, G.E., Osindero, S., Teh, Y.W.: A fast learning algorithm for deep belief nets. Neural Comput. **18**(7), 1527–1554 (2006)

348. Hinton, G.E., Osindero, S., Teh, Y.W.: A fast learning algorithm for deep belief nets. Neural Comput. **18**(7), 1527–1554 (2006). https://doi.org/10.1162/neco.2006.18.7.1527

349. Hinton, G.E., Sabour, S., Frosst, N.: Matrix capsules with em routing. In: ICLR (2018)

350. Hochreiter, S., Schmidhuber, J.: Long short-term memory. Neural Comput. **9**(8), 1735–1780 (1997). https://doi.org/10.1162/neco.1997.9.8.1735

351. Hodan, T., Haluza, P., Obdržálek, S., Matas, J., Lourakis, M.I.A., Zabulis, X.: T-LESS: an RGB-D dataset for 6d pose estimation of texture-less objects. CoRR (2017). arXiv e-prints abs:1701.05498

352. Hoffman, J., Tzeng, E., Park, T., Zhu, J., Isola, P., Saenko, K., Efros, A.A., Darrell, T.: Cycada: Cycle-consistent adversarial domain adaptation. CoRR (2017). arXiv e-prints abs:1711.03213

353. Hoffman, J., Wang, D., Yu, F., Darrell, T.: Fcns in the wild: Pixel-level adversarial and constraint-based adaptation. CoRR (2016). arXiv e-prints abs:1612.02649

354. Hölldobler, S., Kalinke, Y., Lehmann, H.: Designing a counter: Another case study of dynamics and activation landscapes in recurrent networks. In: Brewka, G., Habel, C., Nebel, B. (eds.) KI-97: Advances in Artificial Intelligence, pp. 313–324. Springer, Berlin, Heidelberg (1997)

355. Hong, W., Wang, Z., Yang, M., Yuan, J.: Conditional generative adversarial network for structured domain adaptation. In: 2018 IEEE/CVF Conference on Computer Vision and Pattern Recognition, pp. 1335–1344 (2018). https://doi.org/10.1109/CVPR.2018.00145

356. Hoppen, P., Knieriemen, T., von Puttkamer, E.: Laser-radar based mapping and navigation for an autonomous mobile robot. In: Proceedings of the IEEE International Conference on Robotics and Automation, pp. 948–953 vol. 2 (1990). https://doi.org/10.1109/ROBOT.1990.126113

357. Hou, L., Agarwal, A., Samaras, D., Kurç, T.M., Gupta, R.R., Saltz, J.H.: Unsupervised histopathology image synthesis. CoRR (2017). arXiv e-prints abs:1712.05021

358. Howard, A.G., Zhu, M., Chen, B., Kalenichenko, D., Wang, W., Weyand, T., Andreetto, M., Adam, H.: Mobilenets: Efficient convolutional neural networks for mobile vision applications. CoRR (2017). arXiv e-prints abs:1704.04861

359. Howard, J., Ruder, S.: Fine-tuned language models for text classification. CoRR (2018). arXiv e-prints abs:1801.06146

360. Hu, G., Peng, X., Yang, Y., Hospedales, T.M., Verbeek, J.: Frankenstein: Learning deep face representations using small data. IEEE Trans. Image Process. **27**(1), 293–303 (2018). https://doi.org/10.1109/TIP.2017.2756450

361. Hu, G., Yan, F., Chan, C.H., Deng, W., Christmas, W., Kittler, J., Robertson, N.: Face recognition using a unified 3d morphable model. In: Computer Vision—ECCV 2016: 14th European Conference, Amsterdam, The Netherlands, October 11–14, 2016, Proceedings, Part VIII, Lecture Notes in Computer Science, vol. 9912, pp. 73–89 (2016). https://doi.org/10.1007/978-3-319-46484-8_5

362. Hu, R., Andreas, J., Rohrbach, M., Darrell, T., Saenko, K.: Learning to reason: End-to-end module networks for visual question answering. CoRR (2017). arXiv e-prints abs:1704.05526

363. Hu, R., Rohrbach, M., Andreas, J., Darrell, T., Saenko, K.: Modeling relationships in referential expressions with compositional modular networks. CoRR (2016). arXiv e-prints abs:1611.09978

364. Huang, B., Bayazit, D., Ullman, D., Gopalan, N., Tellex, S.: Flight, camera, action! using natural language and mixed reality to control a drone. In: Proceedings of the IEEE International Conference on Robotics and Automation (2019)

365. Huang, D.: How much did alphago zero cost? (2019)

366. Huang, G.B., Ramesh, M., Berg, T., Learned-Miller, E.: Labeled faces in the wild: A database for studying face recognition in unconstrained environments. Technical Report No. 07-49, University of Massachusetts, Amherst (2007)

367. Huang, H., Guo, H., Ding, Z., Chen, Y., Wu, X.: 3d virtual modeling for crowd analysis. In: 2015 IEEE International Conference on Information and Automation, pp. 2949–2954 (2015). https://doi.org/10.1109/ICInfA.2015.7279793

368. Huang, H., Huang, Q.X., Krähenbühl, P.: Domain transfer through deep activation matching. In: ECCV (2018)

369. Huang, R., Zhang, S., Li, T., He, R.: Beyond face rotation: Global and local perception gan for photorealistic and identity preserving frontal view synthesis. In: 2017 IEEE International Conference on Computer Vision (ICCV), pp. 2458–2467 (2017). https://doi.org/10.1109/ICCV.2017.267

370. Huang, W.R., Emam, Z., Goldblum, M., Fowl, L., Terry, J.K., Huang, F., Goldstein, T.: Understanding generalization through visualizations. CoRR (2019). arXiv e-prints abs:1906.03291

371. Huang, X., Belongie, S.: Arbitrary style transfer in real-time with adaptive instance normalization. In: ICCV (2017)

372. Huang, X., Cheng, X., Geng, Q., Cao, B., Zhou, D., Wang, P., Lin, Y., Yang, R.: The apolloscape dataset for autonomous driving. CoRR (2018). arXiv e-prints abs:1803.06184

373. Huang, X., Liu, M., Belongie, S.J., Kautz, J.: Multimodal unsupervised image-to-image translation. CoRR (2018). arXiv e-prints abs:1804.04732

374. Huang, X., Liu, M.Y., Belongie, S., Kautz, J.: Multimodal unsupervised image-to-image translation. In: ECCV (2018)

375. Huang, Z., Heng, W., Zhou, S.: Learning to paint with model-based deep reinforcement learning (2019). arXiv e-prints abs:1903.04411

376. Hubel, D.H., Wiesel, T.: Receptive fields, binocular interaction, and functional architecture in the cat's visual cortex. J. Physiol. (London) **160**, 106–154 (1962)

377. Hubel, D.H., Wiesel, T.N.: Receptive fields and functional architecture of monkey striate cortex. J. Physiol. **195**(1), 215–243 (1968)

378. Huber, P., Feng, Z., Christmas, W., Kittler, J., Rätsch, M.: Fitting 3d morphable face models using local features. In: 2015 IEEE International Conference on Image Processing (ICIP), pp. 1195–1199 (2015). https://doi.org/10.1109/ICIP.2015.7350989

379. Huffman, D.A.: Impossible object as nonsense sentences. Mach. Intell. **6**, 295–324 (1971)

380. Hurl, B., Czarnecki, K., Waslander, S.L.: Precise synthetic image and lidar (presil) dataset for autonomous vehicle perception (2019). arXiv e-prints abs:1905.00160

381. Hutchins, J.: The georgetown-ibm experiment demonstrated in january 1954. In: Machine translation: from real users to research. 6th Conference of the Association for Machine Translation in the Americas, AMTA 2004, pp. 102–114. Springer (2004)

382. Iandola, F.N., Moskewicz, M.W., Ashraf, K., Han, S., Dally, W.J., Keutzer, K.: Squeezenet: Alexnet-level accuracy with 50x fewer parameters and <1mb model size. CoRR (2016). arXiv e-prints abs:1602.07360

383. Idrees, H., Tayyab, M., Athrey, K., Zhang, D., Al-Máadeed, S., Rajpoot, N.M., Shah, M.: Composition loss for counting, density map estimation and localization in dense crowds. CoRR (2018). arXiv e-prints abs:1808.01050

384. Ilg, E., Mayer, N., Saikia, T., Keuper, M., Dosovitskiy, A., Brox, T.: Flownet 2.0: Evolution of optical flow estimation with deep networks. In: IEEE Conference on Computer Vision and Pattern Recognition (CVPR) (2017)

385. Ilg, E., Saikia, T., Keuper, M., Brox, T.: Occlusions, motion and depth boundaries with a generic network for disparity, optical flow or scene flow estimation. In: European Conference on Computer Vision (ECCV) (2018)

386. Inoue, T., Choudhury, S., De Magistris, G., Dasgupta, S.: Transfer learning from synthetic to real images using variational autoencoders for precise position detection. In: 2018 25th IEEE International Conference on Image Processing (ICIP), pp. 2725–2729 (2018). https://doi.org/10.1109/ICIP.2018.8451064

387. Ioffe, S., Szegedy, C.: Batch normalization: Accelerating deep network training by reducing internal covariate shift. CoRR (2015). arXiv e-prints abs:1502.03167
388. Ionescu, C., Papava, D., Olaru, V., Sminchisescu, C.: Human3.6m: Large scale datasets and predictive methods for 3d human sensing in natural environments. IEEE Trans. Pattern Anal. Mach. Intell. **36**(7), 1325–1339 (2014). https://doi.org/10.1109/TPAMI.2013.248
389. Isola, P., Zhu, J.Y., Zhou, T., Efros, A.A.: Image-to-image translation with conditional adversarial networks. In: 2017 IEEE Conference on Computer Vision and Pattern Recognition (CVPR) pp. 5967–5976 (2017)
390. Itseez: Open source computer vision library (2015). https://github.com/itseez/opencv
391. Izmailov, P., Podoprikhin, D., Garipov, T., Vetrov, D.P., Wilson, A.G.: Averaging weights leads to wider optima and better generalization. CoRR (2018). arXiv e-prints abs:1803.05407
392. Jaderberg, M., Simonyan, K., Vedaldi, A., Zisserman, A.: Synthetic data and artificial neural networks for natural scene text recognition. CoRR (2014). arXiv e-prints abs:1406.2227
393. Jaderberg, M., Simonyan, K., Vedaldi, A., Zisserman, A.: Reading text in the wild with convolutional neural networks. Int. J. Comput. Vision **116**(1), 1–20 (2016). https://doi.org/10.1007/s11263-015-0823-z
394. Jakobi, N., Husbands, P., Harvey, I.: Noise and the reality gap: The use of simulation in evolutionary robotics. In: Morán, F., Moreno, A., Merelo, J.J., Chacón, P. (eds.) Advances in Artificial Life, pp. 704–720. Springer, Berlin, Heidelberg (1995)
395. Jalal, M., Spjut, J., Boudaoud, B., Betke, M.: Sidod: A synthetic image dataset for 3d object pose recognition with distractors. In: 2019 IEEE/CVF Conference on Computer Vision and Pattern Recognition Workshops (CVPRW), pp. 475–477 (2019)
396. James, S., Wohlhart, P., Kalakrishnan, M., Kalashnikov, D., Irpan, A., Ibarz, J., Levine, S., Hadsell, R., Bousmalis, K.: Sim-to-real via sim-to-sim: Data-efficient robotic grasping via randomized-to-canonical adaptation networks (2018). arXiv e-prints abs:1812.07252
397. Jaques, N., Gu, S., Turner, R.E., Eck, D.: Tuning recurrent neural networks with reinforcement learning. CoRR (2016). arXiv e-prints abs:1611.02796
398. Ji, B., Chen, T.: Generative adversarial network for handwritten text. CoRR (2019). arXiv e-prints abs:1907.11845
399. Ji, X., Henriques, J.F., Vedaldi, A.: Invariant information distillation for unsupervised image segmentation and clustering. CoRR(2018). arXiv e-prints abs:1807.06653
400. Ji, Z., Lipton, Z.C., Elkan, C.: Differential privacy and machine learning: a survey and review. CoRR (2014). arXiv e-prints abs:1412.7584
401. Jin, C., Jung, W., Joo, S., Park, E., Ahn, Y.S., Han, I.H., Lee, J., Cui, X.: Deep CT to MR synthesis using paired and unpaired data. CoRR (2018). arXiv e-prints abs:1805.10790
402. Jin, C., Rinard, M.: Learning from context-agnostic synthetic data (2020)
403. Jin, W., Barzilay, R., Jaakkola, T.: Junction Tree Variational Autoencoder for Molecular Graph Generation (2018). arXiv e-prints arXiv:1802.04364
404. Jo, J., Koo, H.I., Soh, J.W., Cho, N.I.: Handwritten text segmentation via end-to-end learning of convolutional neural network. CoRR (2019). arXiv e-prints abs:1906.05229
405. John, V., Mou, L., Bahuleyan, H., Vechtomova, O.: Disentangled representation learning for non-parallel text style transfer. In: Proceedings of the 57th Annual Meeting of the Association for Computational Linguistics, pp. 424–434. Association for Computational Linguistics, Florence, Italy (2019)
406. Johnson, G.R., Donovan-Maiye, R.M., Maleckar, M.M.: Generative Modeling with Conditional Autoencoders: Building an Integrated Cell. arXiv e-prints arXiv:1705.00092 (2017)
407. Johnson, J., Alahi, A., Fei-Fei, L.: Perceptual losses for real-time style transfer and super-resolution. In: ECCV (2016)
408. Johnson, J., Hariharan, B., van der Maaten, L., Fei-Fei, L., Zitnick, C.L., Girshick, R.B.: Clevr: A diagnostic dataset for compositional language and elementary visual reasoning. In: 2017 IEEE Conference on Computer Vision and Pattern Recognition (CVPR) pp. 1988–1997 (2017)
409. Johnson, J., Hariharan, B., van der Maaten, L., Hoffman, J., Li, F.F., Lawrence Zitnick, C., Girshick, R.: Inferring and executing programs for visual reasoning. In: ICCV, pp. 3008–3017 (2017). https://doi.org/10.1109/ICCV.2017.325

410. Johnson, J., Krishna, R., Stark, M., Li, L.J., Shamma, D.A., Bernstein, M.S., Fei-Fei, L.: Image retrieval using scene graphs. In: 2015 IEEE Conference on Computer Vision and Pattern Recognition (CVPR) pp. 3668–3678 (2015)

411. Johnson, S., Everingham, M.: Clustered pose and nonlinear appearance models for human pose estimation. In: Proceedings of the British Machine Vision Conference, pp. 12.1–12.11. BMVA Press (2010). https://doi.org/10.5244/C.24.12

412. Johnson-Roberson, M., Barto, C., Mehta, R., Sridhar, S.N., Rosaen, K., Vasudevan, R.: Driving in the matrix: Can virtual worlds replace human-generated annotations for real world tasks? In: 2017 IEEE International Conference on Robotics and Automation (ICRA), pp. 746–753 (2017). https://doi.org/10.1109/ICRA.2017.7989092

413. Joo, D., Kim, D., Kim, J.: Generating a fusion image: One's identity and another's shape. In: 2018 IEEE/CVF Conference on Computer Vision and Pattern Recognition, pp. 1635–1643 (2018). https://doi.org/10.1109/CVPR.2018.00176

414. Jorg, K.W., Palmes, M., von Puttkamer, E.: Pilot specific realtime world modeling for an autonomous mobile robot using heterogeneous sensor information. In: Fifth Euromicro Workshop on Real-Time Systems, pp. 168–173 (1993). https://doi.org/10.1109/EMWRT.1993.639087

415. Joulin, A., Mikolov, T.: Inferring algorithmic patterns with stack-augmented recurrent nets. In: Proceedings of the 28th International Conference on Neural Information Processing Systems—Volume 1, NIPS'15, pp. 190–198. MIT Press, Cambridge, MA, USA (2015)

416. Joulin, A., Mikolov, T.: Inferring algorithmic patterns with stack-augmented recurrent nets. CoRR (2015). arXiv e-prints abs:1503.01007

417. Justesen, N., Bontrager, P., Togelius, J., Risi, S.: Deep learning for video game playing. IEEE Trans. Games 1–20, (2019). https://doi.org/10.1109/TG.2019.2896986

418. Justesen, N., Torrado, R.R., Bontrager, P., Khalifa, A., Togelius, J., Risi, S.: Procedural level generation improves generality of deep reinforcement learning. CoRR (2018). arXiv e-prints abs:1806.10729

419. Kadurin, A., Aliper, A., Kazennov, A., Mamoshina, P., Vanhaelen, Q., Khrabrov, K., Zhavoronkov, A.: The cornucopia of meaningful leads: Applying deep adversarial autoencoders for new molecule development in oncology. Oncotarget 8, (2016)

420. Kadurin, A., Nikolenko, S.I., Khrabrov, K., Aliper, A., Zhavoronkov, A.: drugan: An advanced generative adversarial autoencoder model for de novo generation of new molecules with desired molecular properties in silico. Mol. Pharm. 14(9), 3098–3104 (2017)

421. Kaiser, L., Sutskever, I.: Neural gpus learn algorithms. CoRR (2016). arXiv e-prints abs:1511.08228

422. Kalashnikov, D., Irpan, A., Pastor, P., Ibarz, J., Herzog, A., Jang, E., Quillen, D., Holly, E., Kalakrishnan, M., Vanhoucke, V., Levine, S.: Scalable deep reinforcement learning for vision-based robotic manipulation. In: A. Billard, A. Dragan, J. Peters, J. Morimoto (eds.) Proceedings of The 2nd Conference on Robot Learning. Proceedings of Machine Learning Research, vol. 87, pp. 651–673. PMLR (2018)

423. Kalchbrenner, N., Grefenstette, E., Blunsom, P.: A convolutional neural network for modelling sentences. CoRR (2014). arXiv e-prints abs:1404.2188

424. Kallweit, S., Müller, T., Mcwilliams, B., Gross, M., Novák, J.: Deep scattering: Rendering atmospheric clouds with radiance-predicting neural networks. ACM Trans. Graph. 36(6) (2017). https://doi.org/10.1145/3130800.3130880

425. Kamnitsas, K., Baumgartner, C., Ledig, C., Newcombe, V., Simpson, J., Kane, A., Menon, D., Nori, A., Criminisi, A., Rueckert, D., Glocker, B.: Unsupervised domain adaptation in brain lesion segmentation with adversarial networks. In: Niethammer, M., Styner, M., Aylward, S., Zhu, H., Oguz, I., Yap, P.T., Shen, D. (eds.) Information Processing in Medical Imaging, pp. 597–609. Springer International Publishing, Cham (2017)

426. Kamnitsas, K., Ledig, C., Newcombe, V.F., Simpson, J.P., Kane, A.D., Menon, D.K., Rueckert, D., Glocker, B.: Efficient multi-scale 3d cnn with fully connected crf for accurate brain lesion segmentation. Med. Image Anal. 36, 61–78 (2017). https://doi.org/10.1016/j.media.2016.10.004

427. Kan, N., Kondo, N., Chinsatit, W., Saitoh, T.: Effectiveness of data augmentation for cnn-based pupil center point detection. In: 2018 57th Annual Conference of the Society of Instrument and Control Engineers of Japan (SICE) pp. 41–46 (2018)

428. Kanellakis, C., Nikolakopoulos, G.: Survey on computer vision for uavs: Current developments and trends. J. Intell. Robot. Syst. (2017). https://doi.org/10.1007/s10846-017-0483-z

429. Kaneva, B., Torralba, A., Freeman, W.T.: Evaluation of image features using a photorealistic virtual world. In: 2011 International Conference on Computer Vision, pp. 2282–2289 (2011). https://doi.org/10.1109/ICCV.2011.6126508

430. Kang, Y., Yin, H., Berger, C.: Test your self-driving algorithm: an overview of publicly available driving datasets and virtual testing environments. IEEE Trans. Intell. Veh. **4**, 171–185 (2019)

431. Kannala, J., Brandt, S.S.: A generic camera model and calibration method for conventional, wide-angle, and fish-eye lenses. IEEE Trans. Pattern Anal. Mach. Intell. **28**(8), 1335–1340 (2006)

432. Kant, N.: Recent advances in neural program synthesis. CoRR (2018). arXiv e-prints abs:1802.02353

433. Kar, A., Prakash, A., Liu, M., Cameracci, E., Yuan, J., Rusiniak, M., Acuna, D., Torralba, A., Fidler, S.: Meta-sim: Learning to generate synthetic datasets. CoRR (2019). arXiv e-prints abs:1904.11621

434. Karis, B.: Real shading in unreal engine 4. Technical Report, Epic Games (2013). http://blog.selfshadow.com/publications/s2013-shading-course/karis/s2013_pbs_epic_notes_v2.pdf

435. Karras, T., Aila, T., Laine, S., Lehtinen, J.: Progressive growing of gans for improved quality, stability, and variation. CoRR (2018). arXiv e-prints abs:1710.10196

436. Karras, T., Aittala, M., Hellsten, J., Laine, S., Lehtinen, J., Aila, T.: Training generative adversarial networks with limited data (2020)

437. Karras, T., Laine, S., Aila, T.: A style-based generator architecture for generative adversarial networks. CoRR (2018). arXiv e-prints abs:1812.04948

438. Karras, T., Laine, S., Aila, T.: A style-based generator architecture for generative adversarial networks. In: Proceedings of the IEEE/CVF Conference on Computer Vision and Pattern Recognition (CVPR) (2019)

439. Karras, T., Laine, S., Aittala, M., Hellsten, J., Lehtinen, J., Aila, T.: Analyzing and improving the image quality of stylegan. In: IEEE/CVF Conference on Computer Vision and Pattern Recognition (CVPR) (2020)

440. Karras, T., Laine, S., Aittala, M., Hellsten, J., Lehtinen, J., Aila, T.: Analyzing and improving the image quality of StyleGAN. In: Proc. CVPR (2020)

441. Kartoun, U.: A methodology to generate virtual patient repositories. CoRR (2016). arXiv e-prints abs:1608.00570

442. Karttunen, J., Kanervisto, A., Hautamäki, V., Kyrki, V.: From video game to real robot: The transfer between action spaces (2019). arXiv e-prints abs:1905.00741

443. Kempka, M., Wydmuch, M., Runc, G., Toczek, J., Jaśkowski, W.: ViZDoom: A Doom-based AI research platform for visual reinforcement learning. In: IEEE Conference on Computational Intelligence and Games, pp. 341–348. IEEE, Santorini, Greece (2016)

444. Kendall, A., Gal, Y., Cipolla, R.: Multi-task learning using uncertainty to weigh losses for scene geometry and semantics. CoRR (2017). arXiv e-prints abs:1705.07115

445. Kendall, K.: A database of computer attacks for the evaluation of intrusion detection systems. M.Sc. Thesis, Massachusetts Institute of Technology (1999)

446. Keskar, N.S., Mudigere, D., Nocedal, J., Smelyanskiy, M., Tang, P.T.P.: On large-batch training for deep learning: Generalization gap and sharp minima. CoRR (2016). arXiv e-prints abs:1609.04836

447. Keskar, N.S., Socher, R.: Improving generalization performance by switching from adam to SGD. CoRR (2017). arXiv e-prints abs:1712.07628

448. Khadka, A.R., Oghaz, M.M., Matta, W., Cosentino, M., Remagnino, P., Argyriou, V.: Learning how to analyse crowd behaviour using synthetic data. In: Proceedings of the 32Nd International Conference on Computer Animation and Social Agents, CASA '19, pp. 11–14. ACM, New York, NY, USA (2019). https://doi.org/10.1145/3328756.3328773

449. Khan, S., Phan, B., Salay, R., Czarnecki, K.: Procsy: Procedural synthetic dataset generation towards influence factor studies of semantic segmentation networks. In: The IEEE Conference on Computer Vision and Pattern Recognition (CVPR) Workshops (2019)

450. Khirodkar, R., Yoo, D., Kitani, K.M.: Vadra: Visual adversarial domain randomization and augmentation (2018). arXiv e-prints arXiv:1812.00491

451. Khodabandeh, M., Joze, H.R.V., Zharkov, I., Pradeep, V.: Diy human action dataset generation. In: 2018 IEEE/CVF Conference on Computer Vision and Pattern Recognition Workshops (CVPRW), pp. 1529–152910 (2018). https://doi.org/10.1109/CVPRW.2018.00194

452. Killian, T.W., Goodwin, J.A., Brown, O.M., Son, S.H.: Kernelized capsule networks (2019). arXiv e-prints arXiv:1906.03164

453. Kim, K., Cheon, Y., Hong, S., Roh, B., Park, M.: PVANET: deep but lightweight neural networks for real-time object detection. CoRR (2016). arXiv e-prints abs:1608.08021

454. Kingma, D.P., Ba, J.: Adam: A method for stochastic optimization. In: Y. Bengio, Y. LeCun (eds.) 3rd International Conference on Learning Representations, ICLR 2015, San Diego, CA, USA, 7–9 May 2015, Conference Track Proceedings (2015)

455. Kingma, D.P., Dhariwal, P.: Glow: Generative flow with invertible 1x1 convolutions. In: S. Bengio, H. Wallach, H. Larochelle, K. Grauman, N. Cesa-Bianchi, R. Garnett (eds.) Advances in Neural Information Processing Systems, vol. 31, pp. 10215–10224. Curran Associates, Inc. (2018)

456. Kingma, D.P., Salimans, T., Jozefowicz, R., Chen, X., Sutskever, I., Welling, M.: Improved variational inference with inverse autoregressive flow. In: D. Lee, M. Sugiyama, U. Luxburg, I. Guyon, R. Garnett (eds.) Advances in Neural Information Processing Systems, vol. 29, pp. 4743–4751. Curran Associates, Inc. (2016)

457. Kingma, D.P., Welling, M.: Auto-Encoding Variational Bayes. In: 2nd International Conference on Learning Representations, ICLR 2014, Banff, AB, Canada, 14–16 April 2014, Conference Track Proceedings (2014)

458. Kingma, D.P., Welling, M.: An introduction to variational autoencoders. CoRR (2019). arXiv e-prints abs:1906.02691

459. Kiran, M., Richmond, P., Holcombe, M., Chin, L.S., Worth, D., Greenough, C.: Flame: Simulating large populations of agents on parallel hardware architectures. In: Proceedings of the 9th International Conference on Autonomous Agents and Multiagent Systems: Volume 1—Volume 1, AAMAS '10, pp. 1633–1636. International Foundation for Autonomous Agents and Multiagent Systems, Richland, SC (2010)

460. Klare, B.F., Klein, B., Taborsky, E., Blanton, A., Cheney, J., Allen, K., Grother, P., Mah, A., Burge, M., Jain, A.K.: Pushing the frontiers of unconstrained face detection and recognition: Iarpa janus benchmark a. In: 2015 IEEE Conference on Computer Vision and Pattern Recognition (CVPR), pp. 1931–1939 (2015). https://doi.org/10.1109/CVPR.2015.7298803

461. Kobayashi, S.: Contextual augmentation: Data augmentation by words with paradigmatic relations. In: Proceedings of the 2018 Conference of the North American Chapter of the Association for Computational Linguistics: Human Language Technologies, Volume 2 (Short Papers), pp. 452–457. Association for Computational Linguistics, New Orleans, Louisiana (2018). https://doi.org/10.18653/v1/N18-2072

462. Kobyzev, I., Prince, S., Brubaker, M.: Normalizing flows: an introduction and review of current methods. IEEE Trans. Patt. Anal. Mach. Intell. (2020). https://doi.org/10.1109/TPAMI.2020.2992934

463. Kocsis, L., Szepesvári, C.: Bandit based monte-carlo planning. In: Fürnkranz, J., Scheffer, T., Spiliopoulou, M. (eds.) Machine Learning: ECML 2006, pp. 282–293. Springer, Berlin, Heidelberg (2006)

464. Koenig, N., Howard, A.: Design and use paradigms for gazebo, an open-source multi-robot simulator. In: 2004 IEEE/RSJ International Conference on Intelligent Robots and Systems

(IROS) (IEEE Cat. No.04CH37566), vol. 3, pp. 2149–2154 vol. 3 (2004). https://doi.org/10.1109/IROS.2004.1389727

465. Kohonen, T.: Self-Organized Formation of Topologically Correct Feature Maps, pp. 509–521. MIT Press, Cambridge, MA, USA (1988)

466. Kolve, E., Mottaghi, R., Gordon, D., Zhu, Y., Gupta, A., Farhadi, A.: AI2-THOR: an interactive 3d environment for visual AI. CoRR (2017). arXiv e-prints abs:1712.05474

467. Korakakis, M., Mylonas, P., Spyrou, E.: A short survey on modern virtual environments that utilize ai and synthetic data. In: MCIS 2018 Proceedings (2018)

468. Koren, Y.: Tutorial on recent progress in collaborative filtering. In: Proceedings of the 2008 ACM Conference on Recommender Systems, RecSys '08, pp. 333–334. Association for Computing Machinery, New York, NY, USA (2008). https://doi.org/10.1145/1454008.1454067

469. Korkinof, D., Rijken, T., O'Neill, M., Yearsley, J., Harvey, H., Glocker, B.: High-resolution mammogram synthesis using progressive generative adversarial networks. CoRR (2018). arXiv e-prints abs:1807.03401

470. Kortylewski, A., Egger, B., Schneider, A., Gerig, T., Morel-Forster, A., Vetter, T.: Empirically analyzing the effect of dataset biases on deep face recognition systems. CoRR (2017). arXiv e-prints abs:1712.01619

471. Kortylewski, A., Egger, B., Schneider, A., Gerig, T., Morel-Forster, A., Vetter, T.: Analyzing and reducing the damage of dataset bias to face recognition with synthetic data. In: The IEEE Conference on Computer Vision and Pattern Recognition (CVPR) Workshops (2019)

472. Kortylewski, A., Schneider, A., Gerig, T., Blumer, C., Egger, B., Reyneke, C., Morel-Forster, A., Vetter, T.: Priming deep neural networks with synthetic faces for enhanced performance. CoRR (2018). arXiv e-prints abs:1811.08565

473. Krasin, I., Duerig, T., Alldrin, N., Ferrari, V., Abu-El-Haija, S., Kuznetsova, A., Rom, H., Uijlings, J., Popov, S., Kamali, S., Malloci, M., Pont-Tuset, J., Veit, A., Belongie, S., Gomes, V., Gupta, A., Sun, C., Chechik, G., Cai, D., Feng, Z., Narayanan, D., Murphy, K.: Openimages: A public dataset for large-scale multi-label and multi-class image classification (2017). Dataset available from https://storage.googleapis.com/openimages/web/index.html

474. Krishna, R., Zhu, Y., Groth, O., Johnson, J., Hata, K., Kravitz, J., Chen, S., Kalantidis, Y., Li, L.J., Shamma, D.A., Bernstein, M.S., Fei-Fei, L.: Visual genome: Connecting language and vision using crowdsourced dense image annotations. Int. J. Comput. Vision **123**(1), 32–73 (2017). https://doi.org/10.1007/s11263-016-0981-7

475. Krishnan, P., Jawahar, C.V.: Hwnet v2: an efficient word image representation for handwritten documents. Int. J. Doc. Anal. Recogn. (IJDAR) (2019). https://doi.org/10.1007/s10032-019-00336-x

476. Krishnan, S., Boroujerdian, B., Fu, W., Faust, A., Reddi, V.J.: Air learning: An AI research platform for algorithm-hardware benchmarking of autonomous aerial robots. CoRR (2019). arXiv e-prints abs:1906.00421

477. Krizhevsky, A., Sutskever, I., Hinton, G.E.: Imagenet classification with deep convolutional neural networks. In: Proceedings of the 25th International Conference on Neural Information Processing Systems—Volume 1, NIPS'12, pp. 1097–1105. Curran Associates Inc., USA (2012)

478. Kuc, R., Siegel, M.W.: Physically based simulation model for acoustic sensor robot navigation. IEEE Trans. Pattern Anal. Mach. Intell. **PAMI-9**(6), 766–778 (1987). https://doi.org/10.1109/TPAMI.1987.4767983

479. Kuipers, B., Byun, Y.T.: A robot exploration and mapping strategy based on a semantic hierarchy of spatial representations. Rob. Auton. Syst. **8**(1), 47–63 (1991). https://doi.org/10.1016/0921-8890(91)90014-C (Special Issue Toward Learning Robots)

480. Kuipers, B., Froom, R., Lee, W.Y., Pierce, D.: The Semantic Hierarchy in Robot Learning, pp. 141–170. Springer US, Boston, MA (1993). https://doi.org/10.1007/978-1-4615-3184-5_6

481. Kuipers, B.J., Byun, Y.T.: A robust, qualitative method for robot spatial learning. In: Proceedings of the Seventh AAAI National Conference on Artificial Intelligence, AAAI'88, pp. 774–779. AAAI Press (1988)

482. Kumar, A., Sattigeri, P., Wadhawan, K., Karlinsky, L., Feris, R., Freeman, W.T., Wornell, G.: Co-regularized alignment for unsupervised domain adaptation. In: Proceedings of the 32Nd International Conference on Neural Information Processing Systems, NIPS'18, pp. 9367–9378. Curran Associates Inc., USA (2018)

483. Kurach, K., Andrychowicz, M., Sutskever, I.: Neural random access machines. ICLR (2016)

484. Kurakin, A., Goodfellow, I.J., Bengio, S.: Adversarial examples in the physical world. CoRR (2016). arXiv e-prints abs:1607.02533

485. Kurnaz, S., Kaynak, O., Konakoglu, E.: Adaptive neuro-fuzzy inference system based autonomous flight control of unmanned air vehicles. In: Proceedings of the 4th International Symposium on Neural Networks: Advances in Neural Networks, ISNN '07, pp. 14–21. Springer-Verlag, Berlin, Heidelberg (2007). https://doi.org/10.1007/978-3-540-72383-7_3

486. Kusner, M.J., Hernández-Lobato, J.M.: GANS for Sequences of Discrete Elements with the Gumbel-softmax Distribution. arXiv e-prints arXiv:1611.04051 (2016)

487. Kusner, M.J., Paige, B., Hernández-Lobato, J.M.: Grammar Variational Autoencoder (2017). arXiv e-prints arXiv:1703.01925

488. Kuzminykh, D., Polykovskiy, D., Kadurin, A., Zhebrak, A., Baskov, I., Nikolenko, S.I., Shayakhmetov, R., Zhavoronkov, A.: 3d molecular representations based on the wave transform for convolutional neural networks. Mol. Pharm. 15(10), 4378–4385 (2018)

489. Kuznetsova, A., Rom, H., Alldrin, N., Uijlings, J., Krasin, I., Pont-Tuset, J., Kamali, S., Popov, S., Malloci, M., Kolesnikov, A., Duerig, T., Ferrari, V.: The open images dataset v4: Unified image classification, object detection, and visual relationship detection at scale. IJCV (2020)

490. Kuznichov, D., Zvirin, A., Honen, Y., Kimmel, R.: Data augmentation for leaf segmentation and counting tasks in rosette plants (2019). arXiv e-prints arXiv:1903.08583

491. Lagani, V., Karozou, A.D., Gomez-Cabrero, D., Silberberg, G., Tsamardinos, I.: A comparative evaluation of data-merging and meta-analysis methods for reconstructing gene-gene interactions. BMC Bioinform. 17(S-5), S194 (2016). https://doi.org/10.1186/s12859-016-1038-1

492. Lai, K.T., Lin, C.C., Kang, C.Y., Liao, M.E., Chen, M.S.: Vivid: Virtual environment for visual deep learning. In: Proceedings of the 26th ACM International Conference on Multimedia, MM '18, pp. 1356–1359. ACM, New York, NY, USA (2018). https://doi.org/10.1145/3240508.3243653

493. Laminar Research: X-plane flight simulator (2018). https://www.x-plane.com/

494. Lample, G., Ballesteros, M., Subramanian, S., Kawakami, K., Dyer, C.: Neural architectures for named entity recognition. In: Proceedings of the 2016 Conference of the North American Chapter of the Association for Computational Linguistics: Human Language Technologies, pp. 260–270. Association for Computational Linguistics, San Diego, California (2016). https://doi.org/10.18653/v1/N16-1030

495. Lample, G., Chaplot, D.S.: Playing fps games with deep reinforcement learning. In: Proceedings of the Thirty-First AAAI Conference on Artificial Intelligence, AAAI'17, pp. 2140–2146. AAAI Press (2017)

496. Landrum, G.: Rdkit: Open-source cheminformatics software (2016). https://github.com/rdkit/rdkit/releases/tag/Release_2016_09_4

497. Larsen, A.B.L., Sønderby, S.K., Larochelle, H., Winther, O.: Autoencoding beyond pixels using a learned similarity metric. In: M.F. Balcan, K.Q. Weinberger (eds.) Proceedings of The 33rd International Conference on Machine Learning. Proceedings of Machine Learning Research, vol. 48, pp. 1558–1566. PMLR, New York, New York, USA (2016)

498. Lategahn, H., Geiger, A., Kitt, B.: Visual slam for autonomous ground vehicles. In: Proceedings—IEEE International Conference on Robotics and Automation, pp. 1732–1737 (2011). https://doi.org/10.1109/ICRA.2011.5979711

499. Learned-Miller, E., Huang, G.B., RoyChowdhury, A., Li, H., Hua, G.: Labeled Faces in the Wild: A Survey, pp. 189–248. Springer International Publishing, Cham (2016). https://doi.org/10.1007/978-3-319-25958-1_8

500. Learned-Miller, G.B.H.E.: Labeled faces in the wild: Updates and new reporting procedures. Technical Report UM-CS-2014-003, University of Massachusetts, Amherst (2014)

501. Lecun, Y., Bottou, L., Bengio, Y., Haffner, P.: Gradient-based learning applied to document recognition. Proc. IEEE **86**(11), 2278–2324 (1998)

502. Lee, B.Y., Liew, L.H., Cheah, W.S., Wang, Y.C.: Simulation videos for understanding occlusion effects on kernel based object tracking. In: Yeo, S.S., Pan, Y., Lee, Y.S., Chang, H.B. (eds.) Computer Science and its Applications, pp. 139–147. Springer, Netherlands, Dordrecht (2012)

503. Lee, D., Ryan, T., Kim, H.J.: Autonomous landing of a vtol uav on a moving platform using image-based visual servoing. In: 2012 IEEE International Conference on Robotics and Automation, pp. 971–976 (2012). https://doi.org/10.1109/ICRA.2012.6224828

504. Lee, K., Moloney, D.: An evaluation of synthetic data for deep learning stereo depth algorithms. In: Proceedings of the International Conference on Watermarking and Image Processing, ICWIP 2017, pp. 42–45. ACM, New York, NY, USA (2017). https://doi.org/10.1145/3150978.3150982

505. Lee, S.: Natural language generation for electronic health records. In: npj Digital Medicine (2018)

506. Lemley, J., Bazrafkan, S., Corcoran, P.: Smart augmentation learning an optimal data augmentation strategy. IEEE Access **5**, 5858–5869 (2017). https://doi.org/10.1109/ACCESS.2017.2696121

507. Lewis, M., Fan, A.: Generative question answering: Learning to answer the whole question. In: International Conference on Learning Representations (2019)

508. Li, C., Li, Y., Wang, K., Rahaman, M.M., Li, X., Sun, C., Chen, H., Wu, X., Zhang, H., Wang, Q.: A comprehensive review for mrf and crf approaches in pathology image analysis (2020)

509. Li, D., Zhao, D., Zhang, Q., Chen, Y.: Reinforcement learning and deep learning based lateral control for autonomous driving. CoRR (2018). arXiv e-prints abs:1810.12778

510. Li, J., Gadde, R., Ginsburg, B., Lavrukhin, V.: Training neural speech recognition systems with synthetic speech augmentation. CoRR (2018). arXiv e-prints abs:1811.00707

511. Li, K., Li, Y., You, S., Barnes, N.: Photo-realistic simulation of road scene for data-driven methods in bad weather. In: 2017 IEEE International Conference on Computer Vision Workshops (ICCVW), pp. 491–500 (2017). https://doi.org/10.1109/ICCVW.2017.65

512. Li, S.Z., Jain, A.K.: Handbook of Face Recognition, 2nd edn. Springer Publishing Company, Incorporated (2011)

513. Li, W., Pan, C., Zhang, R., Ren, J., Ma, Y., Fang, J., Yan, F., Geng, Q., Huang, X., Gong, H., Xu, W., Wang, G., Manocha, D., Yang, R.: Aads: Augmented autonomous driving simulation using data-driven algorithms. CoRR (2019). arXiv e-prints abs:1901.07849

514. Li, Y., Wei, C., Ma, T.: Towards explaining the regularization effect of initial large learning rate in training neural networks. In: H. Wallach, H. Larochelle, A. Beygelzimer, F. d' Alché-Buc, E. Fox, R. Garnett (eds.) Advances in Neural Information Processing Systems 32, pp. 11674–11685. Curran Associates, Inc. (2019)

515. Lian, Q., Lv, F., Duan, L., Gong, B.: Constructing self-motivated pyramid curriculums for cross-domain semantic segmentation: A non-adversarial approach (2019). arXiv e-prints abs:1908.09547

516. Liang, D., Krishnan, R.G., Hoffman, M.D., Jebara, T.: Variational autoencoders for collaborative filtering. In: Proceedings of the 2018 World Wide Web Conference, WWW '18, p. 689–698. International World Wide Web Conferences Steering Committee, Republic and Canton of Geneva, CHE (2018). https://doi.org/10.1145/3178876.3186150

517. Liang, X., Zhang, H., Xing, E.P.: Generative semantic manipulation with contrasting gan. CoRR (2017). arXiv e-prints abs:1708.00315

518. Liebelt, J., Schmid, C.: Multi-view object class detection with a 3d geometric model. In: 2010 IEEE Computer Society Conference on Computer Vision and Pattern Recognition, pp. 1688–1695 (2010). https://doi.org/10.1109/CVPR.2010.5539836

519. Liebelt, J., Schmid, C., Schertler, K.: Viewpoint-independent object class detection using 3d feature maps. In: 2008 IEEE Conference on Computer Vision and Pattern Recognition, pp. 1–8 (2008). https://doi.org/10.1109/CVPR.2008.4587614

520. Ligett, K., Neel, S., Roth, A., Waggoner, B., Wu, S.Z.: Accuracy first: Selecting a differential privacy level for accuracy constrained erm. In: I. Guyon, U.V. Luxburg, S. Bengio, H. Wallach, R. Fergus, S. Vishwanathan, R. Garnett (eds.) Advances in Neural Information Processing Systems 30, pp. 2566–2576. Curran Associates, Inc. (2017)

521. en Lin, C.: Introduction to motion estimation with optical flow (2019). https://nanonets.com/blog/optical-flow/

522. Lin, M., Chen, Q., Yan, S.: Network in network (2013). http://arxiv.org/abs/1312.4400

523. Lin, T., Dollár, P., Girshick, R., He, K., Hariharan, B., Belongie, S.: Feature pyramid networks for object detection. In: 2017 IEEE Conference on Computer Vision and Pattern Recognition (CVPR), pp. 936–944 (2017). https://doi.org/10.1109/CVPR.2017.106

524. Lin, T., Goyal, P., Girshick, R., He, K., Dollár, P.: Focal loss for dense object detection. In: 2017 IEEE International Conference on Computer Vision (ICCV), pp. 2999–3007 (2017)

525. Lin, T., Maire, M., Belongie, S.J., Bourdev, L.D., Girshick, R.B., Hays, J., Perona, P., Ramanan, D., Dollár, P., Zitnick, C.L.: Microsoft COCO: common objects in context. CoRR (2014). arXiv e-prints abs:1405.0312

526. Lin, Z., Gehring, J., Khalidov, V., Synnaeve, G.: STARDATA: A starcraft AI research dataset. CoRR (2017). arXiv e-prints abs:1708.02139

527. Linjordet, T., Balog, K.: Sanitizing synthetic training data generation for question answering over knowledge graphs. Proceedings of the 2020 ACM SIGIR on International Conference on Theory of Information Retrieval (2020). https://doi.org/10.1145/3409256.3409836

528. Lippmann, R., Haines, J.W., Fried, D.J., Korba, J., Das, K.: Analysis and results of the 1999 darpa off-line intrusion detection evaluation. In: Debar, H., Mé, L., Wu, S.F. (eds.) Recent Advances in Intrusion Detection, pp. 162–182. Springer, Berlin, Heidelberg (2000)

529. Little, J.J., Verri, A.: Analysis of differential and matching methods for optical flow. In: Proceedings of the Workshop on Visual Motion, pp. 173–180 (1989). https://doi.org/10.1109/WVM.1989.47107

530. Little, R.: Statistical analysis of masked data. J. Off. Stat. **9**, 407–426 (1993)

531. Liu, A.H., Liu, Y.C., Yeh, Y.Y., Wang, Y.C.F.: A unified feature disentangler for multi-domain image translation and manipulation. In: Proceedings of the 32Nd International Conference on Neural Information Processing Systems, NIPS'18, pp. 2595–2604. Curran Associates Inc., USA (2018)

532. Liu, C., Zoph, B., Shlens, J., Hua, W., Li, L., Fei-Fei, L., Yuille, A.L., Huang, J., Murphy, K.: Progressive neural architecture search. CoRR (2017). arXiv e-prints abs:1712.00559

533. Liu, F., Wang, S., Ding, D., Yuan, Q., Yao, Z., Pan, Z., Li, H.: Retrieving indoor objects: 2d–3d alignment using single image and interactive roi-based refinement. Comput. Graph. **70**, 108–117 (2018). https://doi.org/10.1016/j.cag.2017.07.029 (CAD/Graphics 2017)

534. Liu, G., Siravuru, A., Prabhakar, S., Veloso, M.M., Kantor, G.: Learning end-to-end multimodal sensor policies for autonomous navigation. CoRR (2017). arXiv e-prints abs:1705.10422

535. Liu, M., Guo, Y., Wang, J.: Indoor scene modeling from a single image using normal inference and edge features. Vis. Comput. **33**(10), 1227–1240 (2017). https://doi.org/10.1007/s00371-016-1348-3

536. Liu, M.Y., Breuel, T., Kautz, J.: Unsupervised image-to-image translation networks. In: I. Guyon, U.V. Luxburg, S. Bengio, H. Wallach, R. Fergus, S. Vishwanathan, R. Garnett (eds.) Advances in Neural Information Processing Systems 30, pp. 700–708. Curran Associates, Inc. (2017)

537. Liu, M.Y., Huang, X., Mallya, A., Karras, T., Aila, T., Lehtinen, J., Kautz, J.: Few-shot unsupervised image-to-image translation. In: IEEE International Conference on Computer Vision (ICCV) (2019)

538. Liu, S., Qi, L., Qin, H., Shi, J., Jia, J.: Path aggregation network for instance segmentation. In: 2018 IEEE/CVF Conference on Computer Vision and Pattern Recognition, pp. 8759–8768 (2018)

539. Liu, W., Anguelov, D., Erhan, D., Szegedy, C., Reed, S., Fu, C.Y., Berg, A.C.: Ssd: Single shot multibox detector. In: Leibe, B., Matas, J., Sebe, N., Welling, M. (eds.) Computer Vision–ECCV 2016, pp. 21–37. Springer International Publishing, Cham (2016)

540. Liu, W., Anguelov, D., Erhan, D., Szegedy, C., Reed, S.E., Fu, C., Berg, A.C.: SSD: single shot multibox detector. CoRR (2015). arXiv e-prints abs:1512.02325

541. Liu, X., He, P., Chen, W., Gao, J.: Improving multi-task deep neural networks via knowledge distillation for natural language understanding. CoRR (2019). arXiv e-prints abs:1904.09482

542. Liu, Z., Luo, P., Wang, X., Tang, X.: Deep learning face attributes in the wild. In: Proceedings of International Conference on Computer Vision (ICCV) (2015)

543. Liu, Z., Zhu, J., Bu, J., Chen, C.: A survey of human pose estimation. J. Vis. Comun. Image Represent. **32**(C), 10–19 (2015). https://doi.org/10.1016/j.jvcir.2015.06.013

544. Loiacono, D., Cardamone, L., Lanzi, P.L.: Simulated car racing championship: Competition software manual. CoRR (2013). arXiv e-prints abs:1304.1672

545. Long, M., Cao, Y., Wang, J., Jordan, M.I.: Learning transferable features with deep adaptation networks. In: Proceedings of the 32nd International Conference on International Conference on Machine Learning—Volume 37, ICML'15, pp. 97–105. JMLR.org (2015)

546. Long, M., Zhu, H., Wang, J., Jordan, M.I.: Unsupervised domain adaptation with residual transfer networks. In: Proceedings of the 30th International Conference on Neural Information Processing Systems, NIPS'16, pp. 136–144. Curran Associates Inc., USA (2016)

547. Long, X., Deng, K., Wang, G., Zhang, Y., Dang, Q., Gao, Y., Shen, H., Ren, J., Han, S., Ding, E., Wen, S.: PP-YOLO: An Effective and Efficient Implementation of Object Detector (2020). arXiv e-prints arXiv:2007.12099

548. López, A.M., Xu, J., Gómez, J.L., Vázquez, D., Ros, G.: From Virtual to Real World Visual Perception Using Domain Adaptation—The DPM as Example, pp. 243–258. Springer International Publishing, Cham (2017). https://doi.org/10.1007/978-3-319-58347-1_13

549. Lopez-Rojas, E., Elmir, A., Axelsson, S.: Paysim: A financial mobile money simulator for fraud detection. In: 28th European Modeling and Simulation Symposium, EMSS, Larnaca, pp. 249–255. Dime University of Genoa (2016)

550. Lopez-Rojas, E.A., Axelsson, S.: Money laundering detection using synthetic data. In: The 27th annual workshop of the Swedish Artificial Intelligence Society (SAIS), vol. 71(5), pp. 33–40. Linköping Electronic Conference Proceedings (2012)

551. Lopez-Rojas, E.A., Axelsson, S.: Multi agent based simulation (mabs) of financial transactions for anti money laundering (aml). In: Nordic Conference on Secure IT Systems. Blekinge Institute of Technology (2012)

552. Lopez-Rojas, E.A., Axelsson, S.: A review of computer simulation for fraud detection research in financial datasets. In: 2016 Future Technologies Conference (FTC), pp. 932–935 (2016). https://doi.org/10.1109/FTC.2016.7821715

553. Lopez-Rojas, E.A., Gorton, D., Axelsson, S.: Retsim: A shoestore agent-based simulation for fraud detection. In: 25th European Modeling and Simulation Symposium, EMSS 2013; Athens; Greece, pp. 25–34 (2013)

554. Lopez-Rojas, E.A., Gorton, D., Axelsson, S.: Extending the retsim simulator for estimating the cost of fraud in the retail store domain. In: Proceedings of the European Modeling and Simulation Symposium (2015)

555. Lopez-Rojas, E.A., Gorton, D., Axelsson, S.: Using the retsim simulator for fraud detection research. Int. J. Simul. Process Model. **10**(2), 144–155 (2015). https://doi.org/10.1504/IJSPM.2015.070465

556. Loshchilov, I., Hutter, F.: Fixing weight decay regularization in adam. CoRR (2017). arXiv e-prints abs:1711.05101

557. Loshchilov, I., Hutter, F.: Decoupled weight decay regularization. In: 7th International Conference on Learning Representations, ICLR 2019, New Orleans, LA, USA, May 6-9, 2019. OpenReview.net (2019)

558. Louppe, G., Hermans, J., Cranmer, K.: Adversarial Variational Optimization of Non-Differentiable Simulators. arXiv e-prints arXiv:1707.07113 (2017)

559. Lu, W., Miklau, G., Gupta, V.: Generating private synthetic databases for untrusted system evaluation. In: 2014 IEEE 30th International Conference on Data Engineering, pp. 652–663 (2014). https://doi.org/10.1109/ICDE.2014.6816689

560. Lubashevsky, I., Ando, H.: Intermittent Control Properties of Car Following: Theory and Driving Simulator Experiments. arXiv e-prints arXiv:1609.01812 (2016)

561. Lucas, B.D., Kanade, T.: An iterative image registration technique with an application to stereo vision. In: Proceedings of the 7th International Joint Conference on Artificial Intelligence—Volume 2, IJCAI'81, p. 674–679. Morgan Kaufmann Publishers Inc., San Francisco, CA, USA (1981)

562. Ludl, D., Gulde, T., Thalji, S., Curio, C.: Using simulation to improve human pose estimation for corner cases. In: 2018 21st International Conference on Intelligent Transportation Systems (ITSC), pp. 3575–3582 (2018). https://doi.org/10.1109/ITSC.2018.8569489

563. Lundin, E., Kvarnström, H., Jonsson, E.: A synthetic fraud data generation methodology. In: Deng, R., Bao, F., Zhou, J., Qing, S. (eds.) Information and Communications Security, pp. 265–277. Springer, Berlin, Heidelberg (2002)

564. Luo, M., Tong, Y., Liu, J.: Orthogonal policy gradient and autonomous driving application. CoRR (2018). arXiv e-prints abs:1811.06151

565. Ma, J., Yarats, D.: Quasi-hyperbolic momentum and adam for deep learning. CoRR (2018). arXiv e-prints abs:1810.06801

566. Ma, L., Sun, Q., Schiele, B., Gool, L.V.: A novel bilevel paradigm for image-to-image translation (2019). arXiv e-prints abs:1904.09028

567. Ma, N., Zhang, X., Sun, J.: Activate or Not: Learning Customized Activation. arXiv e-prints arXiv:2009.04759 (2020)

568. Ma, R., Patil, A.G., Fisher, M., Li, M., Pirk, S., Hua, B.S., Yeung, S.K., Tong, X., Guibas, L.J., Zhang, H.: Language-driven synthesis of 3d scenes from scene databases. ACM Trans. Graph. 37, 212:1–212:16 (2018)

569. Ma, X., Hovy, E.: End-to-end sequence labeling via bi-directional LSTM-CNNs-CRF. In: Proceedings of the 54th Annual Meeting of the Association for Computational Linguistics (Volume 1: Long Papers), pp. 1064–1074. Association for Computational Linguistics, Berlin, Germany (2016). https://doi.org/10.18653/v1/P16-1101

570. Maas, A.L., Hannun, A.Y., Ng, A.Y.: Rectifier nonlinearities improve neural network acoustic models. In: in ICML Workshop on Deep Learning for Audio, Speech and Language Processing (2013)

571. Madaan, R., Maturana, D., Scherer, S.: Wire detection using synthetic data and dilated convolutional networks for unmanned aerial vehicles. In: 2017 IEEE/RSJ International Conference on Intelligent Robots and Systems (IROS), pp. 3487–3494 (2017). https://doi.org/10.1109/IROS.2017.8206190

572. Mahmood, F., Chen, R., Durr, N.J.: Unsupervised reverse domain adaptation for synthetic medical images via adversarial training. CoRR (2017). arXiv e-prints abs:1711.06606

573. Mahmood, F., Chen, R.J., Sudarsky, S., Yu, D., Durr, N.J.: Deep learning with cinematic rendering—fine-tuning deep neural networks using photorealistic medical images. CoRR (2018). arXiv e-prints abs:1805.08400

574. Mahoney, M.V., Chan, P.K.: An analysis of the 1999 darpa/lincoln laboratory evaluation data for network anomaly detection. In: Vigna, G., Kruegel, C., Jonsson, E. (eds.) Recent Advances in Intrusion Detection, pp. 220–237. Springer, Berlin, Heidelberg (2003)

575. Mairaj, A., Baba, A.I., Javaid, A.Y.: Application specific drone simulators: Recent advances and challenges. Simul. Model. Pract. Theory 94, 100–117 (2019). https://doi.org/10.1016/j.simpat.2019.01.004

576. Makhzani, A., Shlens, J., Jaitly, N., Goodfellow, I.: Adversarial autoencoders. In: International Conference on Learning Representations (2016)

577. Malik, J., Elhayek, A., Nunnari, F., Varanasi, K., Tamaddon, K., Heloir, A., Stricker, D.: Deephps: End-to-end estimation of 3d hand pose and shape by learning from synthetic depth. In: 2018 International Conference on 3D Vision (3DV), pp. 110–119 (2018). https://doi.org/10.1109/3DV.2018.00023

578. Manna, Z., Waldinger, R.J.: Toward automatic program synthesis. Commun. ACM **14**(3), 151–165 (1971). https://doi.org/10.1145/362566.362568

579. Manolis Savva*, Abhishek Kadian*, Oleksandr Maksymets*, Zhao, Y., Wijmans, E., Jain, B., Straub, J., Liu, J., Koltun, V., Malik, J., Parikh, D., Batra, D.: Habitat: A Platform for Embodied AI Research. arXiv preprint arXiv:1904.01201 (2019)

580. Mao, X., Li, Q., Xie, H., Lau, R.Y.K., Wang, Z., Smolley, S.P.: Least squares generative adversarial networks. In: 2017 IEEE International Conference on Computer Vision (ICCV), pp. 2813–2821 (2017). https://doi.org/10.1109/ICCV.2017.304

581. Marcu, A., Costea, D., Licareţ, V., Pîrvu, M., Sluşanschi, E., Leordeanu, M.: Safeuav: Learning to estimate depth and safe landing areas for uavs from synthetic data. In: Leal-Taixé, L., Roth, S. (eds.) Computer Vision–ECCV 2018 Workshops, pp. 43–58. Springer International Publishing, Cham (2019)

582. Marín, J., Vázquez, D., Gerónimo, D., López, A.M.: Learning appearance in virtual scenarios for pedestrian detection. In: 2010 IEEE Computer Society Conference on Computer Vision and Pattern Recognition, pp. 137–144 (2010). https://doi.org/10.1109/CVPR.2010.5540218

583. Mason, K., Vejdan, S., Grijalva, S.: An "on the fly" framework for efficiently generating synthetic big data sets (2019). arXiv e-prints abs:1903.06798

584. Mautz, R., Tilch, S.: Survey of optical indoor positioning systems. In: 2011 International Conference on Indoor Positioning and Indoor Navigation, IPIN 2011, pp. 1–7 (2011). https://doi.org/10.1109/IPIN.2011.6071925

585. Mayer, N., Ilg, E., Fischer, P., Hazirbas, C., Cremers, D., Dosovitskiy, A., Brox, T.: What makes good synthetic training data for learning disparity and optical flow estimation? Int. J. Comput. Vision **126**(9), 942–960 (2018). https://doi.org/10.1007/s11263-018-1082-6

586. Mayer, N., Ilg, E., Häusser, P., Fischer, P., Cremers, D., Dosovitskiy, A., Brox, T.: A large dataset to train convolutional neural networks for disparity, optical flow, and scene flow estimation. CoRR (2015). arXiv e-prints abs:1512.02134

587. McCarthy, J., Minsky, M., Rochester, N., Shannon, C.E.: A proposal for the dartmouth summer research project on artificial intelligence (1955). http://www-formal.stanford.edu/jmc/history/dartmouth/dartmouth.html

588. McCormac, J., Handa, A., Leutenegger, S., Davison, A.J.: Scenenet rgb-d: Can 5m synthetic images beat generic imagenet pre-training on indoor segmentation? In: 2017 IEEE International Conference on Computer Vision (ICCV), pp. 2697–2706 (2017). https://doi.org/10.1109/ICCV.2017.292

589. McCulloch, W., Pitts, W.: A logical calculus of the ideas immanent in nervous activity. Bull. Math. Biophys. **7**, 115–133 (1943)

590. McHugh, J.: The 1998 lincoln laboratory ids evaluation. In: Debar, H., Mé, L., Wu, S.F. (eds.) Recent Advances in Intrusion Detection, pp. 145–161. Springer, Berlin, Heidelberg (2000)

591. McLachlan, S.: Realism in synthetic data generation. Ph.D. thesis, Massey University, Palmerston North, New Zealand (2017)

592. McLachlan, S., Dube, K., Gallagher, T.: Using the caremap with health incidents statistics for generating the realistic synthetic electronic healthcare record. In: 2016 IEEE International Conference on Healthcare Informatics (ICHI), pp. 439–448 (2016). https://doi.org/10.1109/ICHI.2016.83

593. McMahan, H.B., Ramage, D., Talwar, K., Zhang, L.: Learning differentially private recurrent language models. In: International Conference on Learning Representations (2018)

594. Mdclone: Introducing a new clinical data paradigm (2016). https://www.mdclone.com/

595. Mehta, B., Diaz, M., Golemo, F., Pal, C.J., Paull, L.: Active domain randomization (2019). arXiv e-prints abs:1904.04762

596. Meister, S., Kondermann, D.: Real versus realistically rendered scenes for optical flow evaluation. In: 2011 14th ITG Conference on Electronic Media Technology, pp. 1–6 (2011)

597. Melamud, O., Shivade, C.: Towards automatic generation of shareable synthetic clinical notes using neural language models (2019). arXiv e-prints abs:1905.07002

598. Menze, M., Geiger, A.: Object scene flow for autonomous vehicles. In: Conference on Computer Vision and Pattern Recognition (CVPR) (2015)
599. Merelo Guerv'os, J.J., Fernández-Ares, A., Álvarez-Caballero, A., García-Sánchez, P., Rivas, V.M.: Reddwarfdata: a simplified dataset of starcraft matches. CoRR (2017). arXiv e-prints abs:1712.10179
600. Mikolov, T., Sutskever, I., Chen, K., Corrado, G., Dean, J.: Distributed representations of words and phrases and their compositionality. CoRR (2013). arXiv e-prints abs:1310.4546
601. Milz, S., Arbeiter, G., Witt, C., Abdallah, B., Yogamani, S.: Visual slam for automated driving: Exploring the applications of deep learning. In: The IEEE Conference on Computer Vision and Pattern Recognition (CVPR) Workshops (2018)
602. Mirza, M., Osindero, S.: Conditional generative adversarial nets. CoRR (2014). arXiv e-prints abs:1411.1784
603. Mirzadeh, S.I., Farajtabar, M., Li, A., Ghasemzadeh, H.: Improved knowledge distillation via teacher assistant: Bridging the gap between student and teacher (2019)
604. Misra, D.: Mish: A Self Regularized Non-Monotonic Activation Function. arXiv e-prints arXiv:1908.08681 (2019)
605. Mitra, P., Choudhury, A., Aparow, V.R., Kulandaivelu, G., Dauwels, J.: Towards modeling of perception errors in autonomous vehicles. In: 2018 21st International Conference on Intelligent Transportation Systems (ITSC), pp. 3024–3029 (2018). https://doi.org/10.1109/ITSC.2018.8570015
606. Mnih, V., Badia, A.P., Mirza, M., Graves, A., Lillicrap, T., Harley, T., Silver, D., Kavukcuoglu, K.: Asynchronous methods for deep reinforcement learning. In: M.F. Balcan, K.Q. Weinberger (eds.) Proceedings of The 33rd International Conference on Machine Learning, Proceedings of Machine Learning Research, vol. 48, pp. 1928–1937. PMLR, New York, New York, USA (2016)
607. Mnih, V., Kavukcuoglu, K., Silver, D., Rusu, A.A., Veness, J., Bellemare, M.G., Graves, A., Riedmiller, M., Fidjeland, A.K., Ostrovski, G., Petersen, S., Beattie, C., Sadik, A., Antonoglou, I., King, H., Kumaran, D., Wierstra, D., Legg, S., Hassabis, D.: Human-level control through deep reinforcement learning. Nature **518**(7540), 529–533 (2015). http://dx.doi.org/10.1038/nature14236
608. Mo, K., Zhu, S., Chang, A.X., Yi, L., Tripathi, S., Guibas, L.J., Su, H.: Partnet: A large-scale benchmark for fine-grained and hierarchical part-level 3d object understanding. CoRR (2018). arXiv e-prints arXiv:1812.02713 (2019)
609. Moiseev, B., Konev, A., Chigorin, A., Konushin, A.: Evaluation of traffic sign recognition methods trained on synthetically generated data. In: Blanc-Talon, J., Kasinski, A., Philips, W., Popescu, D., Scheunders, P. (eds.) Advanced Concepts for Intelligent Vision Systems, pp. 576–583. Springer International Publishing, Cham (2013)
610. Moniz, L.J., Buczak, A.L., Hung, L.M., Babin, S., Dorko, M., Lombardo, J.M.: Construction and validation of synthetic electronic medical records. In: Online journal of public health informatics (2009)
611. Moor, J.: The dartmouth college artificial intelligence conference: The next fifty years. AI Mag. **27**(4), 87 (2006). https://doi.org/10.1609/aimag.v27i4.1911
612. Mora, P.B.: Deep 3d pose regression of real objects trained with synthetic data. M.Sc. Thesis, Universitat Politécnica de Catalunya (UPC) (2019)
613. Moravec, H.P.: The Stanford Cart and the CMU Rover, pp. 407–419. Springer New York, New York, NY (1990). https://doi.org/10.1007/978-1-4613-8997-2_30
614. Movshovitz-Attias, Y., Kanade, T., Sheikh, Y.: How useful is photo-realistic rendering for visual learning? In: Hua, G., Jégou, H. (eds.) Computer Vision–ECCV 2016 Workshops, pp. 202–217. Springer International Publishing, Cham (2016)
615. Mualla, Y., Bai, W., Galland, S., Nicolle, C.: Comparison of agent-based simulation frameworks for unmanned aerial transportation applications. Procedia Computer Science **130**, 791 –796 (2018). https://doi.org/10.1016/j.procs.2018.04.137. The 9th International Conference on Ambient Systems, Networks and Technologies (ANT 2018) / The 8th International Conference on Sustainable Energy Information Technology (SEIT-2018) / Affiliated Workshops

616. Nair, V., Hinton, G.E.: Rectified linear units improve restricted boltzmann machines. In: Proceedings of the 27th International Conference on International Conference on Machine Learning, ICML'10, p. 807–814. Omnipress, Madison, WI, USA (2010)

617. Nakano, R.: Neural painters: A learned differentiable constraint for generating brushstroke paintings (2019). arXiv e-prints abs:1904.08410

618. Nakkiran, P., Kaplun, G., Bansal, Y., Yang, T., Barak, B., Sutskever, I.: Deep double descent: Where bigger models and more data hurt. In: International Conference on Learning Representations (2020)

619. Narayanan, A., Shmatikov, V.: Robust de-anonymization of large sparse datasets. In: 2008 IEEE Symposium on Security and Privacy (sp 2008), pp. 111–125 (2008)

620. National Research Council: Language and Machines: Computers in Translation and Linguistics. The National Academies Press, Washington, DC (1966). https://doi.org/10.17226/9547

621. Neff, T.: Data augmentation in deep learning using generative adversarial networks. Ph.D. thesis, Graz University of Technology (2018)

622. Nelson, J., Solawetz, J.: Responding to the controversy about yolov5 (2020). https://blog.roboflow.com/yolov4-versus-yolov5/

623. Nelson, J., Solawetz, J.: Yolov5 is here: State-of-the-art object detection at 140 fps (2020). https://blog.roboflow.com/yolov5-is-here/

624. Nesterov, Y.: A method for unconstrained convex minimization problem with the rate of convergence $o(1/k^2)$. Doklady AN USSR **269**, 543–547 (1983)

625. Nesterov, Y.: Introductory Lectures on Convex Optimization. Springer (2004)

626. Nguyen, T., Miller, I., Cohen, A., Thakur, D., Prasad, S., Guru, A., Taylor, C.J., Chaudrahi, P., Kumar, V.: Pennsyn2real: Training object recognition models without human labeling (2020). arXiv e-prints abs:2009.10292

627. Nie, D., Trullo, R., Lian, J., Wang, L., Petitjean, C., Ruan, S., Wang, Q., Shen, D.: Medical image synthesis with deep convolutional adversarial networks. IEEE Trans. Biomed. Eng. **65**(12), 2720–2730 (2018). https://doi.org/10.1109/TBME.2018.2814538

628. Nie, D., Trullo, R., Petitjean, C., Ruan, S., Shen, D.: Medical image synthesis with context-aware generative adversarial networks. CoRR (2016). arXiv e-prints abs:1612.05362

629. Nie, W., Karras, T., Garg, A., Debnath, S., Patney, A., Patel, A.B., Anandkumar, A.: Semi-supervised stylegan for disentanglement learning (2020)

630. Nielsen, M.A.: Neural Networks and Deep Learning. Determination Press (2018). http://neuralnetworksanddeeplearning.com/

631. Nikolenko, S., Kadurin, A., Arkhangelskaya, E.: Deep Learning. Piter (2017)

632. Nirkin, Y., Keller, Y., Hassner, T.: FSGAN: Subject agnostic face swapping and reenactment. In: Proceedings of the IEEE International Conference on Computer Vision, pp. 7184–7193 (2019)

633. Nocedal, J., Wright, S.J.: Numerical Optimization. Springer Series in Operations Research and Financial Engineering. Springer, New York (2006)

634. North, M.J., Collier, N.T., Ozik, J., Tatara, E.R., Macal, C.M., Bragen, M., Sydelko, P.: Complex adaptive systems modeling with repast simphony. Complex Adapt. Syst. Model. **1**(1), 3 (2013). https://doi.org/10.1186/2194-3206-1-3

635. Nowruzi, F.E., Kapoor, P., Kolhatkar, D., Hassanat, F.A., Laganière, R., Rebut, J.: How much real data do we actually need: Analyzing object detection performance using synthetic and real data (2019). arXiv e-prints abs:1907.07061

636. O'Byrne, M., Pakrashi, V., Schoefs, F., Ghosh, B.: Semantic segmentation of underwater imagery using deep networks trained on synthetic imagery. J. Marine Sci. Eng. **6**(3) (2018). https://doi.org/10.3390/jmse6030093

637. Odena, A., Olah, C., Shlens, J.: Conditional image synthesis with auxiliary classifier gans. In: Proceedings of the 34th International Conference on Machine Learning—Volume 70, ICML'17, pp. 2642–2651. JMLR.org (2017)

638. Oh, J., Chockalingam, V., Singh, S., Lee, H.: Control of memory, active perception, and action in minecraft. In: Proceedings of the 33rd International Conference on International Conference on Machine Learning—Volume 48, ICML'16, pp. 2790–2799. JMLR.org (2016)

639. Oh, K.S., Jung, K.: Gpu implementation of neural networks. Pattern Recogn. **37**(6), 1311–1314 (2004). https://doi.org/10.1016/j.patcog.2004.01.013

640. Olivecrona, M., Blaschke, T., Engkvist, O., Chen, H.: Molecular de-novo design through deep reinforcement learning. J. Cheminform. **9**(1), 48 (2017). https://doi.org/10.1186/s13321-017-0235-x

641. Olson, E.A., Barbalata, C., Zhang, J., Skinner, K.A., Johnson-Roberson, M.: Synthetic data generation for deep learning of underwater disparity estimation. In: OCEANS 2018 MTS/IEEE Charleston pp. 1–6 (2018)

642. van den Oord, A., Dieleman, S., Zen, H., Simonyan, K., Vinyals, O., Graves, A., Kalchbrenner, N., Senior, A., Kavukcuoglu, K.: Wavenet: A generative model for raw audio. In: Arxiv (2016)

643. van den Oord, A., Li, Y., Babuschkin, I., Simonyan, K., Vinyals, O., Kavukcuoglu, K., van den Driessche, G., Lockhart, E., Cobo, L., Stimberg, F., Casagrande, N., Grewe, D., Noury, S., Dieleman, S., Elsen, E., Kalchbrenner, N., Zen, H., Graves, A., King, H., Walters, T., Belov, D., Hassabis, D.: Parallel WaveNet: Fast high-fidelity speech synthesis. In: J. Dy, A. Krause (eds.) Proceedings of the 35th International Conference on Machine Learning, *Proceedings of Machine Learning Research*, vol. 80, pp. 3918–3926. PMLR, Stockholmsmässan, Stockholm Sweden (2018). http://proceedings.mlr.press/v80/oord18a.html

644. van den Oord, A., Vinyals, O., kavukcuoglu, k.: Neural discrete representation learning. In: I. Guyon, U.V. Luxburg, S. Bengio, H. Wallach, R. Fergus, S. Vishwanathan, R. Garnett (eds.) Advances in Neural Information Processing Systems, vol. 30, pp. 6306–6315. Curran Associates, Inc. (2017)

645. Oord, A.V., Kalchbrenner, N., Kavukcuoglu, K.: Pixel recurrent neural networks. In: M.F. Balcan, K.Q. Weinberger (eds.) Proceedings of The 33rd International Conference on Machine Learning. Proceedings of Machine Learning Research, vol. 48, pp. 1747–1756. PMLR, New York, New York, USA (2016). http://proceedings.mlr.press/v48/oord16.html

646. OpenAI, Andrychowicz, M., Baker, B., Chociej, M., Józefowicz, R., McGrew, B., Pachocki, J.W., Pachocki, J., Petron, A., Plappert, M., Powell, G., Ray, A., Schneider, J., Sidor, S., Tobin, J., Welinder, P., Weng, L., Zaremba, W.: Learning dexterous in-hand manipulation. CoRR (2018). arXiv e-prints abs:1808.00177

647. Osa, T., Pajarinen, J., Neumann, G., Bagnell, J.A., Abbeel, P., Peters, J.: An algorithmic perspective on imitation learning. CoRR (2018). arXiv e-prints abs:1811.06711

648. Ostyakov, P., Suvorov, R., Logacheva, E., Khomenko, O., Nikolenko, S.I.: SEIGAN: towards compositional image generation by simultaneously learning to segment, enhance, and inpaint. CoRR (2018). arXiv e-prints abs:1811.07630

649. Paden, B., Čáp, M., Yong, S.Z., Yershov, D., Frazzoli, E.: A survey of motion planning and control techniques for self-driving urban vehicles. IEEE Trans. Intell. Veh. **1**, (2016). https://doi.org/10.1109/TIV.2016.2578706

650. Panchendrarajan, R., Amaresan, A.: Bidirectional LSTM-CRF for named entity recognition. In: Proceedings of the 32nd Pacific Asia Conference on Language, Information and Computation. Association for Computational Linguistics, Hong Kong (2018)

651. Panin, M., Nikolenko, S.I.: Rendering atmospheric clouds with latent space light probes. In: 12th ACM SIGGRAPH Conference and Exhibition on Computer Graphics & Interactive Techniques in Asia, pp. 21–24 (2019)

652. Papamakarios, G., Pavlakou, T., Murray, I.: Masked autoregressive flow for density estimation. In: I. Guyon, U.V. Luxburg, S. Bengio, H. Wallach, R. Fergus, S. Vishwanathan, R. Garnett (eds.) Advances in Neural Information Processing Systems, vol. 30, pp. 2338–2347. Curran Associates, Inc. (2017)

653. Papernot, N., Abadi, M., Erlingsson, Ú., Goodfellow, I.J., Talwar, K.: Semi-supervised knowledge transfer for deep learning from private training data. In: 5th International Conference on Learning Representations, ICLR 2017, Toulon, France, April 24–26, 2017, Conference Track Proceedings (2017)

654. Papert, S.A.: The summer vision project (1966). https://dspace.mit.edu/handle/1721.1/6125

655. Papon, J., Schoeler, M.: Semantic pose using deep networks trained on synthetic rgb-d. In: 2015 IEEE International Conference on Computer Vision (ICCV), pp. 774–782 (2015). https://doi.org/10.1109/ICCV.2015.95

656. Parish, Y.I.H., Müller, P.: Procedural modeling of cities. In: Proceedings of the 28th Annual Conference on Computer Graphics and Interactive Techniques, SIGGRAPH '01, pp. 301–308. ACM, New York, NY, USA (2001). https://doi.org/10.1145/383259.383292

657. Parker, J.R.: Algorithms for Image Processing and Computer Vision, 2nd edn. Wiley Publishing (2010)

658. Pashevich, A., Strudel, R.A.M., Kalevatykh, I., Laptev, I., Schmid, C.: Learning to augment synthetic images for sim2real policy transfer. CoRR (2019). arXiv e-prints abs:1903.07740

659. Patel, S., Kakadiya, A., Mehta, M., Derasari, R., Patel, R., Gandhi, R.: Correlated discrete data generation using adversarial training. CoRR (2018). arXiv e-prints abs:1804.00925

660. Patel, V.M., Gopalan, R., Li, R., Chellappa, R.: Visual domain adaptation: A survey of recent advances. IEEE Signal Process. Mag. **32**(3), 53–69 (2015). https://doi.org/10.1109/MSP.2014.2347059

661. Data-driven policy transfer with imprecise perception simulation: Pecka, M., Zimmermann, K., Petrlx00EDk, M., Svoboda, T. IEEE Robotics and Automation Letters **3**, 3916–3921 (2018)

662. Pendleton, S.D., Andersen, H., Du, X., Shen, X., Meghjani, M., Eng, Y.H., Rus, D., Ang, M.H.: Perception, planning, control, and coordination for autonomous vehicles. Machines **5**(1) (2017). https://doi.org/10.3390/machines5010006

663. Peng, X., Sun, B., Ali, K., Saenko, K.: Exploring invariances in deep convolutional neural networks using synthetic images. CoRR (2014). arXiv e-prints abs:1412.7122

664. Peng, X., Sun, B., Ali, K., Saenko, K.: Learning deep object detectors from 3d models. In: Proceedings of the 2015 IEEE International Conference on Computer Vision (ICCV), ICCV '15, pp. 1278–1286. IEEE Computer Society, Washington, DC, USA (2015). https://doi.org/10.1109/ICCV.2015.151

665. Peng Wang, Xiaohui Shen, Zhe Lin, Cohen, S., Price, B., Yuille, A.: Towards unified depth and semantic prediction from a single image. In: 2015 IEEE Conference on Computer Vision and Pattern Recognition (CVPR), pp. 2800–2809 (2015). https://doi.org/10.1109/CVPR.2015.7298897

666. Pennington, J., Socher, R., Manning, C.: Glove: Global vectors for word representation. In: Proceedings of the 2014 Conference on Empirical Methods in Natural Language Processing (EMNLP), pp. 1532–1543. Association for Computational Linguistics, Doha, Qatar (2014)

667. Pepik, B., Benenson, R., Ritschel, T., Schiele, B.: What is holding back convnets for detection? CoRR (2015). arXiv e-prints abs:1508.02844

668. Pereyra, G., Tucker, G., Chorowski, J., Kaiser, L., Hinton, G.E.: Regularizing neural networks by penalizing confident output distributions. CoRR (2017). arXiv e-prints abs:1701.06548

669. Pérez, P., Gangnet, M., Blake, A.: Poisson image editing. ACM Trans. Graph. **22**(3), 313–318 (2003). https://doi.org/10.1145/882262.882269

670. Perianez-Pascual, J., Rodriguez-Echeverria, R., Burgueño, L., Cabot, J.: Towards the Optical Character Recognition of DSLs, p. 126–132. Association for Computing Machinery, New York, NY, USA (2020)

671. Peris, M., Martull, S., Maki, A., Ohkawa, Y., Fukui, K.: Towards a simulation driven stereo vision system. In: Proceedings of the 21st International Conference on Pattern Recognition (ICPR2012), pp. 1038–1042 (2012)

672. Perot, E., Jaritz, M., Toromanoff, M., d. Charette, R.: End-to-end driving in a realistic racing game with deep reinforcement learning. In: 2017 IEEE Conference on Computer Vision and Pattern Recognition Workshops (CVPRW), pp. 474–475 (2017). https://doi.org/10.1109/CVPRW.2017.64

673. Ping, H., Stoyanovich, J., Howe, B.: Datasynthesizer: Privacy-preserving synthetic datasets. In: Proceedings of the 29th International Conference on Scientific and Statistical Database Management, SSDBM '17, pp. 42:1–42:5. ACM, New York, NY, USA (2017). https://doi.org/10.1145/3085504.3091117

674. Pinheiro, P.H.O., Lin, T., Collobert, R., Dollár, P.: Learning to refine object segments. CoRR (2016). arXiv e-prints abs:1603.08695

675. Pinto, N., Barhomi, Y., Cox, D.D., DiCarlo, J.J.: Comparing state-of-the-art visual features on invariant object recognition tasks. In: 2011 IEEE Workshop on Applications of Computer Vision (WACV), pp. 463–470 (2011). https://doi.org/10.1109/WACV.2011.5711540

676. Pirsiavash, H., Ramanan, D., Fowlkes, C.C.: Globally-optimal greedy algorithms for tracking a variable number of objects. CVPR **2011**, 1201–1208 (2011). https://doi.org/10.1109/CVPR.2011.5995604

677. Planche, B., Wu, Z., Ma, K., Sun, S., Kluckner, S., Lehmann, O., Chen, T., Hutter, A., Zakharov, S., Kosch, H., Ernst, J.: Depthsynth: Real-time realistic synthetic data generation from cad models for 2.5d recognition. In: 2017 International Conference on 3D Vision (3DV), pp. 1–10 (2017). https://doi.org/10.1109/3DV.2017.00011

678. Polykovskiy, D., Zhebrak, A., Sanchez-Lengeling, B., Golovanov, S., Tatanov, O., Belyaev, S., Kurbanov, R., Artamonov, A., Aladinsky, V., Veselov, M., Kadurin, A., Nikolenko, S.I., Aspuru-Guzlik, A., Zhavoronkov, A.: Molecular sets (moses): A benchmarking platform for molecular generation models (2018)

679. Polykovskiy, D., Zhebrak, A., Vetrov, D., Ivanenkov, Y., Aladinskiy, V., Mamoshina, P., Bozdaganyan, M., Aliper, A., Zhavoronkov, A., Kadurin, A.: Entangled conditional adversarial autoencoder for de novo drug discovery. Mol. Pharm. **15**(10), 4398–4405 (2018). https://doi.org/10.1021/acs.molpharmaceut.8b00839. PMID: 30180591

680. Pomerleau, D.A.: Alvinn: An autonomous land vehicle in a neural network. In: Touretzky, D.S. (ed.) Advances in Neural Information Processing Systems 1, pp. 305–313. Morgan Kaufmann Publishers Inc., San Francisco, CA, USA (1989)

681. Popova, M., Isayev, O., Tropsha, A.: Deep reinforcement learning for de novo drug design. Sci. Adv. **4**(7) (2018). https://doi.org/10.1126/sciadv.aap7885

682. Poppe, R.: A survey on vision-based human action recognition. Image Vision Comput. **28**(6), 976–990 (2010). https://doi.org/10.1016/j.imavis.2009.11.014

683. Prabowo, Y.A., Trilaksono, B.R., Triputra, F.R.: Hardware in-the-loop simulation for visual servoing of fixed wing uav. In: 2015 International Conference on Electrical Engineering and Informatics (ICEEI), pp. 247–252 (2015). https://doi.org/10.1109/ICEEI.2015.7352505

684. Prakash, A., Boochoon, S., Brophy, M., Acuna, D., Cameracci, E., State, G., Shapira, O., Birchfield, S.T.: Structured domain randomization: Bridging the reality gap by context-aware synthetic data. CoRR (2018). arXiv e-prints abs:1810.10093

685. Prenger, R., Valle, R., Catanzaro, B.: Waveglow: A flow-based generative network for speech synthesis. CoRR (2018). arXiv e-prints abs:1811.00002

686. Prusinkiewicz, P., Hanan, J.: Lindenmayer systems, fractals, and plants, Springer Science & Business Media, vol. 79. Springer (2013)

687. Qi, S., Zhu, Y., Huang, S., Jiang, C., Zhu, S.: Human-centric indoor scene synthesis using stochastic grammar. In: 2018 IEEE/CVF Conference on Computer Vision and Pattern Recognition, pp. 5899–5908 (2018). https://doi.org/10.1109/CVPR.2018.00618

688. Qi, X., Liao, R., Jia, J., Fidler, S., Urtasun, R.: 3d graph neural networks for rgbd semantic segmentation. In: 2017 IEEE International Conference on Computer Vision (ICCV) pp. 5209–5218 (2017)

689. Qiu, W., Yuille, A.L.: Unrealcv: Connecting computer vision to unreal engine. CoRR (2016). arXiv e-prints abs:1609.01326

690. Qiu, W., Zhong, F., Zhang, Y., Siyuan Qiao, Z.X., Kim, T.S., Wang, Y., Yuille, A.: Unrealcv: Virtual worlds for computer vision. ACM Multimedia Open Source Software Competition (2017)

691. Queiroz, R., Cohen, M., Moreira, J.L., Braun, A., Jacques Junior, J.C., Musse, S.R.: Generating facial ground truth with synthetic faces. In: 2010 23rd SIBGRAPI Conference on Graphics, Patterns and Images, pp. 25–31 (2010). https://doi.org/10.1109/SIBGRAPI.2010.12

692. Quiter, C., Ernst, M.: deepdrive/deepdrive: 2.0 (2018). https://doi.org/10.5281/zenodo.1248998

693. Qureshi, F.Z., Terzopoulos, D.: Surveillance in virtual reality: System design and multi-camera control. In: 2007 IEEE Conference on Computer Vision and Pattern Recognition, pp. 1–8 (2007). https://doi.org/10.1109/CVPR.2007.383071

694. Raczkowsky, J., Mittenbuehler, K.H.: Simulation of cameras in robot applications. IEEE Comput. Graph. Appl. **9**(1), 16–25 (1989). https://doi.org/10.1109/38.20330

695. Rad, M., Oberweger, M., Lepetit, V.: Feature mapping for learning fast and accurate 3d pose inference from synthetic images. CoRR (2017). arXiv e-prints abs:1712.03904

696. Radford, A., Metz, L., Chintala, S.: Unsupervised representation learning with deep convolutional generative adversarial networks. In: 4th International Conference on Learning Representations, ICLR 2016, San Juan, Puerto Rico, 2–4 May 2016, Conference Track Proceedings (2016)

697. Radford, A., Narasimhan, K., Salimans, T., Sutskever, I.: Improving language understanding by generative pre-training (2018). https://s3-us-west-2.amazonaws.com/openai-assets/researchcovers/languageunsupervised/languageunderstandingpaper.pdf

698. Radford, A., Wu, J., Child, R., Luan, D., Amodei, D., Sutskever, I.: Language models are unsupervised multitask learners (2018). https://d4mucfpksywv.cloudfront.net/better-language-models/language-models.pdf

699. Ragheb, H., Velastin, S., Remagnino, P., Ellis, T.: Vihasi: Virtual human action silhouette data for the performance evaluation of silhouette-based action recognition methods. In: 2008 Second ACM/IEEE International Conference on Distributed Smart Cameras, pp. 1–10 (2008). https://doi.org/10.1109/ICDSC.2008.4635730

700. Raghunathan, T., Reiter, J., Rubin, D.: Multiple imputation for statistical disclosure limitation. J. Off. Stat. **19**(1), 1–16 (2003)

701. Rajpura, P.S., Hegde, R.S., Bojinov, H.: Object detection using deep cnns trained on synthetic images. CoRR (2017). arXiv e-prints abs:1706.06782

702. Ramachandran, P., Zoph, B., Le, Q.V.: Searching for activation functions. CoRR (2017). arXiv e-prints abs:1710.05941

703. Ranjan, R., Patel, V.M., Chellappa, R.: Hyperface: A deep multi-task learning framework for face detection, landmark localization, pose estimation, and gender recognition. IEEE Trans. Pattern Anal. Mach. Intell. **41**(1), 121–135 (2019). https://doi.org/10.1109/TPAMI.2017.2781233

704. Rashid, A., Do-Omri, A., Haidar, M.A., Liu, Q., Rezagholizadeh, M.: Bilingual-GAN: A step towards parallel text generation. In: Proceedings of the Workshop on Methods for Optimizing and Evaluating Neural Language Generation, pp. 55–64. Association for Computational Linguistics, Minneapolis, Minnesota (2019). https://doi.org/10.18653/v1/W19-2307

705. Ratcliffe, D.S., Devlin, S., Kruschwitz, U., Citi, L.: Clyde: A deep reinforcement learning doom playing agent. In: AAAI Workshops (2017)

706. Ratner, A.J., Ehrenberg, H., Hussain, Z., Dunnmon, J., Ré, C.: Learning to compose domain-specific transformations for data augmentation. In: I. Guyon, U.V. Luxburg, S. Bengio, H. Wallach, R. Fergus, S. Vishwanathan, R. Garnett (eds.) Advances in Neural Information Processing Systems 30, pp. 3236–3246. Curran Associates, Inc. (2017)

707. Razavi, A., van den Oord, A., Vinyals, O.: Generating diverse high-fidelity images with VQ-VAE-2. CoRR (2019). arXiv e-prints abs:1906.00446

708. Reddi, S.J., Kale, S., Kumar, S.: On the convergence of adam and beyond. CoRR (2019). arXiv e-prints abs:1904.09237

709. Redmon, J., Divvala, S.K., Girshick, R.B., Farhadi, A.: You only look once: Unified, real-time object detection. CoRR (2015). arXiv e-prints abs:1506.02640

710. Redmon, J., Farhadi, A.: YOLO9000: better, faster, stronger. CoRR (2016). arXiv e-prints abs:1612.08242

711. Redmon, J., Farhadi, A.: Yolov3: An incremental improvement. CoRR (2018). arXiv e-prints abs:1804.02767

712. Reiter, J.P.: Releasing multiply imputed, synthetic public use microdata: An illustration and empirical study. J. R. Stat. Soc. Ser. A **168**, 185–205 (2005). https://doi.org/10.1111/j.1467-985X.2004.00343.x

713. Reiter, J.P., Drechsler, J.: Releasing multiply-imputed synthetic data generated in two stages to protect confidentiality. Stat. Sinica **20**, (2007)

714. Reiter, J.P., Raghunathan, T.E.: The multiple adaptations of multiple imputation. J. Am. Stat. Assoc. **102**, 1462–1471 (2007)

715. Rematas, K., Kemelmacher-Shlizerman, I., Curless, B., Seitz, S.: Soccer on your tabletop. In: 2018 IEEE/CVF Conference on Computer Vision and Pattern Recognition, pp. 4738–4747 (2018). https://doi.org/10.1109/CVPR.2018.00498

716. Remez, T., Huang, J., Brown, M.R.: Learning to segment via cut-and-paste. In: ECCV (2018)

717. Ren, J., Hacihaliloglu, I., Singer, E.A., Foran, D.J., Qi, X.: Adversarial domain adaptation for classification of prostate histopathology whole-slide images. In: Medical Image Computing and Computer Assisted Intervention—MICCAI 2018—21st International Conference, Granada, Spain, September 16–20, 2018, Proceedings, Part II, pp. 201–209 (2018). https://doi.org/10.1007/978-3-030-00934-2_23

718. Ren, S., He, K., Girshick, R., Sun, J.: Faster r-cnn: Towards real-time object detection with region proposal networks. In: C. Cortes, N.D. Lawrence, D.D. Lee, M. Sugiyama, R. Garnett (eds.) Advances in Neural Information Processing Systems 28, pp. 91–99. Curran Associates, Inc. (2015)

719. Ren, S., He, K., Girshick, R., Sun, J.: Faster r-cnn: Towards real-time object detection with region proposal networks. IEEE Trans. Pattern Anal. Mach. Intell. **39**(06), 1137–1149 (2017). https://doi.org/10.1109/TPAMI.2016.2577031

720. Ren, Z., Lee, Y.J.: Cross-domain self-supervised multi-task feature learning using synthetic imagery. In: 2018 IEEE/CVF Conference on Computer Vision and Pattern Recognition, pp. 762–771 (2018)

721. Ren, Z., Lee, Y.J., Ryoo, M.S.: Learning to anonymize faces for privacy preserving action detection. In: Ferrari, V., Hebert, M., Sminchisescu, C., Weiss, Y. (eds.) Computer Vision–ECCV 2018, pp. 639–655. Springer International Publishing, Cham (2018)

722. Reymond, J.L., Ruddigkeit, L., Blum, L., van Deursen, R.: The enumeration of chemical space. Wiley Interdisc. Rev. Comput. Mol. Sci. **2**(5), 717–733 (2012). https://doi.org/10.1002/wcms.1104

723. Rezende, D., Mohamed, S.: Variational inference with normalizing flows. In: F. Bach, D. Blei (eds.) Proceedings of the 32nd International Conference on Machine Learning. Proceedings of Machine Learning Research, vol. 37, pp. 1530–1538. PMLR, Lille, France (2015). http://proceedings.mlr.press/v37/rezende15.html

724. Richter, S.R., Hayder, Z., Koltun, V.: Playing for benchmarks. CoRR (2017). arXiv e-prints abs:1709.07322

725. Richter, S.R., Vineet, V., Roth, S., Koltun, V.: Playing for data: Ground truth from computer games. CoRR (2016). arXiv e-prints abs:1608.02192

726. Roberts, M., Paczan, N.: Hypersim: A photorealistic synthetic dataset for holistic indoor scene understanding (2020)

727. Robicquet, A., Sadeghian, A., Alahi, A., Savarese, S.: Learning social etiquette: Human trajectory understanding in crowded scenes. In: Leibe, B., Matas, J., Sebe, N., Welling, M. (eds.) Computer Vision–ECCV 2016, pp. 549–565. Springer International Publishing, Cham (2016)

728. Rohmer, E., Singh, S.P.N., Freese, M.: V-rep: A versatile and scalable robot simulation framework. In: 2013 IEEE/RSJ International Conference on Intelligent Robots and Systems, pp. 1321–1326 (2013). https://doi.org/10.1109/IROS.2013.6696520

729. Rojas-Perez, L.O., Munguia-Silva, R., Martinez-Carranza, J.: Real-time landing zone detection for uavs using single aerial images. In: IMAV2018, 10th International Micro Air Vehicle Conference (IMAV) (2018)

730. Ronneberger, O., P.Fischer, Brox, T.: U-net: Convolutional networks for biomedical image segmentation. In: Medical Image Computing and Computer-Assisted Intervention (MICCAI), *LNCS*, vol. 9351, pp. 234–241. Springer (2015). http://lmb.informatik.uni-freiburg.de/Publications/2015/RFB15a, arXiv:1505.04597 [cs.CV])

731. Ros, G., Sellart, L., Materzynska, J., Vazquez, D., Lopez, A.M.: The synthia dataset: A large collection of synthetic images for semantic segmentation of urban scenes. In: 2016 IEEE Conference on Computer Vision and Pattern Recognition (CVPR), pp. 3234–3243 (2016). https://doi.org/10.1109/CVPR.2016.352

732. Ros, G., Sellart, L., Villalonga, G., Maidanik, E., Molero, F., Garcia, M., Cedeño, A., Perez, F., Ramirez, D., Escobar, E., Gomez, J.L., Vazquez, D., Lopez, A.M.: Semantic Segmentation of Urban Scenes via Domain Adaptation of SYNTHIA, pp. 227–241. Springer International Publishing, Cham (2017). https://doi.org/10.1007/978-3-319-58347-1_12

733. Ros, G., Stent, S., Alcantarilla, P.F., Watanabe, T.: Training constrained deconvolutional networks for road scene semantic segmentation. CoRR (2016). arXiv e-prints abs:1604.01545

734. Rosenberg, A., Patwardhan, R., Shendure, J., Seelig, G.: Learning the sequence determinants of alternative splicing from millions of random sequences. Cell 163(3), 698–711 (2015). https://doi.org/10.1016/j.cell.2015.09.054

735. Rosenblatt, F.: The perceptron: a probabilistic model for information storage and organization in the brain. Psychol. Rev. 65(6), 386 (1958)

736. Rosenblatt, F.: Principles of Neurodynamics. Spartan, New York (1962)

737. Rosique, F., Navarro, P.J., Fernández, C., Padilla, A.: A systematic review of perception system and simulators for autonomous vehicles research. Sensors 19(3) (2019). https://doi.org/10.3390/s19030648

738. Rozantsev, A., Lepetit, V., Fua, P.: On rendering synthetic images for training an object detector. Comput. Vis. Image Underst. 137, 24–37 (2015). https://doi.org/10.1016/j.cviu.2014.12.006

739. Rozantsev, A., Salzmann, M., Fua, P.: Beyond sharing weights for deep domain adaptation. IEEE Trans. Pattern Anal. Mach. Intell. 41(4), 801–814 (2019). https://doi.org/10.1109/TPAMI.2018.2814042

740. Rubin, D.: Discussion: Statistical disclosure limitation. J. Off. Stat. 9, 462–468 (1993)

741. Ruiz, N., Schulter, S., Chandraker, M.: Learning to simulate. In: International Conference on Learning Representations (2019)

742. Rumelhart, D.E., Hinton, G.E., Williams, R.J.: Learning representations by back-propagating errors. Nature 323(6088), 533–536 (1986)

743. Rusu, A.A., Rabinowitz, N.C., Desjardins, G., Soyer, H., Kirkpatrick, J., Kavukcuoglu, K., Pascanu, R., Hadsell, R.: Progressive neural networks. CoRR (2016). arXiv e-prints abs:1606.04671

744. Rusu, A.A., Vecerik, M., Rothörl, T., Heess, N., Pascanu, R., Hadsell, R.: Sim-to-real robot learning from pixels with progressive nets. In: CoRL (2017)

745. Rygaard, L.V.: Using synthesized speech to improve speech recognition for low-resource languages. In: Grace Hopper Celebration 2015, Poster Session (2015)

746. S. Lombardo, J., Moniz, L.: A method for generation and distribution of synthetic medical record data for evaluation of disease-monitoring systems. Johns Hopkins APL Technical Digest (Applied Physics Laboratory) 27 (2008)

747. Saatchi, Y., Wilson, A.G.: Bayesian gan. In: I. Guyon, U.V. Luxburg, S. Bengio, H. Wallach, R. Fergus, S. Vishwanathan, R. Garnett (eds.) Advances in Neural Information Processing Systems 30, pp. 3622–3631. Curran Associates, Inc. (2017)

748. Sabour, S., Frosst, N., Hinton, G.E.: Dynamic routing between capsules (2017). arXiv e-prints abs:1710.09829

749. Sadeghi, F., Levine, S.: Cad2rl: Real single-image flight without a single real image (2017). arXiv e-prints abs:1611.04201

750. Saito, K., Ushiku, Y., Harada, T.: Asymmetric tri-training for unsupervised domain adaptation. In: Proceedings of the 34th International Conference on Machine Learning—Volume 70, ICML'17, pp. 2988–2997. JMLR.org (2017)

751. Sajjan, S.S., Moore, M., Pan, M., Nagaraja, G., Lee, J., Zeng, A., Song, S.: ClearGrasp: 3D Shape Estimation of Transparent Objects for Manipulation. arXiv e-prints arXiv:1910.02550 (2019)

752. Salakhutdinov, R., Hinton, G.: Deep boltzmann machines. In: D. van Dyk, M. Welling (eds.) Proceedings of the Twelth International Conference on Artificial Intelligence and Statistics, *Proceedings of Machine Learning Research*, vol. 5, pp. 448–455. PMLR, Hilton Clearwater Beach Resort, Clearwater Beach, Florida USA (2009). http://proceedings.mlr. press/v5/salakhutdinov09a.html

753. Salakhutdinov, R., Larochelle, H.: Efficient learning of deep boltzmann machines. In: Y.W. Teh, M. Titterington (eds.) Proceedings of the Thirteenth International Conference on Artificial Intelligence and Statistics, *Proceedings of Machine Learning Research*, vol. 9, pp. 693–700. JMLR Workshop and Conference Proceedings, Chia Laguna Resort, Sardinia, Italy (2010). http://proceedings.mlr.press/v9/salakhutdinov10a.html

754. Saleh, F.S., Aliakbarian, M.S., Salzmann, M., Petersson, L., Alvarez, J.M.: Effective use of synthetic data for urban scene semantic segmentation. CoRR (2018). arXiv e-prints abs:1807.06132

755. Salimans, T., Karpathy, A., Chen, X., Kingma, D.P.: Pixelcnn++: Improving the pixelcnn with discretized logistic mixture likelihood and other modifications. CoRR (2017). arXiv e-prints abs:1701.05517

756. Sallab, A.E., Abdou, M., Perot, E., Yogamani, S.: End-to-end deep reinforcement learning for lane keeping assist (2016). arXiv e-prints abs:1612.04340

757. Sallab, A.E., Abdou, M., Perot, E., Yogamani, S.: Deep reinforcement learning framework for autonomous driving (2017). arXiv e-prints abs:1704.02532

758. Santana, E., Hotz, G.: Learning a driving simulator (2016). arXiv e-prints abs:1608.01230

759. Santoro, A., Raposo, D., Barrett, D.G.T., Malinowski, M., Pascanu, R., Battaglia, P.W., Lillicrap, T.P.: A simple neural network module for relational reasoning. In: NIPS (2017)

760. Santoso, K., Kusuma, G.P.: Face recognition using modified openface. Procedia Computer Science **135**, 510–517 (2018). https://doi.org/10.1016/j.procs.2018.08.203. The 3rd International Conference on Computer Science and Computational Intelligence (ICCSCI 2018) : Empowering Smart Technology in Digital Era for a Better Life

761. Savva, M., Chang, A.X., Dosovitskiy, A., Funkhouser, T., Koltun, V.: MINOS: Multimodal indoor simulator for navigation in complex environments (2017). arXiv e-prints arXiv:1712.03931

762. Saxena, A., Driemeyer, J., Ng, A.Y.: Robotic grasping of novel objects using vision. Int. J. Rob. Res. **27**(2), 157–173 (2008). https://doi.org/10.1177/0278364907087172

763. Sayre-McCord, T., Guerra, W., Antonini, A., Arneberg, J., Brown, A., Cavalheiro, G., Fang, Y., Gorodetsky, A., McCoy, D., Quilter, S., Riether, F., Tal, E., Terzioglu, Y., Carlone, L., Karaman, S.: Visual-inertial navigation algorithm development using photorealistic camera simulation in the loop. In: 2018 IEEE International Conference on Robotics and Automation (ICRA), pp. 2566–2573 (2018). https://doi.org/10.1109/ICRA.2018.8460692

764. Scharstein, D., Hirschmüller, H., Kitajima, Y., Krathwohl, G., Nesic, N., Wang, X., Westling, P.: High-resolution stereo datasets with subpixel-accurate ground truth. In: X. Jiang, J. Hornegger, R. Koch (eds.) GCPR, *Lecture Notes in Computer Science*, vol. 8753, pp. 31–42. Springer (2014). http://dblp.uni-trier.de/db/conf/dagm/gcpr2014.html# ScharsteinHKKNWW14

765. Scharstein, D., Szeliski, R., Zabih, R.: A taxonomy and evaluation of dense two-frame stereo correspondence algorithms. In: Proceedings IEEE Workshop on Stereo and Multi-Baseline Vision (SMBV 2001), pp. 131–140 (2001). https://doi.org/10.1109/SMBV.2001.988771

766. Schaul, T., Togelius, J., Schmidhuber, J.: Measuring intelligence through games. CoRR (2011). arXiv e-prints abs:1109.1314

767. Schmidhuber, J.: Deep learning in neural networks: An overview. CoRR (2014). arXiv e-prints abs:1404.7828

768. Schneider, M.J., Abowd, J.M.: A new method for protecting interrelated time series with bayesian prior distributions and synthetic data. J. R. Stat. Soc. A. Stat. Soc. **178**(4), 963–975 (2015). https://doi.org/10.1111/rssa.12100

769. Schneider, P., Schneider, G.: De novo design at the edge of chaos. J. Med. Chem. **59**(9), 4077–4086 (2016). https://doi.org/10.1021/acs.jmedchem.5b01849. PMID: 26881908

770. Schrittwieser, J., Antonoglou, I., Hubert, T., Simonyan, K., Sifre, L., Schmitt, S., Guez, A., Lockhart, E., Hassabis, D., Graepel, T., Lillicrap, T., Silver, D.: Mastering Atari, Go, Chess and Shogi by planning with a learned model (2019). arXiv e-prints arXiv:1911.08265

771. Schroff, F., Kalenichenko, D., Philbin, J.: Facenet: A unified embedding for face recognition and clustering. In: 2015 IEEE Conference on Computer Vision and Pattern Recognition (CVPR), pp. 815–823 (2015)

772. Schulman, J., Wolski, F., Dhariwal, P., Radford, A., Klimov, O.: Proximal policy optimization algorithms. CoRR (2017). arXiv e-prints abs:1707.06347

773. Segler, M.H.S., Kogej, T., Tyrchan, C., Waller, M.P.: Generating focussed molecule libraries for drug discovery with recurrent neural networks. CoRR (2017). arXiv e-prints abs:1701.01329

774. Segler, M.H.S., Kogej, T., Tyrchan, C., Waller, M.P.: Generating focused molecule libraries for drug discovery with recurrent neural networks. ACS Cent. Sci. **4**(1), 120–131 (2018). https://doi.org/10.1021/acscentsci.7b00512. PMID: 29392184

775. Seitz, S.M., Curless, B., Diebel, J., Scharstein, D., Szeliski, R.: A comparison and evaluation of multi-view stereo reconstruction algorithms. In: Proceedings of the 2006 IEEE Computer Society Conference on Computer Vision and Pattern Recognition—Volume 1, CVPR '06, pp. 519–528. IEEE Computer Society, Washington, DC, USA (2006). https://doi.org/10.1109/CVPR.2006.19

776. Sejnova, G., Tesar, M., Vavrecka, M.: Compositional models for vqa: Can neural module networks really count? Proc. Comput. Sci. **145**, 481–487 (2018). https://doi.org/10.1016/j.procs.2018.11.110 (Postproceedings of the 9th Annual International Conference on Biologically Inspired Cognitive Architectures, BICA 2018 (Ninth Annual Meeting of the BICA Society), held August 22–24, 2018 in Prague, Czech Republic)

777. Sela, M., Xu, P., He, J., Navalpakkam, V., Lagun, D.: Gazegan—unpaired adversarial image generation for gaze estimation. CoRR (2017). arXiv e-prints abs:1711.09767

778. Sennrich, R., Haddow, B., Birch, A.: Edinburgh neural machine translation systems for wmt 16. In: Proceedings of the First Conference on Machine Translation, pp. 371–376. Association for Computational Linguistics, Berlin, Germany (2016). https://doi.org/10.18653/v1/W16-2323

779. Sensefly datasets (2019). https://www.sensefly.com/education/datasets/

780. Sepulveda, G., Niebles, J.C., Soto, A.: A deep learning based behavioral approach to indoor autonomous navigation. In: 2018 IEEE International Conference on Robotics and Automation (ICRA), pp. 4646–4653 (2018). https://doi.org/10.1109/ICRA.2018.8460646

781. Serban, A., Poll, E., Visser, J.: Adversarial examples on object recognition: A comprehensive survey. ACM Comput. Surv. **53**(3) (2020). https://doi.org/10.1145/3398394

782. Sermanet, P., Eigen, D., Zhang, X., Mathieu, M., Fergus, R., LeCun, Y.: OverFeat: Integrated Recognition, Localization and Detection using Convolutional Networks. arXiv e-prints arXiv:1312.6229 (2013)

783. Sermanet, P., Eigen, D., Zhang, X., Mathieu, M., Fergus, R., LeCun, Y.: Overfeat: Integrated recognition, localization and detection using convolutional networks. In: Y. Bengio, Y. LeCun (eds.) 2nd International Conference on Learning Representations, ICLR 2014, Banff, AB, Canada, April 14-16, 2014, Conference Track Proceedings (2014)

784. Shaban, M.T., Baur, C., Navab, N., Albarqouni, S.: Staingan: Stain style transfer for digital histological images. CoRR (2018). arXiv e-prints abs:1804.01601

785. Shah, S., Dey, D., Lovett, C., Kapoor, A.: Airsim: High-fidelity visual and physical simulation for autonomous vehicles. In: Field and Service Robotics (2017). arXiv e-prints abs:1705.05065

786. Shaker, N., Togelius, J., Nelson, M.J.: Procedural Content Generation in Games, 1st edn. Springer Publishing Company, Incorporated (2016)

787. Shamir, A.: A survey on mesh segmentation techniques. Comput. Graph. Forum **27**(6), 1539–1556 (2008). https://doi.org/10.1111/j.1467-8659.2007.01103.x

788. Shao, K., Tang, Z., Zhu, Y., Li, N., Zhao, D.: A Survey of Deep Reinforcement Learning in Video Games (2019). arXiv e-prints arXiv:1912.10944

789. Shenbin, I., Alekseev, A., Tutubalina, E., Malykh, V., Nikolenko, S.I.: Recvae: A new varia-
     tional autoencoder for top-n recommendations with implicit feedback. In: 13th International
     Conference on Web Search and Data Mining, pp. 528–536 (2020)
790. Shin, R., Kant, N., Gupta, K., Bender, C., Trabucco, B., Singh, R., Song, D.: Synthetic
     datasets for neural program synthesis. In: International Conference on Learning Represen-
     tations (2019)
791. Shoman, S., Mashita, T., Plopski, A., Ratsamee, P., Uranishi, Y., Takemura, H.: Illumination
     invariant camera localization using synthetic images. In: 2018 IEEE International Sympo-
     sium on Mixed and Augmented Reality Adjunct (ISMAR-Adjunct), pp. 143–144 (2018).
     https://doi.org/10.1109/ISMAR-Adjunct.2018.00053
792. Shorten, C., Khoshgoftaar, T.M.: A survey on image data augmentation for deep learning.
     J. Big Data 6(1), 60 (2019). https://doi.org/10.1186/s40537-019-0197-0
793. Shrivastava, A., Pfister, T., Tuzel, O., Susskind, J., Wang, W., Webb, R.: Learning from
     simulated and unsupervised images through adversarial training. In: 2017 IEEE Conference
     on Computer Vision and Pattern Recognition (CVPR), pp. 2242–2251 (2017)
794. Si, X., Yang, Y., Dai, H., Naik, M., Song, L.: Learning a meta-solver for syntax-guided
     program synthesis. In: International Conference on Learning Representations (2019)
795. Siarohin, A., Lathuilière, S., Tulyakov, S., Ricci, E., Sebe, N.: Animating arbitrary objects
     via deep motion transfer (2018). arXiv e-prints abs:1812.08861
796. Silberman, N., Hoiem, D., Kohli, P., Fergus, R.: Indoor segmentation and support inference
     from rgbd images. In: Fitzgibbon, A., Lazebnik, S., Perona, P., Sato, Y., Schmid, C. (eds.)
     Computer Vision–ECCV 2012, pp. 746–760. Springer, Berlin, Heidelberg (2012)
797. Silva, S., Gutman, B., Romero, E., Thompson, P.M., Altmann, A., Lorenzi, M.: Federated
     learning in distributed medical databases: Meta-analysis of large-scale subcortical brain data
     (2019)
798. Silveira Jacques Junior, J.C., Musse, S.R., Jung, C.R.: Crowd analysis using computer vision
     techniques. IEEE Signal Process. Mag. 27(5), 66–77 (2010). https://doi.org/10.1109/MSP.
     2010.937394
799. Silver, D., Huang, A., Maddison, C.J., Guez, A., Sifre, L., van den Driessche, G., Schrit-
     twieser, J., Antonoglou, I., Panneershelvam, V., Lanctot, M., Dieleman, S., Grewe, D., Nham,
     J., Kalchbrenner, N., Sutskever, I., Lillicrap, T., Leach, M., Kavukcuoglu, K., Graepel, T.,
     Hassabis, D.: Mastering the game of Go with deep neural networks and tree search. Nature
     529(7587), 484–489 (2016). https://doi.org/10.1038/nature16961
800. Silver, D., Hubert, T., Schrittwieser, J., Antonoglou, I., Lai, M., Guez, A., Lanctot, M.,
     Sifre, L., Kumaran, D., Graepel, T., Lillicrap, T., Simonyan, K., Hassabis, D.: A general
     reinforcement learning algorithm that masters chess, shogi, and go through self-play. Science
     362(6419), 1140–1144 (2018). https://doi.org/10.1126/science.aar6404
801. Simard, P.Y., Steinkraus, D., Platt, J.C.: Best practices for convolutional neural networks
     applied to visual document analysis. In: Seventh International Conference on Document
     Analysis and Recognition, 2003. Proceedings., pp. 958–963 (2003). https://doi.org/10.1109/
     ICDAR.2003.1227801
802. Simonyan, K., Zisserman, A.: Very deep convolutional networks for large-scale image recog-
     nition (2014). arXiv e-prints abs:1409.1556
803. Singh, B., Najibi, M., Davis, L.S.: Sniper: Efficient multi-scale training. In: S. Bengio,
     H. Wallach, H. Larochelle, K. Grauman, N. Cesa-Bianchi, R. Garnett (eds.) Advances in
     Neural Information Processing Systems 31, pp. 9310–9320. Curran Associates, Inc. (2018)
804. Sirazitdinov, I., Schulz, H., Saalbach, A., Renisch, S., Dylov, D.V.: Tubular shape aware data
     generation for semantic segmentation in medical imaging (2020)
805. Smith, D.M., Clarke, G.P., Harland, K.: Improving the synthetic data generation process
     in spatial microsimulation models. Environ. Plan. Econ. Space 41(5), 1251–1268 (2009).
     https://doi.org/10.1068/a4147
806. Smith, L.N., Topin, N.: Super-convergence: Very fast training of residual networks using
     large learning rates. CoRR (2017). arXiv e-prints abs:1708.07120

807. Smyth, D.L., Fennell, J., Abinesh, S., Karimi, N.B., Glavin, F.G., Ullah, I., Drury, B., Madden, M.G.: A virtual environment with multi-robot navigation, analytics, and decision support for critical incident investigation (2018). arXiv e-prints abs:1806.04497

808. Smyth, D.L., Glavin, F.G., Madden, M.G.: Using a game engine to simulate critical incidents and data collection by autonomous drones. In: 2018 IEEE Games, Entertainment, Media Conference (GEM), pp. 1–9 (2018). https://doi.org/10.1109/GEM.2018.8516527

809. Sobel, I.: An isotropic 3x3 image gradient operator. Presentation at Stanford A.I. Project 1968 (2014)

810. Socher, R., Ganjoo, M., Manning, C.D., Ng, A.: Zero-shot learning through cross-modal transfer. In: C.J.C. Burges, L. Bottou, M. Welling, Z. Ghahramani, K.Q. Weinberger (eds.) Advances in Neural Information Processing Systems 26, pp. 935–943. Curran Associates, Inc. (2013)

811. Sohn, K., Liu, S., Zhong, G., Yu, X., Yang, M., Chandraker, M.: Unsupervised domain adaptation for face recognition in unlabeled videos. CoRR (2017). arXiv e-prints abs:1708.02191

812. Solovev, P., Aliev, V., Ostyakov, P., Sterkin, G., Logacheva, E., Troeshestov, S., Suvorov, R., Mashikhin, A., Khomenko, O., Nikolenko, S.I.: Learning state representations in complex systems with multimodal data. CoRR (2018). arXiv e-prints abs:1811.11067

813. Song, S., Lichtenberg, S.P., Xiao, J.: Sun rgb-d: A rgb-d scene understanding benchmark suite. In: 2015 IEEE Conference on Computer Vision and Pattern Recognition (CVPR), pp. 567–576 (2015). https://doi.org/10.1109/CVPR.2015.7298655

814. Song, S., Yu, F., Zeng, A., Chang, A.X., Savva, M., Funkhouser, T.: Semantic scene completion from a single depth image. In: Proceedings of 30th IEEE Conference on Computer Vision and Pattern Recognition (2017)

815. d. Souza, C.R., Gaidon, A., Cabon, Y., López, A.M.: Procedural generation of videos to train deep action recognition networks. In: 2017 IEEE Conference on Computer Vision and Pattern Recognition (CVPR), pp. 2594–2604 (2017). https://doi.org/10.1109/CVPR.2017.278

816. Srivastava, N., Hinton, G., Krizhevsky, A., Sutskever, I., Salakhutdinov, R.: Dropout: A simple way to prevent neural networks from overfitting. J. Mach. Learn. Res. **15**, 1929–1958 (2014). http://jmlr.org/papers/v15/srivastava14a.html

817. Stark, M., Goesele, M., Schiele, B.: Back to the future: Learning shape models from 3d cad data. In: BMVC (2010)

818. Stein, G.J., Roy, N.G.: Genesis-rt: Generating synthetic images for training secondary real-world tasks. In: 2018 IEEE International Conference on Robotics and Automation (ICRA), pp. 7151–7158 (2018)

819. Struckmeier, O.: Leagueai: Improving object detector performance and flexibility through automatically generated training data and domain randomization (2019).arXiv e-prints https://arxiv.org/abs/1905.13546

820. Sturm, J., Engelhard, N., Endres, F., Burgard, W., Cremers, D.: A benchmark for the evaluation of rgb-d slam systems. In: 2012 IEEE/RSJ International Conference on Intelligent Robots and Systems, pp. 573–580 (2012)

821. Su, H., Maji, S., Kalogerakis, E., Learned-Miller, E.: Multi-view convolutional neural networks for 3d shape recognition. In: 2015 IEEE International Conference on Computer Vision (ICCV), pp. 945–953 (2015). https://doi.org/10.1109/ICCV.2015.114

822. Su, H., Qi, C.R., Li, Y., Guibas, L.J.: Render for cnn: Viewpoint estimation in images using cnns trained with rendered 3d model views. In: 2015 IEEE International Conference on Computer Vision (ICCV), pp. 2686–2694 (2015). https://doi.org/10.1109/ICCV.2015.308

823. Su, Z., Ye, M., Zhang, G., Dai, L., Sheng, J.: Improvement multi-stage model for human pose estimation. CoRR (2019). arXiv e-prints abs:1902.07837

824. Sulkowski, T., Bugiel, P., Izydorczyk, J.: In search of the ultimate autonomous driving simulator. In: 2018 International Conference on Signals and Electronic Systems (ICSES), pp. 252–256 (2018). https://doi.org/10.1109/ICSES.2018.8507288

825. The 2019 sumo workshop 360° indoor scene understanding and modeling (2019). https://sumochallenge.org/2019-sumo-workshop.html

826. Sun, B., Saenko, K.: From virtual to reality: Fast adaptation of virtual object detectors to real domains. In: Proceedings of the British Machine Vision Conference. BMVA Press (2014)

827. Sun, C., Shrivastava, A., Singh, S., Gupta, A.: Revisiting unreasonable effectiveness of data in deep learning era. In: 2017 IEEE International Conference on Computer Vision (ICCV), pp. 843–852 (2017). https://doi.org/10.1109/ICCV.2017.97

828. Sun, K., Xiao, B., Liu, D., Wang, J.: Deep high-resolution representation learning for human pose estimation. CoRR (2019). arXiv e-prints abs:1902.09212

829. Supervisely: Ai assisted labeling (2019). https://docs.supervise.ly/

830. Sutton, C., McCallum, A.: An introduction to conditional random fields. Found. Trends Mach. Learn. **4**(4), 267–373 (2012). https://doi.org/10.1561/2200000013

831. Sutton, R.S., Barto, A.G.: Reinforcement Learning: An Introduction, 2nd ed. edn. MIT Press, Cambridge, MA (2018)

832. Sweeney, L.: Maintaining patient confidentiality when sharing medical data requires a symbiotic relationship between technology and policy. MIT A.I. Working Paper No. AIWP-WP344b (1997)

833. Sweeney, L.: Weaving technology and policy together to maintain confidentiality. J. Law Med. Ethics **25**(2–3), 98–110 (1997). https://doi.org/10.1111/j.1748-720X.1997.tb01885.x

834. Synnaeve, G., Nardelli, N., Auvolat, A., Chintala, S., Lacroix, T., Lin, Z., Richoux, F., Usunier, N.: Torchcraft: a library for machine learning research on real-time strategy games. CoRR (2016). arXiv e-prints abs:1611.00625

835. Szegedy, C., Ioffe, S., Vanhoucke, V.: Inception-v4, inception-resnet and the impact of residual connections on learning. CoRR (2016). arXiv e-prints abs:1602.07261

836. Szegedy, C., Liu, W., Jia, Y., Sermanet, P., Reed, S.E., Anguelov, D., Erhan, D., Vanhoucke, V., Rabinovich, A.: Going deeper with convolutions (2014). arXiv e-prints abs:1409.4842

837. Szegedy, C., Vanhoucke, V., Ioffe, S., Shlens, J., Wojna, Z.: Rethinking the inception architecture for computer vision (2015). arXiv e-prints abs:1512.00567

838. Szegedy, C., Vanhoucke, V., Ioffe, S., Shlens, J., Wojna, Z.: Rethinking the inception architecture for computer vision. In: 2016 IEEE Conference on Computer Vision and Pattern Recognition (CVPR), pp. 2818–2826 (2016). https://doi.org/10.1109/CVPR.2016.308

839. Szeliski, R.: Computer Vision: Algorithms and Applications. Texts in Computer Science. Springer, London (2011)

840. Sánchez, M., Martínez, J.L., Morales, J., Robles, A., Morán, M.: Automatic generation of labeled 3d point clouds of natural environments with gazebo. In: 2019 IEEE International Conference on Mechatronics (ICM), vol. 1, pp. 161–166 (2019). https://doi.org/10.1109/ICMECH.2019.8722866

841. Tai, L., Liu, M.: Deep-learning in mobile robotics—from perception to control systems: A survey on why and why not. CoRR (2016). arXiv e-prints abs:1612.07139

842. Tai, L., Paolo, G., Liu, M.: Virtual-to-real deep reinforcement learning: Continuous control of mobile robots for mapless navigation. In: 2017 IEEE/RSJ International Conference on Intelligent Robots and Systems (IROS), pp. 31–36 (2017). https://doi.org/10.1109/IROS.2017.8202134

843. Taigman, Y., Yang, M., Ranzato, M., Wolf, L.: Deepface: Closing the gap to human-level performance in face verification. In: 2014 IEEE Conference on Computer Vision and Pattern Recognition, pp. 1701–1708 (2014). https://doi.org/10.1109/CVPR.2014.220

844. Tan, B., Xu, N., Kong, B.: Autonomous driving in reality with reinforcement learning and image translation. CoRR (2018). arXiv e-prints abs:1801.05299

845. Tan, B., Xu, N., Kong, B.: Autonomous driving in reality with reinforcement learning and image translation (2018). arXiv e-prints abs:1801.05299

846. Tan, M., Chen, B., Pang, R., Vasudevan, V., Le, Q.V.: Mnasnet: Platform-aware neural architecture search for mobile. CoRR (2018). arXiv e-prints abs:1807.11626

847. Tan, M., Le, Q.V.: Efficientnet: Rethinking model scaling for convolutional neural networks. CoRR (2019). arXiv e-prints abs:1905.11946

848. Tan, M., Pang, R., Le, Q.V.: Efficientdet: Scalable and efficient object detection. In: Proceedings of the IEEE/CVF Conference on Computer Vision and Pattern Recognition (CVPR) (2020)

849. Tang, Y., Tang, Y., Xiao, J., Summers, R.M.: Xlsor: A robust and accurate lung segmentor on chest x-rays using criss-cross attention and customized radiorealistic abnormalities generation. CoRR (2019). arXiv e-prints abs:1904.09229

850. Tang, Y.B., Oh, S., Tang, Y.X., Xiao, J., Summers, R.M.: Ct-realistic data augmentation using generative adversarial network for robust lymph node segmentation. In: SPIE Medical Imaging 2019: Computer-Aided Diagnosis, vol. 10950 (2019). https://doi.org/10.1117/12.2512004

851. Tang, Z., Shao, K., Zhu, Y., Li, D.M., Zhao, D., Huang, T.: A review of computational intelligence for starcraft ai. In: 2018 IEEE Symposium Series on Computational Intelligence (SSCI), pp. 1167–1173 (2018)

852. Taylor, C.J., Cowley, A.: Parsing indoor scenes using rgb-d imagery. In: Robotics: Science and Systems (2012)

853. Taylor, G.R., Chosak, A.J., Brewer, P.C.: Ovvv: Using virtual worlds to design and evaluate surveillance systems. In: 2007 IEEE Conference on Computer Vision and Pattern Recognition, pp. 1–8 (2007). https://doi.org/10.1109/CVPR.2007.383518

854. The SPRINT Research Group: A randomized trial of intensive versus standard blood-pressure control. N. Engl. J. Med. 373(22), 2103–2116 (2015). https://doi.org/10.1056/NEJMoa1511939. PMID: 26551272

855. Thieling, J., Roßmann, J.: Highly-scalable and generalized sensor structures for efficient physically-based simulation of multi-modal sensor networks. In: 2018 12th International Conference on Sensing Technology (ICST), pp. 202–207 (2018). https://doi.org/10.1109/ICSensT.2018.8603563

856. Tian, Y., Gong, Q., Shang, W., Wu, Y., Zitnick, L.: ELF: an extensive, lightweight and flexible research platform for real-time strategy games. CoRR (2017). arXiv e-prints abs:1707.01067

857. Tian, Y., Li, X., Wang, K., Wang, F.Y.: Training and testing object detectors with virtual images. IEEE/CAA J. Autom. Sinica 5, 539–546 (2018). https://doi.org/10.1109/JAS.2017.7510841

858. Times, N.Y.: New navy device learns by doing; psychologist shows embryo of computer designed to read and grow wiser (1958). https://www.nytimes.com/1958/07/08/archives/new-navy-device-learns-by-doing-psychologist-shows-embryo-of.html

859. To, T., Tremblay, J., McKay, D., Yamaguchi, Y., Leung, K., Balanon, A., Cheng, J., Hodge, W., Birchfield, S.: NDDS: NVIDIA deep learning dataset synthesizer (2018). https://github.com/NVIDIA/Dataset_Synthesizer

860. Tobin, J., Biewald, L., Duan, R., Andrychowicz, M., Handa, A., Kumar, V., McGrew, B., Ray, A., Schneider, J., Welinder, P., Zaremba, W., Abbeel, P.: Domain randomization and generative models for robotic grasping. In: 2018 IEEE/RSJ International Conference on Intelligent Robots and Systems (IROS), pp. 3482–3489 (2018). https://doi.org/10.1109/IROS.2018.8593933

861. Tobin, J., Fong, R., Ray, A., Schneider, J., Zaremba, W., Abbeel, P.: Domain randomization for transferring deep neural networks from simulation to the real world. In: 2017 IEEE/RSJ International Conference on Intelligent Robots and Systems (IROS), pp. 23–30 (2017). https://doi.org/10.1109/IROS.2017.8202133

862. Todorov, E., Erez, T., Tassa, Y.: Mujoco: A physics engine for model-based control. In: 2012 IEEE/RSJ International Conference on Intelligent Robots and Systems, pp. 5026–5033 (2012)

863. Togelius, J., Kastbjerg, E., Schedl, D., Yannakakis, G.N.: What is procedural content generation?: Mario on the borderline. In: Proceedings of the 2Nd International Workshop on Procedural Content Generation in Games, PCGames '11, pp. 3:1–3:6. ACM, New York, NY, USA (2011). https://doi.org/10.1145/2000919.2000922

864. Tommasi, T., Patricia, N., Caputo, B., Tuytelaars, T.: A Deeper Look at Dataset Bias, pp. 37–55. Springer International Publishing, Cham (2017). https://doi.org/10.1007/978-3-319-58347-1_2

865. Tran, A.T., Hassner, T., Masi, I., Medioni, G.G.: Regressing robust and discriminative 3d morphable models with a very deep neural network. In: 2017 IEEE Conference on Computer Vision and Pattern Recognition (CVPR), pp. 1493–1502 (2017)

866. Tran, L., Yin, X., Liu, X.: Disentangled representation learning gan for pose-invariant face recognition. In: 2017 IEEE Conference on Computer Vision and Pattern Recognition (CVPR), pp. 1283–1292 (2017). https://doi.org/10.1109/CVPR.2017.141

867. Tremblay, J., Prakash, A., Acuna, D., Brophy, M., Jampani, V., Anil, C., To, T., Cameracci, E., Boochoon, S., Birchfield, S.T.: Training deep networks with synthetic data: Bridging the reality gap by domain randomization. In: 2018 IEEE/CVF Conference on Computer Vision and Pattern Recognition Workshops (CVPRW), pp. 1082–10828 (2018)

868. Tremblay, J., To, T., Birchfield, S.: Falling things: A synthetic dataset for 3d object detection and pose estimation. In: 2018 IEEE/CVF Conference on Computer Vision and Pattern Recognition Workshops (CVPRW), pp. 2119–21193 (2018). https://doi.org/10.1109/CVPRW.2018.00275

869. Tremblay, J., To, T., Sundaralingam, B., Xiang, Y., Fox, D., Birchfield, S.T.: Deep object pose estimation for semantic robotic grasping of household objects. In: CoRL (2018)

870. Triastcyn, A., Faltings, B.: Generating Artificial Data for Private Deep Learning (2018). arXiv e-prints arXiv:1803.03148

871. Triastcyn, A., Faltings, B.: Generating differentially private datasets using gans. CoRR (2018). arXiv e-prints abs:1803.03148

872. Trieb, R., Von Puttkamer, E.: The 3d7-simulation environment: a tool for autonomous mobile robot development. In: Proceedings of International Workshop on Modeling, Analysis and Simulation of Computer and Telecommunication Systems, pp. 358–361 (1994). https://doi.org/10.1109/MASCOT.1994.284398

873. Triyonoputro, J.C., Wan, W., Harada, K.: Quickly inserting pegs into uncertain holes using multi-view images and deep network trained on synthetic data (2019). arXiv e-prints abs:1902.09157

874. Trumble, M., Gilbert, A., Malleson, C., Hilton, A., Collomosse, J.: Total capture: 3d human pose estimation fusing video and inertial sensors. BMVC (2017). https://doi.org/10.5244/C.31.14

875. Tsirikoglou, A., Eilertsen, G., Unger, J.: A survey of image synthesis methods for visual machine learning. Comput. Graph. Forum **39**(6), 426–451 (2020). https://doi.org/10.1111/cgf.14047

876. Tsirikoglou, A., Kronander, J., Wrenninge, M., Unger, J.: Procedural modeling and physically based rendering for synthetic data generation in automotive applications. CoRR (2017). arXiv e-prints abs:1710.06270

877. Turing, A.M.: Computing machinery and intelligence. Mind **59**(236), 433–460 (1950). http://www.jstor.org/stable/2251299

878. Turing, A.M.: Computing Machinery and Intelligence, pp. 11–35. MIT Press, Cambridge, MA, USA (1995)

879. Turk, M.A., Morgenthaler, D.G., Gremban, K.D., Marra, M.: Vits-a vision system for autonomous land vehicle navigation. IEEE Trans. Pattern Anal. Mach. Intell. **10**(3), 342–361 (1988)

880. Tzeng, E., Devin, C., Hoffman, J., Finn, C., Peng, X., Levine, S., Saenko, K., Darrell, T.: Towards adapting deep visuomotor representations from simulated to real environments. CoRR (2015). arXiv e-prints abs:1511.07111

881. Tzeng, E., Hoffman, J., Darrell, T., Saenko, K.: Simultaneous deep transfer across domains and tasks. In: Proceedings of the 2015 IEEE International Conference on Computer Vision (ICCV), ICCV '15, pp. 4068–4076. IEEE Computer Society, Washington, DC, USA (2015). https://doi.org/10.1109/ICCV.2015.463

882. Tzeng, E., Hoffman, J., Zhang, N., Saenko, K., Darrell, T.: Deep domain confusion: Maximizing for domain invariance. CoRR (2014). arXiv e-prints abs:1412.3474

883. Ubbens, J., Cieslak, M., Prusinkiewicz, P., Stavness, I.: The use of plant models in deep learning: an application to leaf counting in rosette plants. Plant Methods **14**(1), 6 (2018). https://doi.org/10.1186/s13007-018-0273-z

884. Uijlings, J., van de Sande, K., Gevers, T., Smeulders, A.: Selective search for object recognition. Int. J. Comput. Vision (2013)

885. Ullah, I., Abinesh, S., Smyth, D.L., Karimi, N.B., Drury, B., Glavin, F.G., Madden, M.G.: A virtual testbed for critical incident investigation with autonomous remote aerial vehicle surveying, artificial intelligence, and decision support. In: Alzate, C., Monreale, A., Assem, H., Bifet, A., Buda, T.S., Caglayan, B., Drury, B., García-Martín, E., Gavaldà, R., Koprinska, I., Kramer, S., Lavesson, N., Madden, M., Molloy, I., Nicolae, M.I., Sinn, M. (eds.) ECML PKDD 2018 Workshops, pp. 216–221. Springer International Publishing, Cham (2019)
886. Ulyanov, D., Lebedev, V., Vedaldi, A., Lempitsky, V.: Texture networks: Feed-forward synthesis of textures and stylized images. In: Proceedings of the 33rd International Conference on International Conference on Machine Learning—Volume 48, ICML'16, pp. 1349–1357. JMLR.org (2016)
887. Uria, B., Côté, M.A., Gregor, K., Murray, I., Larochelle, H.: Neural autoregressive distribution estimation. J. Mach. Learn. Res. **17**(205), 1–37 (2016). http://jmlr.org/papers/v17/16-272.html
888. Vahdat, A., Kautz, J.: Nvae: A deep hierarchical variational autoencoder (2021)
889. Varol, G., Romero, J., Martin, X., Mahmood, N., Black, M.J., Laptev, I., Schmid, C.: Learning from synthetic humans. CoRR (2017). arXiv e-prints abs:1701.01370
890. Vaswani, A., Bengio, S., Brevdo, E., Chollet, F., Gomez, A.N., Gouws, S., Jones, L., Kaiser, L., Kalchbrenner, N., Parmar, N., Sepassi, R., Shazeer, N., Uszkoreit, J.: Tensor2tensor for neural machine translation. CoRR (2018). arXiv e-prints abs:1803.07416
891. Vaswani, A., Shazeer, N., Parmar, N., Uszkoreit, J., Jones, L., Gomez, A.N., Kaiser, L.u., Polosukhin, I.: Attention is all you need. In: I. Guyon, U.V. Luxburg, S. Bengio, H. Wallach, R. Fergus, S. Vishwanathan, R. Garnett (eds.) Advances in Neural Information Processing Systems 30, pp. 5998–6008. Curran Associates, Inc. (2017)
892. Vázquez, D., López, A.M., Marín, J., Ponsa, D., Gerónimo, D.: Virtual and real world adaptation for pedestrian detection. IEEE Trans. Pattern Anal. Mach. Intell. **36**(4), 797–809 (2014). https://doi.org/10.1109/TPAMI.2013.163
893. Veach, E., Guibas, L.J.: Metropolis light transport. In: Proceedings of the 24th Annual Conference on Computer Graphics and Interactive Techniques, SIGGRAPH '97, pp. 65–76. ACM Press/Addison-Wesley Publishing Co., New York, NY, USA (1997). https://doi.org/10.1145/258734.258775
894. Vijayakumar, A.J., Mohta, A., Polozov, O., Batra, D., Jain, P., Gulwani, S.: Neural-guided deductive search for real-time program synthesis from examples. CoRR (2018). arXiv e-prints abs:1804.01186
895. Vincent, P., Hugo, L., Bengio, Y., Manzagol, P.A.: Extracting and composing robust features with denoising autoencoders. In: Proceedings of the 25th international conference on Machine learning, ICML '08, pp. 1096–1103. ACM, New York, NY, USA (2008). https://doi.org/10.1145/1390156.1390294
896. Vinyals, O., Ewalds, T., Bartunov, S., Georgiev, P., Vezhnevets, A.S., Yeo, M., Makhzani, A., Küttler, H., Agapiou, J., Schrittwieser, J., Quan, J., Gaffney, S., Petersen, S., Simonyan, K., Schaul, T., van Hasselt, H., Silver, D., Lillicrap, T.P., Calderone, K., Keet, P., Brunasso, A., Lawrence, D., Ekermo, A., Repp, J., Tsing, R.: Starcraft II: A new challenge for reinforcement learning. CoRR (2017). arXiv e-prints abs:1708.04782
897. Vuong, Q.V., Vikram, S., Su, H., Gao, S., Christensen, H.I.: How to pick the domain randomization parameters for sim-to-real transfer of reinforcement learning policies? (2019). arXiv e-prints abs:1903.11774
898. Walonoski, J., Kramer, M., Nichols, J., Quina, A., Moesel, C., Hall, D., Duffett, C., Dube, K., Gallagher, T., McLachlan, S.: Synthea: An approach, method, and software mechanism for generating synthetic patients and the synthetic electronic health care record. J. Am. Med. Inform. Assoc. **25**(3), 230–238 (2017)
899. Wandarosanza, R., Trilaksono, B.R., Hidayat, E.: Hardware-in-the-loop simulation of uav hexacopter for chemical hazard monitoring mission. In: 2016 6th International Conference on System Engineering and Technology (ICSET), pp. 189–193 (2016). https://doi.org/10.1109/ICSEngT.2016.7849648

900. Wang, C.Y., Liao, H.Y.M., Yeh, I.H., Wu, Y.H., Chen, P.Y., Hsieh, J.W.: CSPNet: A New Backbone that can Enhance Learning Capability of CNN. arXiv e-prints arXiv:1911.11929 (2019)

901. Wang, F., Zhuang, Y., Gu, H., Hu, H.: Automatic generation of synthetic lidar point clouds for 3-d data analysis. IEEE Trans. Instrum. Meas. **68**(7), 2671–2673 (2019). https://doi.org/10.1109/TIM.2019.2906416

902. Wang, H., Kang, B., Kim, D.: Pfw: A face database in the wild for studying face identification and verification in uncontrolled environment. In: 2013 2nd IAPR Asian Conference on Pattern Recognition, pp. 356–360 (2013). https://doi.org/10.1109/ACPR.2013.53

903. Wang, H., Wang, Q., Yang, F., Zhang, W., Zuo, W.: Data augmentation for object detection via progressive and selective instance-switching (2019). arXiv e-prints abs:1906.00358

904. Wang, K., Shi, F., Wang, W., Nan, Y., Lian, S.: Synthetic data generation and adaption for object detection in smart vending machines. CoRR (2019). arXiv e-prints abs:1904.12294

905. Wang, K., Zhao, R., Ji, Q.: A hierarchical generative model for eye image synthesis and eye gaze estimation. In: 2018 IEEE/CVF Conference on Computer Vision and Pattern Recognition, pp. 440–448 (2018). https://doi.org/10.1109/CVPR.2018.00053

906. Wang, K., Zheng, J., Seah, H.S., Ma, Y.: Triangular mesh deformation via edge-based graph. Comput. Aided Des. Appl. **9**(3), 345–359 (2012). https://doi.org/10.3722/cadaps.2012.345-359

907. Wang, Q., Gao, J., Lin, W., Yuan, Y.: Learning from synthetic data for crowd counting in the wild. CoRR (2019). arXiv e-prints abs:1903.03303

908. Wang, S., Jia, D., Weng, X.: Deep reinforcement learning for autonomous driving. CoRR (2018). arXiv e-prints abs:1811.11329

909. Wang, T., Liu, M., Zhu, J., Tao, A., Kautz, J., Catanzaro, B.: High-resolution image synthesis and semantic manipulation with conditional gans. In: 2018 IEEE/CVF Conference on Computer Vision and Pattern Recognition, pp. 8798–8807 (2018). https://doi.org/10.1109/CVPR.2018.00917

910. Wang, T., Wu, D.J., Coates, A., Ng, A.Y.: End-to-end text recognition with convolutional neural networks. In: Proceedings of the 21st International Conference on Pattern Recognition (ICPR2012), pp. 3304–3308 (2012)

911. Wang, W., Zheng, V.W., Yu, H., Miao, C.: A survey of zero-shot learning: Settings, methods, and applications. ACM Trans. Intell. Syst. Technol. **10**(2) (2019). https://doi.org/10.1145/3293318

912. Wang, W.Y., Yang, D.: That's so annoying!!!: A lexical and frame-semantic embedding based data augmentation approach to automatic categorization of annoying behaviors using #petpeeve tweets. In: EMNLP (2015)

913. Wang, X., Pham, H., Dai, Z., Neubig, G.: Switchout: an efficient data augmentation algorithm for neural machine translation. In: Proceedings of the 2018 Conference on Empirical Methods in Natural Language Processing, pp. 856–861. Association for Computational Linguistics, Brussels, Belgium (2018)

914. Wang, X., Wang, K., Lian, S.: A survey on face data augmentation. CoRR (2019). arXiv e-prints abs:1904.11685

915. Wang, X., Wang, S., Zhang, S., Fu, T., Shi, H., Mei, T.: Support vector guided softmax loss for face recognition. CoRR (2018). arXiv e-prints abs:1812.11317

916. Wang, Y., Yao, Q.: Few-shot learning: A survey. CoRR (2019). arXiv e-prints abs:1904.05046

917. Wang, Y., Yao, Q., Kwok, J.T., Ni, L.M.: Generalizing from a few examples: A survey on few-shot learning. ACM Comput. Surv. **53**(3) (2020). https://doi.org/10.1145/3386252

918. Wang, Y.X., Fienberg, S.E., Smola, A.J.: Privacy for free: Posterior sampling and stochastic gradient monte carlo. In: Proceedings of the 32Nd International Conference on International Conference on Machine Learning—Volume 37, ICML'15, pp. 2493–2502. JMLR.org (2015)

919. Wang, Z., Chai, J., Xia, S.: Combining recurrent neural networks and adversarial training for human motion synthesis and control. CoRR (2018). arXiv e-prints abs:1806.08666

920. Ward, D., Moghadam, P., Hudson, N.: Deep leaf segmentation using synthetic data. In: BMVC (2018)

921. Warren, M., Upcroft, B.: Robust scale initialization for long-range stereo visual odometry. In: 2013 IEEE/RSJ International Conference on Intelligent Robots and Systems, pp. 2115–2121 (2013). https://doi.org/10.1109/IROS.2013.6696652

922. Weber, M.: Where to? a history of autonomous vehicles (2014). https://computerhistory.org/blog/where-to-a-history-of-autonomous-vehicles/

923. Weininger, D.: Smiles, a chemical language and information system. 1. introduction to methodology and encoding rules. J. Chem. Inf. Comput. Sci. **28**(1), 31–36 (1988). https://doi.org/10.1021/ci00057a005

924. Werbos, P.J.: Applications of advances in nonlinear sensitivity analysis. In: Proceedings of the 10th IFIP Conference, 31.8–4.9, NYC, pp. 762–770 (1981)

925. Wiesel, D.H., Hubel, T.N.: Receptive fields of single neurones in the cat's striate cortex. J. Physiol. **148**, 574–591 (1959)

926. Wiles, J., Elman, J.L.: Learning to count without a counter: A case study of dynamics and activation landscapes in recurrent networks. In: ACCSS (1995)

927. Wilson, A.C., Roelofs, R., Stern, M., Srebro, N., Recht, B.: The marginal value of adaptive gradient methods in machine learning. In: I. Guyon, U.V. Luxburg, S. Bengio, H. Wallach, R. Fergus, S. Vishwanathan, R. Garnett (eds.) Advances in Neural Information Processing Systems 30, pp. 4148–4158. Curran Associates, Inc. (2017)

928. Wilson, A.G.: The Case for Bayesian Deep Learning (2020). arXiv e-prints arXiv:2001.10995

929. Wilson, A.G., Izmailov, P.: Bayesian Deep Learning and a Probabilistic Perspective of Generalization (2020). arXiv e-prints arXiv:2002.08791

930. Wistuba, M., Rawat, A., Pedapati, T.: A survey on neural architecture search. CoRR (2019). arXiv e-prints abs:1905.01392

931. Wohlhart, P., Lepetit, V.: Learning descriptors for object recognition and 3d pose estimation. 2015 IEEE Conference on Computer Vision and Pattern Recognition (CVPR) pp. 3109–3118 (2015)

932. Wolterink, J.M., Dinkla, A.M., Savenije, M.H.F., Seevinck, P.R., van den Berg, C.A.T., Isgum, I.: Deep MR to CT synthesis using unpaired data. CoRR (2017). arXiv e-prints abs:1708.01155

933. Wood, E., Baltruaitis, T., Zhang, X., Sugano, Y., Robinson, P., Bulling, A.: Rendering of eyes for eye-shape registration and gaze estimation. In: 2015 IEEE International Conference on Computer Vision (ICCV), pp. 3756–3764 (2015). https://doi.org/10.1109/ICCV.2015.428

934. Wood, E., Baltrušaitis, T., Morency, L.P., Robinson, P., Bulling, A.: Learning an appearance-based gaze estimator from one million synthesised images. In: Proceedings of the Ninth Biennial ACM Symposium on Eye Tracking Research & Applications, ETRA '16, pp. 131–138. ACM, New York, NY, USA (2016). https://doi.org/10.1145/2857491.2857492

935. Wrenninge, M., Unger, J.: Synscapes: A photorealistic synthetic dataset for street scene parsing. CoRR (2018). arXiv e-prints abs:1810.08705

936. Wu, B., Wan, A., Yue, X., Keutzer, K.: Squeezeseg: Convolutional neural nets with recurrent CRF for real-time road-object segmentation from 3d lidar point cloud. CoRR (2017). arXiv e-prints abs:1710.07368

937. Wu, B., Zhou, X., Zhao, S., Yue, X., Keutzer, K.. Squeezesegv2: Improved model structure and unsupervised domain adaptation for road-object segmentation from a lidar point cloud. CoRR (2018). arXiv e-prints abs:1809.08495

938. Wu, F., Liu, J., Wu, C., Huang, Y., Xie, X.: Neural chinese named entity recognition via cnn-lstm-crf and joint training with word segmentation. In: The World Wide Web Conference, WWW '19, p. 3342–3348. Association for Computing Machinery, New York, NY, USA (2019). https://doi.org/10.1145/3308558.3313743

939. Wu, H., Zhang, J., Huang, K.: Msc: A dataset for macro-management in starcraft ii (2017). arXiv e-prints abs:1710.03131

940. Wu, S., Huang, H., Portenier, T., Sela, M., Cohen-Or, D., Kimmel, R., Zwicker, M.: Specular-to-diffuse translation for multi-view reconstruction. CoRR (2018). arXiv e-prints abs:1807.05439

941. Wu, X., He, R., Sun, Z., Tan, T.: A light cnn for deep face representation with noisy labels. IEEE Trans. Inf. Forensics Secur. **13**, 2884–2896 (2018)

942. Wu, Y., Tian, Y.: Training agent for first-person shooter game with actor-critic curriculum learning. In: ICLR (2017)

943. Wu, Y., Wu, Y., Gkioxari, G., Tian, Y.: Building generalizable agents with a realistic and rich 3d environment. CoRR (2018). arXiv e-prints abs:1801.02209

944. Wu, Z., Song, S., Khosla, A., Tang, X., Xiao, J.: 3d shapenets for 2.5d object recognition and next-best-view prediction (2014). arXiv e-prints abs:1406.5670

945. Wulfmeier, M., Bewley, A., Posner, I.: Addressing appearance change in outdoor robotics with adversarial domain adaptation. 2017 IEEE/RSJ International Conference on Intelligent Robots and Systems (IROS) pp. 1551–1558 (2017)

946. Wymann, B., Espié, E., Guionneau, C., Dimitrakakis, C., Coulom, R., Sumner, A.: TORCS, The Open Racing Car Simulator (2014). http://www.torcs.org

947. Xia, F., Zamir, A.R., He, Z., Sax, A., Malik, J., Savarese, S.: Gibson env: Real-world perception for embodied agents. CoRR (2018). arXiv e-prints abs:1808.10654

948. Xia, X., Kulis, B.: W-net: A deep model for fully unsupervised image segmentation. CoRR (2017). arXiv e-prints abs:1711.08506

949. Xian, Y., Lampert, C.H., Schiele, B., Akata, Z.: Zero-shot learning—A comprehensive evaluation of the good, the bad and the ugly. CoRR (2017). arXiv e-prints abs:1707.00600

950. Xiang, Y., Alahi, A., Savarese, S.: Learning to track: Online multi-object tracking by decision making. In: 2015 IEEE International Conference on Computer Vision (ICCV), pp. 4705–4713 (2015). https://doi.org/10.1109/ICCV.2015.534

951. Xiao, J., Owens, A., Torralba, A.: Sun3d: A database of big spaces reconstructed using sfm and object labels. In: 2013 IEEE International Conference on Computer Vision, pp. 1625–1632 (2013). https://doi.org/10.1109/ICCV.2013.458

952. Xiao, S., Feng, J., Xing, J., Lai, H., Yan, S., Kassim, A.: Robust facial landmark detection via recurrent attentive-refinement networks. In: Leibe, B., Matas, J., Sebe, N., Welling, M. (eds.) Computer Vision—ECCV 2016, pp. 57–72. Springer International Publishing, Cham (2016)

953. Xiao, X., Ganguli, S., Pandey, V.: Vae-info-cgan: Generating synthetic images by combining pixel-level and feature-level geospatial conditional inputs. In: Proceedings of the 13th ACM SIGSPATIAL International Workshop on Computational Transportation Science, IWCTS '20. Association for Computing Machinery, New York, NY, USA (2020). https://doi.org/10.1145/3423457.3429361

954. Xiaoxia Hu, Xuefeng Liu, Zhenming He, Jiahua Zhang: Batch modeling of 3d city based on esri cityengine. In: IET International Conference on Smart and Sustainable City 2013 (ICSSC 2013), pp. 69–73 (2013). https://doi.org/10.1049/cp.2013.1979

955. Xie, L., Lin, K., Wang, S., Wang, F., Zhou, J.: Differentially private generative adversarial network. CoRR (2018). arXiv e-prints abs:1802.06739

956. Xie, Q., Luong, M.T., Hovy, E., Le, Q.V.: Self-training with Noisy Student improves ImageNet classification (2019). arXiv e-prints arXiv:1911.04252

957. Xie, X., Liu, H., Zhang, Z., Qiu, Y., Gao, F., Qi, S., Zhu, Y., Zhu, S.C.: Vrgym: A virtual testbed for physical and interactive ai. In: Proceedings of the ACM Turing Celebration Conference—China, ACM TURC '19, pp. 100:1–100:6. ACM, New York, NY, USA (2019). https://doi.org/10.1145/3321408.3322633

958. Xie, Z., Wang, S.I., Li, J., Lévy, D., Nie, A., Jurafsky, D., Ng, A.Y.: Data noising as smoothing in neural network language models. CoRR (2017). arXiv e-prints abs:1703.02573

959. Xiong, H.Y., Alipanahi, B., Lee, L.J., Bretschneider, H., Merico, D., Yuen, R.K.C., Hua, Y., Gueroussov, S., Najafabadi, H.S., Hughes, T.R., Morris, Q., Barash, Y., Krainer, A.R., Jojic, N., Scherer, S.W., Blencowe, B.J., Frey, B.J.: The human splicing code reveals new insights into the genetic determinants of disease. Science **347**(6218) (2015). https://doi.org/10.1126/science.1254806

960. Xiong, X., Wang, J., Zhang, F., Li, K.: Combining deep reinforcement learning and safety based control for autonomous driving. CoRR (2016). arXiv e-prints abs:1612.00147

961. Xu, D., Ouyang, W., Wang, X., Sebe, N.: Pad-net: Multi-tasks guided prediction-and-distillation network for simultaneous depth estimation and scene parsing. In: 2018 IEEE/CVF Conference on Computer Vision and Pattern Recognition, pp. 675–684 (2018)

962. Xu, G.Y., Zhang, Q., Liu, D., Lin, G., Wang, J., Zhang, Y.: Adversarial adaptation from synthesis to reality in fast detector for smoke detection. IEEE Access **7**, 29471–29483 (2019)

963. Xu, H., Ma, Y., Liu, H., Deb, D., Liu, H., Tang, J., Jain, A.K.: Adversarial attacks and defenses in images, graphs and text: A review. CoRR (2019). arXiv e-prints abs:1909.08072

964. Xu, J., Vázquez, D., Ramos, S., López, A.M., Ponsa, D.: Adapting a pedestrian detector by boosting lda exemplar classifiers. In: 2013 IEEE Conference on Computer Vision and Pattern Recognition Workshops, pp. 688–693 (2013). https://doi.org/10.1109/CVPRW.2013.104

965. Xu, X., Dou, P., Le, H.A., Kakadiaris, I.A.: When 3d-aided 2d face recognition meets deep learning: An extended UR2D for pose-invariant face recognition. CoRR (2017). arXiv e-prints abs:1709.06532

966. Yan, M., Zhao, M., Xu, Z., Zhang, Q., Wang, G., Su, Z.: Vargfacenet: An efficient variable group convolutional neural network for lightweight face recognition. CoRR (2019). arXiv e-prints abs:1910.04985

967. Yan, P., Xu, S., Rastinehad, A.R., Wood, B.J.: Adversarial image registration with application for MR and TRUS image fusion. CoRR (2018). arXiv e-prints abs:1804.11024

968. Yang, G.R., Ganichev, I., Wang, X.J., Shlens, J., Sussillo, D.: A dataset and architecture for visual reasoning with a working memory. In: ECCV (2018)

969. Yang, L., Liang, X., Xing, E.P.: Unsupervised real-to-virtual domain unification for end-to-end highway driving. CoRR (2018). arXiv e-prints abs:1801.03458

970. Yang, W., Li, S., Ouyang, W., Li, H., Wang, X.: Learning feature pyramids for human pose estimation. CoRR (2017). arXiv e-prints abs:1708.01101

971. Yi, L., Guibas, L., Hertzmann, A., Kim, V.G., Su, H., Yumer, E.: Learning hierarchical shape segmentation and labeling from online repositories. ACM Trans. Graph. **36**(4), 70:1–70:12 (2017). https://doi.org/10.1145/3072959.3073652

972. Yi, L., Kim, V.G., Ceylan, D., Shen, I.C., Yan, M., Su, H., Lu, C., Huang, Q., Sheffer, A., Guibas, L.: A scalable active framework for region annotation in 3d shape collections. ACM Trans. Graph. **35**(6), 210:1–210:12 (2016). https://doi.org/10.1145/2980179.2980238

973. Yi, L., Shao, L., Savva, M., Huang, H., Zhou, Y., Wang, Q., Graham, B., Engelcke, M., Klokov, R., Lempitsky, V.S., Gan, Y., Wang, P., Liu, K., Yu, F., Shui, P., Hu, B., Zhang, Y., Li, Y., Bu, R., Sun, M., Wu, W., Jeong, M., Choi, J., Kim, C., Geetchandra, A., Murthy, N., Ramu, B., Manda, B., Ramanathan, M., Kumar, G., Preetham, P., Srivastava, S., Bhugra, S., Lall, B., Häne, C., Tulsiani, S., Malik, J., Lafer, J., Jones, R., Li, S., Lu, J., Jin, S., Yu, J., Huang, Q., Kalogerakis, E., Savarese, S., Hanrahan, P., Funkhouser, T.A., Su, H., Guibas, L.J.: Large-scale 3d shape reconstruction and segmentation from shapenet core55. CoRR (2017). arXiv e-prints abs:1710.06104

974. Yi, X., Walia, E., Babyn, P.: Generative adversarial network in medical imaging: A review. CoRR (2018). arXiv e-prints abs:1809.07294

975. Yogamani, S., Hughes, C., Horgan, J., Sistu, G., Varley, P., O'Dea, D., Uricár, M., Milz, S., Simon, M., Amende, K., Witt, C., Rashed, H., Chennupati, S., Nayak, S., Mansoor, S., Perroton, X., Perez, P.: Woodscape: A multi-task, multi-camera fisheye dataset for autonomous driving. CoRR (2019). arXiv e-prints abs:1905.01489

976. Yoon, J., Jordon, J., van der Schaar, M.: PATE-GAN: Generating synthetic data with differential privacy guarantees. In: International Conference on Learning Representations (2019)

977. You, Y., Pan, X., Wang, Z., Lu, C.: Virtual to real reinforcement learning for autonomous driving. CoRR (2017). arXiv e-prints abs:1704.03952

978. Yu, B., Fan, Z.: A comprehensive review of conditional random fields: variants, hybrids and applications. Artif. Intell. Rev. **53**, 1–45 (2020). https://doi.org/10.1007/s10462-019-09793-6

979. Yu, B., Zhou, L., Wang, L., Fripp, J., Bourgeat, P.: 3d cgan based cross-modality mr image synthesis for brain tumor segmentation. In: 2018 IEEE 15th International Symposium on Biomedical Imaging (ISBI 2018), pp. 626–630 (2018). https://doi.org/10.1109/ISBI.2018.8363653

980. Yu, F., Koltun, V.: Multi-Scale Context Aggregation by Dilated Convolutions. arXiv e-prints arXiv:1511.07122 (2015)

981. Yu, F., Xian, W., Chen, Y., Liu, F., Liao, M., Madhavan, V., Darrell, T.: BDD100K: A diverse driving video database with scalable annotation tooling. CoRR (2018). arXiv e-prints abs:1805.04687

982. Yu, L., Zhang, W., Wang, J., Yu, Y.: Seqgan: Sequence generative adversarial nets with policy gradient. In: Proceedings of the Thirty-First AAAI Conference on Artificial Intelligence, AAAI'17, pp. 2852–2858. AAAI Press (2017)

983. Yue, X., Wu, B., Seshia, S.A., Keutzer, K., Sangiovanni-Vincentelli, A.L.: A lidar point cloud generator: from a virtual world to autonomous driving. In: ICMR (2018)

984. Yue, X., Zhang, Y., Zhao, S., Sangiovanni-Vincentelli, A., Keutzer, K., Gong, B.: Domain randomization and pyramid consistency: Simulation-to-real generalization without accessing target domain data (2019). arXiv e-prints abs:1909.00889

985. Yılmaz, B., Amasyalı, M.F., Balcılar, M., Uslu, E., Yavuz, S.: Impact of artificial dataset enlargement on performance of deformable part models. In: 2016 24th Signal Processing and Communication Application Conference (SIU), pp. 193–196 (2016). https://doi.org/10.1109/SIU.2016.7495710

986. Zakharov, S., Kehl, W., Ilic, S.: Deceptionnet: Network-driven domain randomization. CoRR (2019). arXiv e-prints abs:1904.02750

987. Zaremba, W., Sutskever, I.: Learning to execute. CoRR (2014). arXiv e-prints abs:1410.4615

988. Zavershynskyi, M., Skidanov, A., Polosukhin, I.: NAPS: natural program synthesis dataset. CoRR (2018). arXiv e-prints abs:1807.03168

989. Zeiler, M.D.: ADADELTA: an adaptive learning rate method. CoRR (2012). arXiv e-prints abs:1212.5701

990. Zellers, R., Holtzman, A., Rashkin, H., Bisk, Y., Farhadi, A., Roesner, F., Choi, Y.: Defending against neural fake news. CoRR (2019). arXiv e-prints abs:1905.12616

991. Zhang, B., Zhao, W., Liu, J., Wu, R., Tang, X.: Character recognition in natural scene images using local description. In: Zhang, Y., Zhou, Z.H., Zhang, C., Li, Y. (eds.) Intelligent Science and Intelligent Data Engineering, pp. 193–200. Springer, Berlin, Heidelberg (2012)

992. Zhang, C., Bengio, S., Hardt, M., Recht, B., Vinyals, O.: Understanding deep learning requires rethinking generalization. CoRR (2016). arXiv e-prints abs:1611.03530

993. Zhang, H., Cissé, M., Dauphin, Y.N., Lopez-Paz, D.: mixup: Beyond empirical risk minimization. CoRR (2017). arXiv e-prints abs:1710.09412

994. Zhang, H., Xu, T., Li, H., Zhang, S., Huang, X., Wang, X., Metaxas, D.N.: Stackgan: Text to photo-realistic image synthesis with stacked generative adversarial networks. CoRR (2016). arXiv e-prints abs:1612.03242

995. Zhang, H.B., Lei, Q., Zhong, B.N., Du, J.X., Peng, J.: A survey on human pose estimation. Intell. Autom. Soft Comput. 22(3), 483–489 (2016). https://doi.org/10.1080/10798587.2015.1095419

996. Zhang, H.B., Zhang, Y.X., Zhong, B., Lei, Q., Yang, L., Du, J.X., Chen, D.S.: A comprehensive survey of vision-based human action recognition methods. Sensors 19, 1005 (2019). https://doi.org/10.3390/s19051005

997. Zhang, J., Cormode, G., Procopiuc, C.M., Srivastava, D., Xiao, X.: Privbayes: Private data release via bayesian networks. ACM Trans. Database Syst. 42(4), 25:1–25:41 (2017). https://doi.org/10.1145/3134428

998. Zhang, J., Li, W., Ogunbona, P., Xu, D.: Recent advances in transfer learning for cross-dataset visual recognition: A problem-oriented perspective. ACM Comput. Surv. 52(1), 7:1–7:38 (2019). https://doi.org/10.1145/3291124

999. Zhang, J., Tai, L., Xiong, Y., Liu, M., Boedecker, J., Burgard, W.: VR goggles for robots: real-to-sim domain adaptation for visual control. CoRR (2018). arXiv e-prints abs:1802.00265

1000. Zhang, K., Zhang, Z., Li, Z., Qiao, Y.: Joint face detection and alignment using multitask cascaded convolutional networks. IEEE Signal Process. Lett. 23(10), 1499–1503 (2016). https://doi.org/10.1109/LSP.2016.2603342

1001. Zhang, S.H., Zhang, S.J., Liang, Y., Hall, P.: A survey of 3d indoor scene synthesis. J. Comput. Sci. Technol. **34**, 594–608 (2019)

1002. Zhang, X., Ji, S., Wang, T.: Differentially private releasing via deep generative model. CoRR (2018). arXiv e-prints abs:1801.01594

1003. Zhang, X., Sugano, Y., Fritz, M., Bulling, A.: Appearance-based gaze estimation in the wild. In: 2015 IEEE Conference on Computer Vision and Pattern Recognition (CVPR), pp. 4511–4520 (2015). https://doi.org/10.1109/CVPR.2015.7299081

1004. Zhang, X., Zhao, J., LeCun, Y.: Character-level convolutional networks for text classification. In: C. Cortes, N.D. Lawrence, D.D. Lee, M. Sugiyama, R. Garnett (eds.) Advances in Neural Information Processing Systems 28, pp. 649–657. Curran Associates, Inc. (2015)

1005. Zhang, Y., David, P., Foroosh, H., Gong, B.: A curriculum domain adaptation approach to the semantic segmentation of urban scenes. CoRR (2018). arXiv e-prints abs:1812.09953

1006. Zhang, Y., David, P., Gong, B.: Curriculum domain adaptation for semantic segmentation of urban scenes. CoRR (2017). arXiv e-prints abs:1707.09465

1007. Zhang, Y., Gan, Z., Fan, K., Chen, Z., Henao, R., Shen, D., Carin, L.: Adversarial feature matching for text generation (2017). arXiv e-prints arXiv:1706.03850

1008. Zhang, Y., Qiu, W., Chen, Q., Hu, X.C., Yuille, A.L.: Unrealstereo: A synthetic dataset for analyzing stereo vision (2016). arXiv e-prints arXiv:1612.04647

1009. Zhang, Y., Song, S., Yumer, E., Savva, M., Lee, J., Jin, H., Funkhouser, T.A.: Physically-based rendering for indoor scene understanding using convolutional neural networks. CoRR (2016). arXiv e-prints abs:1612.07429

1010. Zhang, Y., Wallace, B.C.: A sensitivity analysis of (and practitioners' guide to) convolutional neural networks for sentence classification. CoRR (2015). arXiv e-prints abs:1510.03820

1011. Zhang, Y., Yang, L., Chen, J., Fredericksen, M., Hughes, D.P., Chen, D.Z.: Deep adversarial networks for biomedical image segmentation utilizing unannotated images. In: Descoteaux, M., Maier-Hein, L., Franz, A., Jannin, P., Collins, D.L., Duchesne, S. (eds.) Medical Image Computing and Computer Assisted Intervention (MICCAI 2017), pp. 408–416. Springer International Publishing, Cham (2017)

1012. Zhang, Y., Zhou, D., Chen, S., Gao, S., Ma, Y.: Single-image crowd counting via multi-column convolutional neural network. In: 2016 IEEE Conference on Computer Vision and Pattern Recognition (CVPR), pp. 589–597 (2016). https://doi.org/10.1109/CVPR.2016.70

1013. Zhang, Z., Yang, L., Zheng, Y.: Translating and segmenting multimodal medical volumes with cycle- and shape-consistency generative adversarial network. In: 2018 IEEE/CVF Conference on Computer Vision and Pattern Recognition, pp. 9242–9251 (2018)

1014. Zhao, H., Li, H., Cheng, L.: Synthesizing filamentary structured images with gans. CoRR (2017). arXiv e-prints abs:1706.02185

1015. Zhao, H., Zhang, S., Wu, G., Costeira, J.A.P., Moura, J.M.F., Gordon, G.J.: Adversarial multiple source domain adaptation. In: Proceedings of the 32Nd International Conference on Neural Information Processing Systems, NIPS'18, pp. 8568–8579. Curran Associates Inc., USA (2018)

1016. Zhao, J., Xiong, L., Cheng, Y., Cheng, Y., Li, J., Zhou, L., Xu, Y., Karlekar, J., Pranata, S., Shen, S., Xing, J., Yan, S., Feng, J.: 3d-aided deep pose-invariant face recognition. In: Proceedings of the Twenty-Seventh International Joint Conference on Artificial Intelligence, IJCAI-18, pp. 1184–1190. International Joint Conferences on Artificial Intelligence Organization (2018). https://doi.org/10.24963/ijcai.2018/165

1017. Zhao, J., Xiong, L., Li, J., Xing, J., Yan, S., Feng, J.: 3d-aided dual-agent gans for unconstrained face recognition. IEEE Trans. Patt. Anal. Mach. Intell. **2380–2394**, (2018). https://doi.org/10.1109/TPAMI.2018.2858819

1018. Zhao, J.J., Kim, Y., Zhang, K., Rush, A.M., LeCun, Y.: Adversarially regularized autoencoders for generating discrete structures. CoRR (2017). arXiv e-prints abs:1706.04223

1019. Zhao, J.J., Mathieu, M., LeCun, Y.: Energy-based generative adversarial network. CoRR (2016). arXiv e-prints abs:1609.03126

1020. Zheng, C., Cham, T.J., Cai, J.: T2net: Synthetic-to-realistic translation for solving single-image depth estimation tasks. In: Ferrari, V., Hebert, M., Sminchisescu, C., Weiss, Y.

(eds.) Computer Vision–ECCV 2018, pp. 798–814. Springer International Publishing, Cham (2018)

1021. Zheng, N., Jiang, Y., jiang Huang, D.: Strokenet: A neural painting environment. In: ICLR (2019)

1022. Zheng, S., Jayasumana, S., Romera-Paredes, B., Vineet, V., Su, Z., Du, D., Huang, C., Torr, P.H.S.: Conditional random fields as recurrent neural networks. In: 2015 IEEE International Conference on Computer Vision (ICCV), pp. 1529–1537 (2015). https://doi.org/10.1109/ICCV.2015.179

1023. Zhou, F.: A survey of mainstream indoor positioning systems. J. Phys. Conf. Ser. **910**, 012069 (2017). https://doi.org/10.1088/1742-6596/910/1/012069

1024. Zhou Wang, Bovik, A.C., Sheikh, H.R., Simoncelli, E.P.: Image quality assessment: from error visibility to structural similarity. IEEE Transactions on Image Processing **13**(4), 600–612 (2004). https://doi.org/10.1109/TIP.2003.819861

1025. Zhu, J., Park, T., Isola, P., Efros, A.A.: Unpaired image-to-image translation using cycle-consistent adversarial networks. In: 2017 IEEE International Conference on Computer Vision (ICCV), pp. 2242–2251 (2017)

1026. Zhu, S.C., Mumford, D.: A stochastic grammar of images. Found. Trends. Comput. Graph. Vis. **2**(4), 259–362 (2006). https://doi.org/10.1561/0600000018

1027. Zhu, X., Yan, J., Yi, D., Lei, Z., Li, S.Z.: Discriminative 3d morphable model fitting. In: 2015 11th IEEE International Conference and Workshops on Automatic Face and Gesture Recognition (FG), vol. 1, pp. 1–8 (2015). https://doi.org/10.1109/FG.2015.7163096

1028. Zhu, Y., Aoun, M., Krijn, M., Vanschoren, J.: Data augmentation using conditional generative adversarial networks for leaf counting in arabidopsis plants. In: BMVC (2018)

1029. Zhu, Y., Elhoseiny, M., Liu, B., Peng, X., Elgammal, A.: A generative adversarial approach for zero-shot learning from noisy texts. In: 2018 IEEE/CVF Conference on Computer Vision and Pattern Recognition, pp. 1004–1013 (2018). https://doi.org/10.1109/CVPR.2018.00111

1030. Zimmer, U.R., von Puttkamer, E.: Realtime-learning on an autonomous mobile robot with neural networks. In: Proceedings Sixth Euromicro Workshop on Real-Time Systems, pp. 40–44 (1994). https://doi.org/10.1109/EMWRTS.1994.336867

1031. Zoph, B., Le, Q.V.: Neural architecture search with reinforcement learning. CoRR (2016). arXiv e-prints abs:1611.01578

1032. Zoph, B., Vasudevan, V., Shlens, J., Le, Q.V.: Learning transferable architectures for scalable image recognition. CoRR (2017). arXiv e-prints abs:1707.07012

1033. Zou, D., Xu, P., Gu, Q.: Stochastic gradient hamiltonian monte carlo methods with recursive variance reduction. In: H. Wallach, H. Larochelle, A. Beygelzimer, F. d' Alché-Buc, E. Fox, R. Garnett (eds.) Advances in Neural Information Processing Systems, vol. 32. Curran Associates, Inc. (2019)

Printed in the United States
by Baker & Taylor Publisher Services